The Volterra Chronicles

The Life and Times
of an Extraordinary Mathematician
1860–1940

The Volterra Chronicles

The Life and Times
of an Extraordinary Mathematician
1860–1940

Judith R. Goodstein

Cover art: Anonymous, Portrait of Vito Volterra, Pisa, ca. 1879, The Burndy Library, Cambridge, Massachusetts. Oil on canvas.

2000 *Mathematics Subject Classification.* Primary 01A60, 01A70.

For additional information and updates on this book, visit
www.ams.org/bookpages/hmath-31

Library of Congress Cataloging-in-Publication Data

Goodstein, Judith R.
The Volterra chronicles : the life and times of an extraordinary mathematician, 1860–1940 / Judith R. Goodstein.
    p. cm. — (History of mathematics, ISSN 0899-2428 ; v. 31)
Includes bibliographical references and index.
ISBN-13: 978-0-8218-3969-0 (alk. paper)
ISBN-10: 0-8218-3969-1 (alk. paper)
1. Volterra, Vito, 1860–1940. 2. Mathematicians—Italy—Biography. 3. Mathematics—Italy—History—19th century. 4. Mathematics—Italy—History—20th century. 5. Jews—Italy—Social conditions—19th century. 6. Jews—Italy—Social conditions—20th century. 7. Italy—Politics and government—19th century. 8. Italy—Politics and government—20th century. I. Title.

QA29.V64G66   2007
510.92—dc22

                                                                        2006051752

With love

to David, Marcia, and Mark

and

the memory of Fanny E. Winagrad Koral

December 18, 1913 – August 29, 1991

Contents

Illustrations

1. Map of Italy, 1870.
Credit: from George Holmes, ed., The Oxford History of Italy (1997). By permission of Oxford University Press.

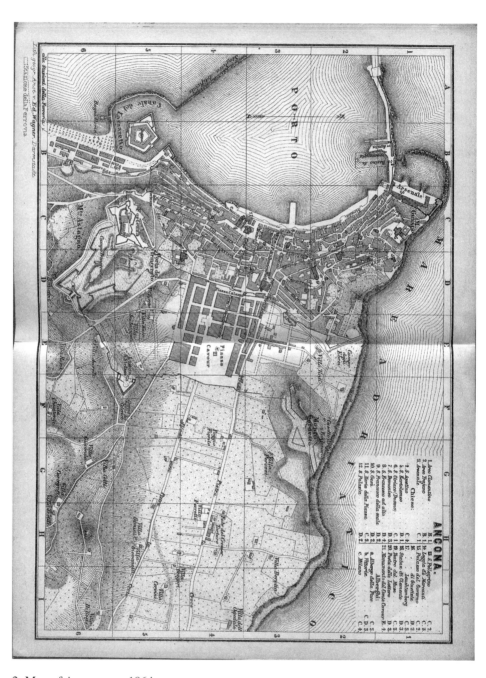

2. Map of Ancona, ca. 1864.
Credit: Baedeker's Central Italy, 1874 edition/courtesy of the Department of Special Collections, Charles E. Young Research Library, UCLA.

3. The Ancona ghetto, ca. 1813. The ghetto was abolished in 1860, and the construction of a new street, named for King Vittorio Emanuele, running from Piazza Cavour to Piazza del Teatro, effectively opened up Via Lata and Via Bagno, two of the principal streets of the old ghetto.

Credit: E. Sori, "Una comunità crepuscolare: Ancona tra Otto e Novecento," Proposte e Ricerche, n. 14.

4. Portrait by an unknown artist of Saul Almagià and his son Edoardo, ca. 1852.
Credit: Edoardo Achille Almagià.

5. Alfonso Almagià (seated, second from right) and co-workers in the central office of
the Italian National Bank (today the Bank of Italy) in Florence, Italy, 1868.
*Credit: Banca d'Italia, Archivio storico della Banca d'Italia, Album del personale della
Banca Nazionale del Regno d'Italia, Ritratto di gruppo dei dirigenti e degli impiegati
della Sezione I – Divisione I della Direzione Generale, 1868.*

6. Cesare Arzelà (1847–1912)
*Credit: Biblioteca
dell'Accademia Nazionale dei
Lincei e Corsiniana, Fondo
Volterra.*

7. Enrico Betti (1823–1892)
*Credit: Biblioteca dell'Accademia Nazionale dei
Lincei e Corsiniana, Fondo Volterra.*

8. Guido Castelnuovo
(1865–1952)
Credit: Fonti iconografiche,
Biblioteca Matematica
Giuseppe Peano, Department
of Mathematics, University
of Turin.

9. Ulisse Dini (1845–1918)
Credit: Institut Mittag–Leffler.

10. Griffith Evans (1887–1973)
Credit: Biblioteca dell'Accademia Nazionale dei Lincei e Corsiniana, Fondo Volterra.

11. Giovanni Battista Guccia (1855–1914)
Credit: Biblioteca dell'Accademia Nazionale dei Lincei e Corsiniana, Fondo Volterra.

12. Sofia Kovalevskaya (1850–1891)
Credit: Institut Mittag–Leffler.

13. Tullio Levi–Civita (1873–1941)
Credit: Institut Mittag–Leffler.

14. Antonio Ròiti (1843–1921)
Credit: Biblioteca dell'Accademia Nazionale dei Lincei e Corsiniana, Fondo Volterra.

15. Giovanni Vailati (1863–1909)
Credit: M. Calderoni, U. Ricci, and G. Vacca, eds. Scritti di G. Vailati (Leipzig, Johann A. Barth, 1911).

16. Vito Volterra (left) and Carlo Somigliana, Pisa, 1881.
Credit: Pontificia Academia Scientiarum, 1942.

17. Vito Volterra's Train Pass, Pisa, ca. 1884.
Credit: from a private collection.

18. Mathematics graduates, Scuola Normale Superiore di Pisa, 1888. The dedication on the reverse reads: "Al chiarissimo prof. Vito Volterra i Normalisti laureandi in matematica dell'anno 1887–1888. Pisa, 31 Maggio 1888."
Credit: Biblioteca dell'Accademia Nazionale dei Lincei e Corsiniana, Fondo Volterra.

19. Via Po, Turin, ca. 1890. The University Building is situated in the middle of Via Po (the main thoroughfare, on the left side of the street). The Mole Antonelliana, visible in the upper left corner, held the title of tallest brick construction in Europe for many years.
Credit: Fonti iconografiche, Biblioteca Matematica Giuseppe Peano, Department of Mathematics, University of Turin.

20. Angelica Volterra, 1900.
*Credit: The Burndy Library,
Cambridge, Massachusetts.*

21. Amelia Almagià Ambron's portrait of Edoardo Almagià, ca. 1900.
Credit: from a private collection.

22. Virginia Almagià, ca. 1900.
Credit: from a private collection.

23. Vito Volterra and Virginia Almagià, 1900.
Credit: from a private collection.

24. The Palazzo Fiano–Almagià, at the corner of Piazza in Lucina and the Via del Corso in Rome, where Volterra lived from 1900 to 1940. The palazzo runs the length of the block on the Corso, with another entrance on Via in Lucina, on the south side of the building. Standing in the heart of the city's historic center, the building had the added advantage of being around the corner from the Chamber of Deputies and a short walk to the Senate.
Credit: Edizioni Quasar s.r.l. di Severino Tognon.

25. The professor and his wife in his library at home, ca. 1901. During World War II, the janitor of the building walled over the entrance to the room, and Volterra's rare book collection, prized for its many unique items, survived the war intact.
Credit: Biblioteca dell'Accademia Nazionale dei Lincei e Corsiniana, Fondo Volterra.

26. Leading physicists and mathematicians met at Clark University in 1909 to celebrate the school's twentieth anniversary. Speakers and Clark participants included: Front row: Robert Williams Wood (first on the left), Eliakins Hastings Moore (second from left), Vito Volterra (third from left), A. A. Michelson (fourth from left), William Edward Story (second from right). Second row: Arthur Gordon Webster (second from left), Edward Burr Van Vleck (fourth from left), Ernest Fox Nichols (third from right). Third row: Ernest Rutherford (third from left), Carl Barus (fifth from left, with beard). Fourth row: Robert Hutchings Goddard (far left).
Credit: Clark University Archives.

27. After establishing Italy's Office for War Inventions in 1917, Volterra made many trips to France and England to promote military cooperation among the Allies. In his application to visit these countries in May 1918, he states that the purpose of the journey is an official mission in connection with the War Ministry.
Credit: from a private collection.

28. Albert Einstein (in profile) and Federigo Enriques, Bologna, 1921.
Credit: Federico Enriques.

29. On a postcard bearing his own picture, Volterra wrote his epitaph for Mussolini's
Italy: "Empires die, but Euclid's theorems keep their youth forever," ca. 1931.
Credit: The Burndy Library, Cambridge, Massachusetts.

"The Jewish Mathematician"

Vito Volterra, born in 1860, died in Rome in 1940, having lived a tumultuous life that spanned the period from the unification of Italy to the outbreak of World War II. Born into a Jewish family of modest means, he became a world-renowned mathematician, developing a powerful mathematical language and theories whose influence would be felt in fields ranging from physics and applied mathematics to biology and economics. The literature of modern mathematics and applied mathematics is shot through with his name, the hallmark of immortality for a scientist. *Volterra processes* are widely used in materials science and control theory, and *Volterra integral equations* and *Volterra integro-differential equations* inform the theory of elastic media. The *Lotka-Volterra equations* appear in the first few pages of any contemporary treatise on theoretical ecology, sometimes disguised as the *Volterra population equation.* Elsewhere we find *Volterra systems, Volterra series, Volterra operators, Volterra kernels,* and *Volterra functionals.* Few mathematicians of any time and place have left behind such a legacy. By the age of twenty-three, Vito Volterra was a tenured university professor at Pisa, and he scaled the academic ladder quickly, becoming a professor at the University of Rome in 1900. Five years later, he was a central figure in international academic, political, and intellectual circles, and that same year he was appointed a senator of the Kingdom of Italy, making him a member for life of the upper house of Parliament.

For the next twenty years, Vito Volterra was the undisputed head of the Italian school of mathematics, which included a formidable group of Jewish university professors, among them Tullio Levi-Civita, Federigo Enriques, and Guido Castelnuovo. An aristocratic and imposing figure, Volterra zealously promoted the cause of science within and outside his country. He became president of the faltering Italian Physical Society and breathed new life into it. In an era when scientists in northern and central Europe were presiding over the birth of modern physics, Volterra and his circle of mathematicians filled the lacuna created by the absence of such physicists in Italy, a vacuum that would persist until the emergence of Enrico Fermi.

It did not take long for the fruits of Einstein's 1905 "annus mirabilis," particularly his special theory of relativity, to claim the attention of these articulate, versatile, and consummate mathematicians. In Italy, they alone had the training, temperament, and breadth of interest to embrace the new theory, and the story of relativity in the Italian Peninsula belongs to them.

Quick to grasp the profound importance of Einstein's ideas, Italy's mathematicians quickly set about making the theory known at home. Volterra organized discussion groups within the Italian Physical Society. Guido Castelnuovo—who, along with Federigo Enriques, took a keen interest in the philosophical implications of the theory—encouraged the translation and publication of Hermann Minkowski's famous 1908 address on "Space and Time" in the physics journal *Nuovo Cimento*. He also lectured and wrote about relativity and helped organize, in 1919, the first series of seminar talks on general relativity at the University of Rome. Enriques enlisted his own journal, *Scientia*, in the campaign. Between 1908 and 1914, a number of relativity articles, both favorable and unfavorable to relativity, appeared in his journal. ("To be sure," Einstein wrote to a friend in 1914, after reading an article contributed by Max Abraham, an outspoken critic of Einstein's efforts to generalize the special theory of relativity, "he fulminates against all relativity in *Scientia*, but he does it with understanding.")[1]

The contributions of Italy's foremost mathematicians to the intellectual migration and reception of relativity theory, both special and general, culminated in an extraordinary correspondence in 1915 between Einstein in Berlin and Tullio Levi-Civita in Padua, in which Levi-Civita took Einstein to task for not being rigorous enough in a particular proof. On April 14, 1915, after a flurry of letters, Einstein conceded that his proof failed in the special case treated by Levi-Civita and offered to "include the corrections I have learned from our memorable correspondence," if the occasion to do so arose. He continued in a more personal vein: "I shall take pleasure in striving to allow our acquaintance by letter [to] turn into a personal one: one more reason at last to cross the Alps again one day. It is to be hoped that our fatherlands will not rebel against each other as well!"[2] But Italy's decision later that year to take up arms against its one-time treaty partners Austria-Hungary and Germany forced Einstein to postpone his visit to the Italian mathematical community.

These dedicated mathematicians were also proud Italian patriots. When Italy entered World War I in 1915, the fifty-five-year-old Volterra volunteered as a lieutenant in the Italian Army Corps of Engineers. He spent the next two years working at the Aeronautics Institute in Rome, carrying out aerial warfare experiments on airships in Tuscany and testing various phototelemetric devices, first along the Italian-Austrian front between Verona and Gorizia and later on the French front from Reims to Soisson. While in the army, he also made several trips to England and France to see how Italy's allies had organized and recruited scientists to work on military problems, developed his own plans for an organization to support such work in Italy, and participated in the founding of the International Research Council in London; in all, he did not don civilian clothes again until the Armistice of November 1918.

When the war ended, Volterra returned to the University of Rome. There he resumed his teaching and administrative duties as the dean of the science faculty, and continued his rise, which, ominously in retrospect, can be seen to have paralleled that of the dictator. By 1925, when Mussolini had completed his takeover of Italy, Vito Volterra was president of the Accademia dei Lincei, the world's oldest scientific society, to which Galileo had been elected in 1611. He had also by that time founded Italy's National Research Council, which was charged with organizing and promoting scientific and industrial development and national defense. He acted as its president while also serving as vice president of its parent International Research Council, the new multinational federation of research councils in fifteen Allied and neutral countries. Indeed, Volterra's importance loomed even larger internationally than it did in Italy, where, according to his American friend the astronomer George Ellery Hale, he was "the leading spirit"—Mr. Italian Science, if you will—of his embattled nation.[3]

The rise of Fascism, however, brought a halt to the golden age of Italian mathematics. To be sure, age may well have been a factor: By 1925, the members of Volterra's influential circle, all born between 1860 and 1873, were well into, if not past, their scientific prime. Be that as it may, Mussolini's regime was actively hostile to the circle of Jewish Italian mathematicians who, in effect, had dominated the old scientific order. The records of the Accademia d'Italia—the most prominent of Mussolini's new institutions of culture—shows only too clearly that the Fascist state did not seek an accommodation with Volterra's circle, and in fact used the new academy as a means of isolating them and marginalizing their scientific contributions. The selection of candidates both for membership in the new academy and for its major prizes reflects a pronounced anti-Semitism, despite the regime's initial disavowals. Matters had already deteriorated badly by the time Anti-Semitism became an official component of the Fascist state in 1938. That year, the government issued the *Manifesto of Italian Racism*, which declared in part that Jews could not be considered members of a so-called Italian "race," and put into effect the infamous racial laws, which among other measures banished Italy's Jewish professors from their universities. The racial laws exacted a heavy toll on Italy's scientific community, in which Italian Jews were represented far out of proportion to their percentage of the population.[*]

Seven years earlier, in 1931, Vito Volterra had refused to sign the oath of allegiance to the Fascist government, which all university professors were required to sign. For this courageous act—he was one of only twelve Italian university professors who refused—he was drummed out of Italy's scientific establishment and ostracized by his profession and his nation. Upon learning the news, George Ellery Hale sent him a letter of support. "I wish I could also tell you personally of... my admiration of your attitude and courage,"

---

[*]In the early thirties there were only about 39,000 Jews in a country of some 40 million people. Although they figured as only 0.1 percent of Mussolini's subjects, ninety-seven of them were professors, 7 percent of the total. (1931 census figures reported in Renzo De Felice, *Storia degli ebrei italiani sotto il fascismo*, 1st ed., Turin: Einaudi, 1961, p. 6.)

he wrote. In his answer, Volterra, by now all but banished from public view, made it clear that he harbored no illusions. "If you no longer see my name among the membership of the Accademia dei Lincei," he wrote to Hale in 1934, "do not think I am dead."[4]

Vito Volterra had become an invisible man. Late in 1938, Enrico Fermi, that year's Nobel laureate in physics and the new Mr. Italian Science, left Mussolini's Italy for good. Accompanied by his wife Laura, their two children, and a nanny, he headed for America, stopping in Stockholm to collect his prize before continuing on to Columbia University in New York. Two years later Volterra died—in such obscurity that when the Nazis occupied Rome in 1943, German soldiers knocked on the door of his house in Via in Lucina 17 confidently expecting to arrest him and deport him to a concentration camp.

Volterra's life exemplifies the post-unification rise of Italian mathematics, its prominence in the first quarter of the twentieth century, and its precipitous decline under Mussolini. This intellectual history in turn parallels the rise of Italian Jewry in the latter half of the nineteenth century and its travails during the Second World War. The meteoric rise and tragic fall of Volterra and his circle thus constitutes a lens through which we may examine in intimate detail the fortunes of Italian science in an epic scientific age.

* * * * * *

With the proclamation of the Kingdom of Italy in 1861, Victor Emmanuel II of the House of Savoy took his place among other European monarchs. Before the Risorgimento, the Italian peninsula's campaign for national unification, Italy had been a patchwork of states governed, protected, and fought over by foreign rulers. Between 1859 and 1861 Camillo Cavour, the prime minister of Piedmont, in the northwest corner of the country, engineered the annexation of Lombardy to the east and the provinces of Parma, Modena, and Tuscany to the south, and sent the King's troops into the Papal States. When the fighting ceased, the Adriatic coast from Ancona to Bologna and the lands of Umbria and the Marches had been annexed to the Kingdom of Italy as well. The charismatic Risorgimento general Giuseppe Garibaldi and his army of volunteers had in the meantime conquered the island of Sicily and now moved north and chased the Bourbons out of Naples. Garibaldi planned to march on Rome, but under pressure from Cavour he surrendered southern Italy to the King in October 1860. Only Venice (until 1866), Rome (until 1870), and Trieste (until 1919) remained outside the fold.

In September 1858, inspired by the unification fever that gripped their country, a trio of Italian mathematicians packed their bags and set out on a scientific expedition across the Alps. Motivated in part by a sense of scientific isolation at home, in part by the lament of colleagues that their work was unknown abroad, Enrico Betti, Francesco Brioschi, and Felice Casorati headed for Germany and France, home to some of Europe's finest mathematicians. Their trip was a huge success. For the thirty-five-year-old Betti, the newly anointed professor of algebra at Pisa, the trip's high point was meeting the great analyst, geometer, and number theorist Georg Friedrich Bernhard Riemann at Göttingen. Riemann, who would later visit Pisa, had a lasting influence on Betti's approach to mathematics and on Italian mathematics in general. When Betti returned to Pisa, he would translate into Italian Riemann's doctoral dissertation of 1851, "Foundations for a General Theory of Functions of a Complex Variable," for the *Annali di Matematica pura e applicata,* an ambitious new journal intended not just to publish original papers by Italian authors but also to introduce important results by German and English mathematicians to a new generation of Italian mathematicians. Betti's later work would focus almost entirely on problems of mathematical physics, ranging from electrostatics and magnetism to the theory of potentials and drawing inspiration from Riemann's ideas regarding mathematical descriptions of the physical world. Mathematicians today are familiar with Betti numbers from algebraic topology, the study of geometric shapes.[†]

For Betti's colleagues, Brioschi and Casorti, the Alpine crossing had an equally salutary outcome. Brioschi, then thirty-four and a professor at Pavia, plunged into animated conversations first with Leopold Kronecker in Berlin and later with Charles Hermite in Paris, and returned home reinvigorated and ready to adapt their methods to his own work on the theory of the transformation of elliptic and Abelian functions. Casorati, then Brioschi's assistant, also came under Riemann's spell and would return to Berlin in 1864 for further productive discussions with Karl Weierstrass and other mathematicians there. More than a mere trip, that foray across the Alps by Betti and his companions represented an important stage in Italy's evolution into a modern scientific country. Speaking for the first generation of post-unification mathematicians, Vito Volterra told an international assembly of mathematicians in Paris in 1900 that "the scientific existence of Italy as a nation" dated from this continental trip.[5]

By the time Rome had become the capital of the newly created state of Italy in 1870, the Italian mathematical community was undergoing a rebirth of its own. Mathematicians like Betti and Luigi Cremona, both veterans of the 1848 War of Independence, were now members of Parliament. Cremona had been a student of Brioschi and later became the director of the school of engineering at Rome, and his voice carried great weight in the

---

[†]The question is, How many times can you cut up the surface of a solid shape without separating it into two objects? For example, if you cut through one side of a doughnut-shaped torus, you can straighten it and it's a cylinder. Then if you cut the cylinder's surface parallel to its axis, you can lay it flat and it's a sheet, still one object. Thus a torus has a Betti number two, and a cylinder has Betti number one. Any closed cut of the surface of a sphere separates it into two pieces, so a sphere has Betti number zero.

Senate on educational matters. Betti had recently become the director of the Scuola Normale Superiore at Pisa, a select teachers college connected to the University of Pisa, and there he set about building a paramount school of mathematics whose roster of distinguished alumni (Volterra was among them) grew by the decade. Brioschi, who combined his university position with a ministerial appointment in the new government, helped to establish the Milan Polytechnic Institute, where he was both teacher and administrator. As soldiers, teachers, and politicians, these scientists not only assisted in the building of their state but after 1870 put Italian mathematics on the scientific map of Europe.

From around the turn of the century until Benito Mussolini became dictator in 1925, mathematics was Italy's foremost scientific discipline, and a close-knit circle of largely Jewish Italian mathematicians dominated science in Italy. As Jews, they shared a history and common traditions, and as scientists and Italian citizens, they enjoyed an international reputation. Nor were their interests confined to mathematics. Volterra combined a love of French literature with a passion for rare books and Roman antiquities. He also developed a boundless enthusiasm for international travel: In 1909 he represented Italy at the twentieth anniversary of the founding of Clark University, in Worcester, Massachusetts, and a few years later participated in the inauguration of the Rice Institute, in Houston. After Mussolini came to power, Paris would become, for a time, his second home. Enriques, an authority on algebraic geometry who wrote extensively on the philosophy and history of science, founded and coedited *Scientia*, a culturally grounded international review of scientific ideas. When the end of World War I finally enabled Einstein to make that long-anticipated and, in the event, highly celebrated, trip to Italy, it was Enriques who made the arrangements. Levi-Civita's taste in mathematical problems ranged from the pure to the applied, with a special interest in the general theory of relativity. Castelnuovo, for his part, spent considerable time building up a school of mathematics at the University of Rome, bringing Levi-Civita there from the University of Padua in 1918; Enriques followed from Bologna in 1923. By then, Rome had become an international mecca for mathematicians, ranging from the Russian-born Oscar Zariski to Rockefeller fellows Dirk Struik and Szolem Mandelbrojt to the American Griffith Evans—all at the start of their careers. Vito Volterra's circle of university professors were in the vanguard of what Struik would describe many years later, in a celebrated talk on the history of mathematics in Europe, as "a new phenomenon—namely, the Jewish mathematician."[6]

In the course of compiling brief biographical profiles of more than 300 Italian mathematicians who died between 1861 and 1960, the late Francesco Tricomi, professor of mathematics at the University of Turin, assembled a vast quantity of statistical information: the relatively high number who died a violent death, the dearth of women in their ranks, the length of mathematical careers. (Volterra's is among the longest, since he became a tenured university professor at an unusually young age.) Many of the published obituaries and commemorations that Tricomi examined glossed over certain matters. He writes, "[O]ne almost never says, even during the pre-Fascist period, whether the person about whom one is writing is Jewish. Why? Particularly since, in our science, that would be almost a title of honor, considering the number and above all the merit of these Jewish mathematicians!" Not himself Jewish, Tricomi supplied the missing data himself, using the letter "E" (for *Ebreo*) to identify the Jewish mathematicians on his list. Of the 371 mathematicians included in his census, 29 (8 percent of the group) have an "E" before their name—an impressively high number, given that Italian Jews had never represented more than a tenth of a percent of the country's citizens.[7]

Before unification, Italian Jews paid a heavy price for being Jewish. Citizens of a Roman Republic and Empire long before Rome became the sacred seat of Christendom, Italy's Jews saw their fortunes rise and fall as emperors, kings, and papal rulers governed the land. During the Counter Reformation and the Inquisition, Jewish books were confiscated and burned. Roman Jews were burned alive in Campo de' Fiori, on the site where a statue of Giordano Bruno, who suffered the same fate in 1600 for his belief in a plurality of worlds, stands today. By the late sixteenth century, most Italian cities had segregated the Jews from the rest of the population, forcing them to live in ghettos. (The very word derives from the Italian *getto*, meaning metal casting, and refers to the location in Venice of the original ghetto, which was set up in 1516 near the site of an iron foundry.) The confinement of Italian Jews in ghettos persisted, with few exceptions, for 300 years.

The nineteenth-century German historian Ferdinand Gregorovius came upon the narrow streets and alleys of the Rome ghetto in the mid-nineteenth century, while working on his monumental history of medieval Rome. He describes it in an essay as "the dreariest quarter in Rome, a corner of filth and poverty." Some 3,800 Jews lived there, a few steps from the Tiber River, in the shadow of the Theater of Marcellus and the Porticus of Octavia, in row houses, "tower-like masses of bizarre design, with numerous flowerpots in the windows and countless household utensils hanging on the walls." When the banks of the Tiber overflowed, the basements of houses built at the river's edge filled with water. "Those who live beneath take refuge in the upper floors, which are intolerably crowded and tainted by pestilential atmosphere," he wrote. "The stoppage of food supply and of work increase the misfortune, and the flood ruins everything that cannot be removed." Prosperous ghetto Jews lived in houses situated on higher land; the wealthiest, who had the means to pay the heavy taxes and tributes levied on Jews in Italy, often bought their way out and emigrated north, to Lombardy and Tuscany—provinces belonging to rulers other than the pope.[8]

Ghetto life was harsh and degrading. Gregorovius observed men and women who would sit

in their doorways or outdoors, which affords scarcely more
light than their damp and dismal rooms, and tend their
ragged merchandise or industriously patch and sew. The
chaos of patching and mending of rags (called *cenci* in Italian) is indescribable. All the world seems to lie about, transformed into Jewish trash, tattered and torn, in countless rags
and scraps. Pieces of junk of every kind and color are heaped
high before the doors: scraps of golden fringe, pieces of silk
brocade, rags of velvet, patches of red, scraps of blue, orange,
yellow, black, white, old, torn, threadbare, badly worn scraps
and tatters.[9]

The glaring contrast between the dark and foul-smelling streets of the
Rome ghetto and the majestic views of ancient Roman ruins on the Palatine
and bustling Trastevere just across the river puzzled Gregorovius: How had
the Jews managed to survive while the civilizations that had conquered them
had vanished? He was still living in Rome in 1870 when the troops of King
Victor Emmanuel II entered the city and emancipated its Jews.

It was in the cities under papal rule that the Jews of Italy had endured
the most repressive civil, political, and religious restrictions. To facilitate the
conversion of Jews to Christianity, houses of catechumens were established
in the sixteenth century, where Jews could be held indefinitely against their
will, while their "conversions" were monitored. Jews were forced to listen
to evangelical sermons. They could not attend public schools at any level,
with the exception of the faculty of medicine (so that they could practice on
fellow Jews). They were prohibited from owning property, maintaining shops
outside the ghetto, remaining outside the ghetto after sundown, or hiring
Christian servants—which Jewish families sometimes did anyway, often to
their distinct disadvantage.

In Ancona, Modena, Reggio Emilia, and other cities with Jewish populations, the surreptitious baptism of Jewish children by Christian servants
was a commonplace event well into the nineteenth century. A case that
drew the attention of Cavour, Garibaldi, and other Risorgimento leaders
and caused an uproar in England, France, and the United States involved
the kidnapping of six-year-old Edgardo Mortara in Bologna in June of 1858
by papal guards. A servant charged with caring for Edgardo during a childhood illness claimed to have baptized him in secret. Under canon law, once
a Jewish child was baptized in the Catholic faith, the church had an unim-
peachable right to raise him as a Catholic, regardless of the parents' wishes.
Accordingly, little Edgardo was taken to Rome, placed in the house of the
catechumens there, and adopted by Pius IX, whose refusal to return him to
his family only intensified the anticlerical feelings of liberals and democrats
in Piedmont and accelerated the downfall of the Papal States.[10] The following year, Bologna, a papal city for more than two centuries, overthrew
the papal government and pledged its allegiance to the War of Independence

and the House of Savoy. In the process, Italian Jews in Romagna gained the same civil rights as Christians.

In Piedmont, where the Risorgimento began, Massimo D'Azeglio and other figures active in the movement had linked Jewish emancipation to the political emancipation of the Italian people up and down the peninsula. In 1848 Piedmont became the first state to extend full civil rights to Jews and other non-Catholics living within its borders; Romagna and the other states that were annexed to Piedmont to form the United Kingdom of Italy followed suit in 1859 and 1860. Slowly but surely, the emancipation of Jews spread throughout the country. Italy's Jews firmly believed that they owed their political, economic, and social rebirth to the House of Savoy, the Risorgimento, and revolutionaries like Garibaldi. Ghetto Jews, who saw Garibaldi's cause as their own, fought alongside Garibaldi's volunteers in the Rome campaign of 1849, in the Sicilian expedition of 1860, and in the Trentino in 1868. After 1860, some Jewish families began to choose patriotic names for their newborn children—for the girls, Italia (coupled with a traditional name); for the boys, Victor Emmanuel, Umberto, or Carlo Alberto—and the practice ("the double identity," one writer called it) of coupling traditional Jewish names with an Italian Christian name became widespread.[11]

The Risorgimento culminated in equal rights for all Italians, including the Italian Jews. They could—and did—live anyplace, stay out all night, buy and sell property in their own names, attend public schools, choose a trade, become university professors. "And all this," the writer Primo Levi once remarked about the Jews of his native Turin, "at a time when in Italy the great majority of the population was illiterate." Noting that Jews had always prized culture, education, and literacy, he added, "Therefore emancipation did not catch them unprepared [and] within one or two generations the Jews out of the ghetto easily moved from crafts and small commerce to the newly born industries, to administration, high public office, the armed forces, and the universities."[12] Levi's characterization of the Jewish communities in Piedmont applies to many Jewish communities in Italy at that time, including that of Ancona, where Vito Volterra was born in 1860.

CHAPTER 1

"A New Era Is Dawning," 1860

Second only to Venice as the largest commercial port on the Adriatic coast, Ancona was settled in Greek times by colonists from Syracuse, who called it Ancon ("elbow" in Greek), after the shape of Monte Conero, a promontory protecting the harbor, which was one of the best in Italy. Under the Romans, the city, situated in the northeast region known as the Marches, became a bustling port and a favorite of the Emperor Trajan, who enlarged its harbor in 114 C. E. The city and its wharves have changed hands many times since then. After the fall of Rome in 410, the Goths and others overran Ancona; in the Middle Ages the rulers of Rimini, up the Adriatic coast, governed the city. In the meantime, Ancona's small Jewish community had settled in and become money lenders and maritime traders, carrying out commerce with Turkey, Armenia, and other countries across the Aegean and Adriatic. Town authorities had given the Jews permission to live in a quarter of the city near the port, on and between Monte Guasco, with its towering medieval cathedral, and Monte Astagno to the south, where a citadel stood guard. The fall of Constantinople in 1453, the expulsion of the Jews from Spain and Sicily in 1492, the sack of Rome in 1527, and Pius V's decree of 1569—which expelled Jews from every city state under papal rule except Rome and Ancona—all brought fresh waves of new Jewish immigrants into Ancona's Jewish quarter.

By 1555, the quarter had been transformed into a walled city whose gates were opened in the morning and locked at night. Curfew began at sunset. As elsewhere, Jews were required to wear a yellow badge and were not allowed to own land. By then, Ancona had become a papal fiefdom—one more small Italian state controlled by the church, whose political, economic, and territorial ambitions beyond Rome matched those of the kings of Spain and France and Italy's other foreign rulers. Not until the French Revolution and Napoleon's subsequent military victories in Italy would papal rule come to an end, and with it the ghettoization of the Jews of Ancona.

At the start of the nineteenth century, about 1,400 Jews lived in this city of some 17,000 people. In 1811, while serving in the French Army in Milan, the writer Stendhal visited Ancona. "The shoreline is nothing but arid cliffs," he wrote in his journal. "In Ancona, one goes up and down continuously, and this greatly limits the use of carriages. The houses are built of bricks, rather tall, the streets very narrow....There are no trees in Ancona. One goes for a walk around the French harbor, on the barren beach, or in

the area where new fortifications [are under construction]." He noted that
the wealthy citizens often spent their evenings at the Casino Dorico, a private club founded in 1806, where dances, poetry readings, singing recitals,
concerts, and receptions took place in a large ornate ballroom with brocade
curtains draped over the windows; candelabras affixed to the walls and a
chandelier of candles suspended from a two-story domed ceiling brightened
the room. A new opera house, inaugurated in 1827 with productions of
Rossini's *Aureliano in Palmira* and *Riccardo and Zoraide*, also figured prominently in the cultural life of the city, as did several popular coffee houses.
The extravagant religious and civic celebrations that took place throughout
the year, accompanied by musical bands, fireworks displays, and colorful
banners did not impress the visiting Frenchman, who dismissed the provincial city as "small, where you see very few people and you die of boredom."
He did not visit the ghetto.[1]

Ancona's Jewish ghetto, a city within a city, teemed with people. A
map from the Napoleonic era shows five major arteries—Via Lata (later
Via Astagno), the most important, with thirty-six passageways leading to
a maze of apartment buildings; Via del Traffico and Via del Bagno, each
with fifteen such entryways; and Via dello Speziale and Via dei Banchieri,
with eleven and eight respectively—and a small number of alleys, lanes, and
footpaths.[2] The names of certain streets reflect the religious requirements
of the community: Via delle Azzimelle refers to the ovens used to bake
unleavened bread; Via del Macello, the kosher meat stores; Via del Bagno,
the special bathing establishments for women. Via dei Banchieri and Via
dello Speziale denote the banking and shopping districts. At the beginning
of the nineteenth century, more than half the Jewish community lived there
below the poverty level; and only 26 percent of its members could afford to
pay the various assessments levied on the entire community by the Università
Israelitica, the ghetto's governing body. By 1830, the ghetto consisted of
391 families, of whom 48 enjoyed the luxury of one or more Jewish servants.
By mid-century, the population in the ghetto had risen to 1,800—still a
very small fraction of the entire city's population of 25,000. The French
diplomat Michel Mangourit, who toured the ghetto shortly after Napoleon
entered Ancona in 1797, paints an appalling picture: "[T]he alleys, covered
with garbage. . . look like the tunnels of a mine. Their houses rise up four-five
floors; they are bilges of filth. There one climbs a gloomy spiral staircase;
and you will find under their roofs and in the arches of the walls up to 100,000
Spanish piastras time and again."[3] Whether Mangourit had actually seen
such caches of coins—money that would have represented one of the few
forms of tangible security that these persecuted and historically tormented
people possessed—or was just indulging in reflexive anti-Semitism remains
unclear.

Vito Volterra was named after his mother's father, Vito Almagià, who
was born in the Ancona ghetto in 1797, the same year that Napoleon's troops
entered the city. A schoolteacher, Almagià taught in the Jewish elementary

school housed in the Italian synagogue in the center of the ghetto. (Situated at an angle between Via Lata and Vicolo Strettore, the building complex was demolished in the 1930s to make way for the Corso Stamira.) In the first half of the nineteenth century, primary and middle schools like the one where Vito Almagià taught fell under the jurisdiction of the community's Talmud Torà [Torah], an organization run by an autonomous board of governors, reporting to and financially dependent on the Università Israelitica. In the ghetto even the poorest children went to school, and illiteracy, which was otherwise endemic throughout Italy, was seldom encountered inside the ghetto. A network of yeshivas—there were ten in Ancona, some completely secular—offered adults an opportunity to continue their Hebrew and, in some cases, scriptural studies, after work and functioned both as social and cultural associations.

Like other families in the ghetto, the Almagiàs did not own the roof over their heads; rather, they rented in perpetuity an apartment from a Christian landlord, whose ancestors had probably inhabited it in pre-ghetto times. Almagià roots in Ancona can be traced at least as far back as 1737, when an official act granted one of the family's patriarchs a permanent dwelling at a fixed rent on what is now the Corso Giuseppe Garibaldi. The law governing these rental agreements, known as the *jus gazzagà*, permitted the Jewish tenant to alter the rental property, to transfer it to relatives or others, and to sublet a portion of the space to other Jews.

Bearing the revolutionary message of equal rights for all citizens on the tips of their bayonets, the French Army tore down the two gates of the ghetto. Ancona, now no longer the property of the Catholic Church, was recast as an Italian republic, and Jews who could afford it were now permitted to live outside the ghetto—if they wanted to. Initially, few took advantage of the opportunity. The daily lives of most of the Jews still revolved around their religious institutions, their synagogues, and kosher butcher shops, and the proximity of friends and family. The fear of living among people who might turn on them unexpectedly also kept many families from moving out. The Almagià family did not rush to leave the ghetto.

For Vito Almagià and the other Jews who lived in Ancona, the presence of French troops opened civic doors as well; when a new city council was convened, it included several Jews. But by the time Vito Almagià had turned eighteen, Napoleon's empire no longer existed, and the Congress of Vienna had restored the pope to his throne and returned the Papal States, including Ancona, to him. In 1826, Pope Leo XII ordered new gates to be built for the city's Jewish quarter, but they were taken down once again in 1831. In 1848, in the wake of the War of Independence, the church permanently abolished the ghetto. Despite the abolition, the Jewish quarter still continued to house the city's Jews; even after unification, the ghetto persisted for years in fact, if not in law. In 1865 the conflict between the state's *Cassa ecclesiastica*, which was eager to transform the old property arrangements into simple rental agreements, and the Jewish community, many of whose members

wanted to disassociate themselves from the dwellings they had been forced
to occupy under the rule of *jus gazzagà*, would be widely reported in the
local press. The community's lawyer, Annibale Ninchi, remarked that "it's
enough to have eyes to notice that in Ancona above all a place exists where
the Jews live. For some Catholics to be in the ghetto and for some Jews to
live outside the ghetto does not undo the existence of the ghetto."[4] Within
four decades, however, no more than forty-six families would still be living on
the ghetto's main street: Emigration, the rise of new communities near the
ghetto, conversions to Catholicism, name changes, and growing numbers of
secular Jews would temper much of Ninchi's fear that the ghetto had become
a permanent state of mind.

Pope Gregory XVI visited Ancona in 1841. Seven of the city's prominent traders, nearly all Jewish and among them an Almagià, had offered to
share with the papacy the cost of restoring the port's ancient wharf. The
gesture was an act of desperation. The merchants and bankers whose livelihood depended on buying and selling coffee and spices, dyes and chemicals,
and textiles were doing less business, in part because of the recent years of
political turmoil in Italy but also because the port's infrastructure—credit,
shipping insurance, ground transportation, and other services related to the
importation and exportation of goods—had been neglected. Whatever the
Jewish community's private feelings toward Gregory may have been, its
leaders made a concerted effort to get him to spend papal funds on the de-
teriorating dock area. The pope was escorted to an obelisk that had been
erected on Ancona's principal street, especially for his visit, and taken to
look at the façade of Vito Almagià's school, festooned with gauzy decorations for the occasion. Almagià, a learned man with a gift for poetry,
had been persuaded to compose a sonnet for Gregory whose first couplet
concluded, "I, too, long to quench myself in your spring." He signed this
effusion—one of several special greetings composed, printed, and presented
to the pontiff upon his arrival—"Your most respectful and faithful Israelite
subject, Vito Almagià."[5] Not long after his visit to Ancona, Gregory agreed
to the building of a new dockyard.

Vito Almagià died penniless two years later, in 1843, leaving his wife,
Fortunata Basman, and two small children, Alfonso and Angelica, ages nine
and seven. Now the man of the house, Alfonso dutifully recorded the sad
milestone in sloping handwriting in the family's Italian-Hebrew psalm book,
using the Jewish lunar calendar, 17 Sivan 5603 (corresponding to June 15,
1843), to mark the date. In the Jewish community, well-off relatives were
expected to pitch in and help their less fortunate family members; Vito's
brother Saul, one of Ancona's most prosperous Jews, stepped forward quietly
and without fuss (Saul Almagià was "a man with a big heart," his son
Edoardo once remarked) and looked after the bereaved family.[6]

Unlike his penurious brother, Saul lived in a well-appointed house on
the ghetto's main street, close by one of its two synagogues. A painting by
the city's leading artist, Francesco Podesti, adorned his ceiling, and from

his window, he could see the gray stone cathedral of S. Ciriaco, built on the summit of Monte Guasco in the form of a Greek cross and crowned at the center by a 300-foot-high, twelve-sided dome. The source of Saul's prosperity was his ownership of one of five Jewish banks in Ancona. Apprenticed in 1822 at the age of fourteen as a junior clerk to the Pereira Bank, he had quickly risen in the banking business. In 1830, having gone to work for another local Jewish bank, Gallico & Costantini, he attained the rank of attorney and ended up running the bank, which made loans and accepted deposits for foreign investment. Family lore has it that the papal authorities in Ancona looked the other way when Saul, a pillar in the Jewish community and among the first Jews to be invited to join the city's elite Casino Dorico, would stay out late at the club and return to his home in the ghetto after curfew.

Saul was the father of three sons and a daughter—Roberto, Edoardo, Vito, and Virginia. Thrown together by circumstances from childhood on, they and Vito Almagià's fatherless children, their first cousins, became life-long friends. In particular, Edoardo had a soft spot in his heart for his cousin Angelica; they stayed in touch as the years passed, and this association ultimately led to the marriage of Edoardo's daughter Virginia to Angelica's son, Vito Volterra.

Recognized early as a highly gifted young man, Edoardo Almagià was the first member of his family to go to college. His elementary education took place in a private school in Florence, where he was sent in 1852, at the age of eleven. Although Jews were no longer compelled to live in the ghetto, Ancona's public schools were still off-limits to young people from Jewish families, but in Florence, the capital of the grand duchy of Tuscany, a more tolerant attitude prevailed. Ordinarily, Jews in the Papal States were not allowed to relocate to another city; it took a special traveler's permit. Not everyone could manage to pull that off. In Ancona, as family tradition reports it, people would say, "Who do you think you are? An Almagià?"

On April 5, 1852, the day of his departure, Edoardo rose before dawn. "I was fighting between my desire to experience new things and my sorrow at leaving my father's home," he wrote in his diary. "I find myself again in front of a window at 5 in the morning, looking mechanically toward the Cathedral, but with my mind absorbed in doubts and in uncertainties never experienced before. [My] eyes filling with tears which erupted later—at the last moment, when I kissed my grandfather for the last time and exchanged affectionate embraces with my mother."[7] There is a portrait that seems to date from this period, by an unknown artist, of Saul and Edoardo. It shows Saul seated at a table with short wavy hair, and a broad, well-groomed mustache that hides his upper lip. Edoardo leans against him, in the shelter of his arm. A large map of Italy is on the table, resting on an easel, and Saul is pointing to "TUSCANIA," neatly spelled out in bold capitals. Father and son are wearing well-cut clothes—crisp white shirts and vests, discreet bow ties, somber business jackets, and contrasting striped pants. Saul's eyes are fixed on his son's face while Edoardo, one hand on his hip, the other resting on his father's lap, is looking impassively off into the distance. He would continue his studies in Florence under the guidance of Cesare Scartabelli, professor of literature at the University of Florence; later he would enroll in the University of Pisa, from which he received his doctorate in applied mathematics in 1861, one year after the birth of Vito Volterra.

Edoardo was still a student in Pisa when his cousin Angelica became engaged to Abramo Sabato Volterra. In the tradition of the day it was an arranged union, whose details were spelled out in an elegantly penned nuptial agreement negotiated between the two families, signed and notarized in Ancona in 1857. The wedding itself took place two years later, on March 14, 1859, in a religious ceremony. The bride brought to the marriage a dowry of cash, clothing, sheets, and gold jewelry; every garment, bath accessory, and earring was listed and described in an inventory appended to the nuptial agreement. Her brother, Alfonso, agreed to pay for her trousseau and pledged additional cash, to be given to the groom in four equal installments, starting with the first wedding anniversary. These business arrangements uniting the couple were concluded the day before the wedding when in front of the notary, Angelica's uncles Saul and Leone handed the couple the sum of 1,000 *scudi* in silver and gold coins (a *scudo* being roughly equal to a dollar). Abramo added a sack of Jewish coins—an old Jewish custom in Ancona— plus cash equal to twenty percent of the bride's dowry and bath linen for the bride's trousseau valued at seventy *scudi*. This was no small sacrifice on his part. Born into a poor family, Abramo, like his father before him, made his living buying and selling cloth, one of the few trades permitted to Jews who lived in cities under papal rule. With a cash dowry of 1,400 *scudi* in his pocket, an educated young wife at his side, and the pious wedding verses of his extended family ringing in his ears, Abramo faced the future. It would be brief: He would die suddenly, of unknown causes, in 1862. But, on the auspicious occasion of his marriage, his older brother Mosès summed up the family's hopes for him in a letter, in which he recalled the many financial hardships endured by their family and counseled Abramo to make Angelica happy, for she had "abandoned the name of her father to take that of our Family." He concluded by expressing the collective hope that "you give life to new intelligent children."[8]

Samuel Giuseppe Vito Volterra was born a little over a year later, on May 3, 1860, in Via Astagno 19. At birth, the newborn's name was recorded in the annals of the Jewish community's archive. The house that Angelica and Abramo lived in, which was later deeded to the Jewish community, belonged to Saul Almagià. Later that year when the Italian soldiers laid siege to Ancona under General Cialdini, a shell shattered the wall of the house, destroying Vito's cradle. Fortunately, mother and infant were else-where at the time. Angelica, a deeply religious woman, would remember it as a miraculous escape. Soon afterward, Victor Emmanuel's army over-whelmed the pope's troops, and by the end of September the Italians had entered Ancona. Angelica's Uncle Saul, recently elected deputy (the official representative) of the city's Jewish community, was glad to see the end of papal authority. "I have already presented myself to all the authorities, and one of these days I will have to pay a visit to our King, who comforts us for several days by his presence," he wrote to his son Roberto, in Trieste. Saul saved the best piece of news for last: "The decree for our emancipation is already fixed, and for we Jews therefore a new era is dawning."[9] Unlike the generations of his family before him, Samuel Giuseppe Vito Volterra would start life with the same rights and privileges as other Italians.

Volterra's full name reflects his Italian-Jewish heritage. Among Italians of Jewish descent, the custom of taking their city of origin as the family name originated in the second half of the sixteenth century, when Pope Pius V expelled the Jews from all the cities in the Papal States except Rome and Ancona. From the ancient Etruscan city of Volterra in Tuscany, where they had lived for more than a century, Volterra's paternal ancestors fanned out to other cities across the Italian peninsula, including Ancona. After the death of her husband Abramo in 1862, Angelica instructed the court to transfer her inheritance to their infant son, "Samuel Giuseppe known as Vito Volterra." From then on, he went through life as Vito Volterra. While some Jewish traditions, like naming a baby after a relative who has died, were subsequently exported around the world, Volterra's grave marker does not follow the venerable custom. In the local cemetery in Ariccia, where he is buried, the only name chiseled on the tombstone is "Vito Volterra," nothing more.

Shortly after the end of World War II, Volterra's colleagues and family, and Ancona officials organized the first commemoration there in honor of the city's illustrious native son. The first step was to find out whether the house in which Volterra was born still existed so that a commemorative plaque could be mounted on the exterior. A cousin by marriage, Vito Terni, consulted Volterra's eldest son, Edoardo, who was a law professor in Bologna. On the copy of the birth certificate that Edoardo Volterra sent Terni, yet another traditional name appears—one that posed a problem for Edoardo, as his accompanying letter reveals: "I beg you, most emphatically not to use the name of Isacar on the plaque" Edoardo wrote. He issued similar instructions to his mother, Virginia: "I enclose the two copies of the [birth] document....But it's absolutely essential that the name Isacar not appear on the plaque."[10] He need not have worried. By then, nothing was left of the old Ancona Jewish quarter except a synagogue—now boarded up—on once vibrant Via Astagno, itself now no more than a dark, narrow, ghostly lane off a busy downtown thoroughfare. Volterra's birthplace, no. 19, had been completely destroyed by aerial bombing during the war. In the end, the city fathers named the extension of Via Maratta, as it approaches the sea, Via Vito Volterra. Several years later the provincial government named several technical schools in the area after Volterra.

The irony is that Vito Volterra never thought of Ancona as his home-town, and in fact he lived there, off and on, only for the first five years of his life. After his father died, Vito and his mother were penniless. A year later, they moved in with Angelica's brother and in 1863, they left Ancona with Alfonso, who eventually settled in Florence. It was this exquisite Renaissance city on the Arno that would come to hold a special place of honor in Volterra's heart.

CHAPTER 2

"This, Above All, I Promise," 1863-1870

Vito was three and fatherless when he and his mother, Angelica, moved in with her brother, Alfonso, and her mother, Fortunata. At twenty-nine, Alfonso was single, his fiancée having died earlier that year. The outlines of a new domestic arrangement quickly took shape: Fortunata and Angelica kept house, while Alfonso was the breadwinner and de facto head of the household. He paid the bills, provided his sister and mother with spending money, and loomed large in his nephew Vito's life.

In the fall of 1863, Alfonso moved the family from Ancona to Terni, a town in Umbria, where he had started working for a railroad construction company. The firm, one of many private companies armed with economic concessions and guarantees from the new state to lay railroad tracks, in 1864 dispatched two of its engineers to work on train station projects elsewhere. This enabled Alfonso to move his family into the rooms formerly occupied by one of the men. Angelica was delighted with her new quarters, situated next to Alfonso's office. While not the best apartment in the building (the windows did not face the town's main street), it did come with a terrace, where Vito could play. More important, Angelica could now "distance [herself] from the rather coarse people" in whose midst she had been obliged to live up until then. Her new neighbors, she confided by letter to her wealthy cousin Virginia back in Ancona, were "much more educated."[1]

In January 1865, Alfonso moved to Piedmont to work in the central office of the Italian National Bank (today the Bank of Italy), headquartered in Turin; and Vito, his mother, and grandmother returned temporarily to Ancona. Angelica had already taken Vito's education in hand, having taught him to write when he was three. Soon he was signing his name and adding a brief message ("Vito gives you many kisses" or "Uncle, I'm being good") to his mother's letters. "What do you think?" Angelica wrote to Alfonso and her cousin Virginia. "It's a little scribbled. But what do you want" from a four-year-old? Angelica was every son's Italian-Jewish mother, and Vito learned at an early age what it took to please her. Nevertheless, he was not the most willing of pupils, and Angelica confessed that she found it very tiring to make him pay attention to her. She was certain that if a teacher had instructed him instead, he would have made much greater progress, for Vito had what his mother called a "rare intelligence and a great facility for learning."[2]

As with many gifted children, Vito evinced special abilities early. As a small child, he effortlessly learned poems by heart and quickly advanced to word games that involved remembering and continuing the thread of a conversation as it traveled from person to person around the room. At a party one evening in Ancona, Vito, then not quite five, joined other young people in one such game. The adults who had gathered round to listen were astonished at his proficiency. "You think it such a small matter to compete against twelve-year-old boys?" Angelica proudly wrote her brother afterward, and being a pious woman, she added a small prayer: "May God keep him always for our consolation."[3]

That July, Alfonso was transferred to the Italian Bank's new central headquarters in Florence, which had replaced Turin as the provisional capital of the Kingdom of Italy. Angelica, Vito, and Fortunata soon joined him there. Unlike Turin—which looked the part of a national capital, with broad, straight streets neatly laid out at right angles—Florence in 1865 was still an untidy medieval city, hemmed in by its stout and ancient walls. Save for the main railroad station, constructed in 1847 in the orchard behind the church of S. Maria Novella, the buildings, squares, twisting passageways, and bustling market situated in the heart of the city would not have looked unfamiliar to Michelangelo, Machiavelli, or Galileo. To the politicians and French diplomats who had orchestrated the transfer of the seat of government from Piedmont to Tuscany, Florence was a temporary but necessary stopping point on the way to Rome.

Vito's early residence in the city coincided with a visit that Mark Twain made there in the 1860s in the course of a round-the-world trip. America's premier author characterized his stay in the newly anointed capital of Italy as "chiefly unpleasant" and perhaps not surprisingly, given his life-long attachment to the Mississippi River, dismissed the river Arno, which flows through the city, as "a great historical creek. They [the Florentines] even help out the delusion by building bridges over it. I do not see why they are too good to wade." Instead of appreciating the engineering skill involved, Twain also complained about the three miles of railroad tunnels that deprived him of a view of the surrounding hills as he entered the city by train. To top it off, he got lost in Florence one night. As he later related in *Innocents Abroad*, he "staid lost in that labyrinth of narrow streets and long rows of vast buildings that look all alike, until toward three o'clock in the morning, I grew accustomed to prowling about mysterious drifts and tunnels and astonishing and interesting myself with coming around corners expecting to find the hotel staring me in the face, and not finding it doing any thing of the kind. At last I came unexpectedly to one of the city gates. The soldiers thought I wanted to leave the city, and they sprang up and barred the way with their muskets. I said: 'Hotel d'Europe!'"—the only foreign words Twain knew. The King's soldiers barely understood him. By the time he got to sleep in his hotel bed, Twain had come to appreciate the simple joy "of comprehending every single word that is said, and knowing that every word one says in return will be understood as well!"[4]

Angelica's adjustment to Florence was not easy either. Between 1865 and 1870, more than 30,000 workers like Alfonso Almagià would descend on the city with the King, all needing to find housing for themselves and their families right away. Prices rose sharply, from the cost of food to the monthly rent on an apartment. To accommodate the newcomers, thousands of new dwellings sprang up, eating away at much of the undeveloped land available within the city's walls. Older buildings were divided, with floors and stairs added, to shelter whole families in spaces once reserved for a few people. Florence was crowded, dusty, and noisy—cold in the winter, hot in the summer. Angelica, just turning thirty, found herself homesick for Ancona.

One Sunday in August of 1866, to escape the suffocating heat, the whole family went up to Fiesole, a village in the hills just north of the city. Walking in the open fields and filling her lungs with fresh air rejuvenated Angelica. It "seemed like I could re-create myself," she wrote to her cousin Virginia, thanks to the fresh smells, "which are not easily enjoyed by anyone living in a big city." Florence, she reported, seemed cramped, like "a hen house in an abyss," when viewed from on high. The villas surrounding Fiesole captivated her, however—in part because some of them, as she noted proudly, belonged to fellow Jews.[5]

Several days later, mother and son went with friends on another excursion, this time a picnic, in a field not very far from where their friends lived. "I imagine you are pleased to hear that I am having some fun!" she wrote to Virginia, who had a summer house outside Ancona. "Here"—she wrote with perhaps a touch of envy—"the wives are returning from the seashore and preparing now to go on their country holidays." From time to time, friends and family from Ancona visited Angelica in Florence, filling her ears with hometown stories and gossip. Such news was better than nothing, but it only made her feel more isolated. "[E]ven that is not enough to satisfy the intense desire that I feel to see you, dear Virginia, and my other relatives, and to savor everyone's presence," she blurted out in a letter written in the fall of 1866. "I dream always that this wish (God willing a favorable opportunity) will come true soon."[6]

In the summer of 1869, Angelica, her mother, and son went back to Ancona for the longed-for visit. They stayed with Cousin Virginia and her family, and Alfonso remained in Florence. "My Vito is very well and enjoying himself immensely," Angelica wrote to her brother. "[H]is appetite is something to behold and his coloring has improved."[7] Vito could do no wrong that summer. He was well behaved, eager to please, polite to his younger cousins. Virginia's husband, Peppino Terni, took Vito under his wing. He gave him columns of figures to add up and monitored Vito's schoolwork—compositions and arithmetic problems that Vito's teacher in Florence had given him to do over the summer. Virginia read and corrected

Vito's translation of a French story into Italian. The boy also worked on his French dictation, read books ranging from Greek history to Dante, and wrote verse. He was nine years old.

Angelica couldn't wait to show off her son. Everyone Angelica visited, it seemed, wanted to hear Vito speak Italian the way it was spoken in Florence. "Read to us," Angelica would say. Vito willingly obliged. Book in hand, he would begin reading aloud a short story, the words rolling off his tongue in a pure Tuscan accent. The adults loved it. "If Vito continues to be like this," Angelica commented in yet another letter to her brother, "what satisfaction awaits me ahead."[8]

Meanwhile, Vito was blossoming into a gregarious, witty, and observant correspondent, eager to try out his newly attained literary skills. Whereas Angelica's letters to Alfonso that summer largely deal with domestic matters in Florence—instructions for the maid, Matilde ("Tell her to make sure the peppers don't spoil. I'm trusting that we won't find anything missing when we return"); messages for the neighbors ("Send our greetings to Mr. Federico. Ask Mrs. Anielli how her little grandson is doing"); and tasks for Alfonso ("I'm entrusting you to choose a nice present for Virginia, whose hospitality is unbounded")—Vito's letters to his uncle provide a vivid picture of a day in the life of the Almagià family in Ancona seen through the eyes of a sharp and receptive nine-year-old. In one of his longer letters, Vito describes in detail a morning spent strolling through the city's annual watermelon and straw fair, the stalls overflowing with fruit, baskets, hats, and toys. The adults in the party—Virginia, her father Saul, and her brothers Roberto and Edoardo—showered small gifts (dolls, construction paper, noisy tin drums) on all the children, including Vito. They returned home for lunch, followed some hours later, according to Vito, by "a wonderful dinner," capped off with a dessert he had never before tasted "called 'Strudel,' consisting," he explained to his uncle, of thin layers of dough filled with pears and raisins.[9]

More happy surprises lay in store for Vito in the coming weeks. He took his first sea voyage. The occasion was a visit to one of the town's offshore bathing establishments, known as *bagni galleggianti* (floating baths), houselike structures protruding from the water. To get to them, Vito and the family took a boat at the Marinelli seashore bathing concession. The ride lasted some time. "When we stopped in the middle of the sea," he later reported to his uncle, he was overcome with "joy to see the sea quiet like oil, and to feel a freshness that one feels only at sea." Then everyone got off the boat and inspected the bathing facilities. People would go bathing inside these little huts, which sheltered them from prying eyes. Apparently Vito did not go for a swim. Still, the day ended on a splendid note—the family returned to Cousin Virginia's house in a horse-drawn carriage.[10]

Like the rest of the family, young Vito mostly got around town on foot. A hike to a scenic spot above the city brought out the nature lover in the impressionable boy—"the serene sky and the beautiful fields and the sea always at the bottom," he exclaimed in a letter to Alfonso—and a taste for the outdoors that he never lost. Continuing with his description of how the day was spent, Vito once again displayed what can only be called a precocious enthusiasm for the local fare:

> As usual, we walked there the whole way. I was dead tired when we arrived at the villa and I sat down and had a lemonade, then I had to go see the garden. Then the open fields. At the end we found ourselves in a square where we sat down. I asked the watermelon man for a watermelon large enough for 24 people. After we returned home I ate a little more watermelon.[11]

When there was no room on the tissue-thin paper for Vito to write about how he spent his days, he joked about it. "My dearest Uncle," he wrote on one occasion, squeezing in a couple of lines at the end of his mother's letter, "You see how little space mama has left for me to write you? And I wanted to tell you so many things! Patience! Another time I will recount to you everything I wanted to tell you. . . ."[12]

Watching her son enjoy himself and relishing the hospitality that her family lavished on her in Ancona, Angelica found reasons to prolong their stay. To leave the port city of her birth so soon after arriving there, she wrote her brother in mid-August, struck her as a bit hasty, particularly when there were so many new things to see and just when "it seems to me I'm becoming another person." But in the very next line she worried aloud about abusing the generosity of their relatives; they really should return to Florence at the end of the month, as agreed upon.[13]

Her cousin Virginia thought otherwise. In a letter to Alfonso on August 22, 1869, Virginia asserted that "our dearest Angelica has improved, really improved since she arrived here," adding, "The change in her is so noticeable as to be perceived not only by those who saw her again after the first days, but also to us who are always with her. To me it seems that she is, both in her physical appearance, and in her spirits, as she was in those sweet years in which we were neighbors."[14] In other words, could Alfonso have forgotten already that the cousins had been playmates who had practically grown up together after the death in 1843 of Alfonso's father? Surely Alfonso could appreciate the advantages that would flow from his sister's remaining in Ancona for several more weeks. Alfonso was not appreciative. The Jewish New Year was approaching, and his sister's absence meant that for the first time he would be spending this most important holiday apart from his immediate family. He responded by giving them all the silent treatment for a while.

"Mama tells me that we will remain here another 15 days," Vito wrote in haste to his uncle at the end of that summer. "I can't write any more now because they're all waiting for me to go on a walk." In their notes to Alfonso, his mother and sister tried to put the best face on Cousin Virginia's small victory. It wasn't their idea, Fortunata wrote to her son, to stay on in Ancona—their hosts had forced their hand. She continued, "You cannot believe how sorry it makes me to be far away from you especially in the first days of our New Year." Troubled by Alfonso's silence, Angelica begged him to write "as often as you can if you want us to be calm." At the end of this particular letter, Cousin Roberto joined the chorus: "You think it's nice to cause your family pain? I can't believe that you have not written today."[15]

Relations had thawed enough by early September that Alfonso sent Virginia a saddle for her horse and enclosed a letter. In her thank-you note, she lamented how little she had done to merit such a splendid gift. "Of our sacrifices (as you would call it) we have been hugely compensated by the pleasure of the company of your beloved family that I had been struggling to embrace again for such a long time." Like everyone else in the family, Virginia wished her cousin many years of long life and happiness in the New Year. Vito followed suit with his own New Year's letters. To Alfonso he wished "a thousand years of life," and asked his uncle to forgive him all the imperfections of the past year, promising in the New Year to correct his mistakes and "to compensate you for all that you do for me with my good behavior." In his holiday greeting to his mother, Vito swore to be always good, obedient, and studious—"This above all I promise."[16]

The whole family, Vito included, spent the day before their departure making wine; that evening they pressed the grapes, and Vito, with Angelica's permission ("Mama says it's good for chapped skin"), did his share of stomping in the huge tub. At 9:30 the next morning, they took the train to Bologna, where they stayed with friends for a few days, before making their way back to Florence.

Life changed dramatically for Vito after that summer. In April 1870, Alfonso married a young woman from Florence named Ester Supino. The following spring he rented a place outside the city for the family, which now included Ester's kin, to use during the summer, creating one less reason for Vito and his mother to vacation in Ancona. In the meantime, Alfonso and Ester started raising a family. Seven children were eventually born, four of whom lived, and the family's lifestyle became increasingly peripatetic. Between the years 1870 and 1880, Vito and his relatives lived in three houses: Via dei Conti 13, a corner building in central Florence overlooking Piazza San Lorenzo; Piazza degli Adimari 7, not far from the Palazzo Vecchio and the Duomo; and Via delle Farine 1, a handsome building on a small street directly opposite the Palazzo Vecchio. Meanwhile, the ongoing cost of feeding, clothing, and educating Vito had emerged as an issue. In 1877, with Alfonso's fourth child on the way and his strong-minded nephew now in his last year of high school, a pitched and shrill battle of words took place in the Via delle Farine apartment over the boy's future. The referee was none other than Angelica's beloved cousin Edoardo Almagià, now a seasoned and tough-minded civil engineer and the family's esteemed college graduate.

CHAPTER 3

"That Damned Passion," 1874-1877

The war of words over Vito's future raged unabated for more than three years, starting in 1874, when he became a student at the local technical high school, and grew more heated with each passing year. His announcement at about the age of twelve that he was going to be a mathematician when he grew up sparked endless bickering at home. Angelica badly wanted him to choose a more practical line of work. Her brother, Alfonso, working long hours to support not only his wife and children, but his mother, sister, and nephew, thought that Vito needed a basic lesson in economics, not a degree from the University of Pisa, where he had his heart set on going after high school. Alfonso and Angelica first brought their cousin Edoardo Almagià into the picture in the summer of 1875, hoping to enlist him on their side. "I make Vito come to the bank every day where I give him some work to do," Alfonso confided in a letter to Edoardo that August. But Vito took no pleasure in that line of work, his uncle added. "God grant that he might overcome his disgust for anything that takes him away from his beloved science. If at least he would show an inclination to apply himself to some technical branch, such as engineering, mechanics, land surveying and the like, I would try to please him, but that damned passion for the study of pure mathematics discourages me."[1]

"That damned passion," as Alfonso called it, had deep roots. His first taste of science, Volterra recalled many years later, came from reading a children's book by Jean Macé, a prominent educational reformer in mid-nineteenth-century France. Organized as a series of letters to a young girl and dealing with the vital processes in humans and animals, Macé's *Histoire d' une bouchée de pain* ("Story of a mouthful of bread") inspired the intro-spective youngster, then nine, to examine more closely the natural world around him and deduce, if possible, the laws underlying the phenomena he observed. According to a story published years later, it was around this time that he discovered that the oscillations produced by the torsion of a string are isochronous, like those of a pendulum—that is, each swing takes the same amount of time independent of its amplitude. Only later, when he took a course in physics, did he learn that his discovery had been made long before.[2] This account unfortunately provides no further details about the circumstances that might have helped a young boy to this conclusion (it appears in an anonymously authored memoir of Volterra that was published in 1941 in Argentina, most likely by the French mathematician Joseph Pérès.)*

Vito read Macé's book in French for the simple reason that such works were seldom translated into Italian. Right up to the end of the nineteenth century, only about 600,000 Italians in a country of 25 million spoke and wrote the language commonly regarded today as Italian. (Judging by their correspondence, this minority included Vito and his family.) The great majority of the population was illiterate—compulsory elementary-school attendance was not established until 1879—and most Italians spoke one of a multitude of local dialects. Moreover, the whole cultural orientation of the day (from philosophy to history to science) revolved around France and Germany, with England not far behind. Even the menus at banquets, from the Florentine aristocratic court to routine city hall functions, were printed in French and featured French cuisine. Growing up in this provincial intellectual milieu, Vito and other scientifically minded students had little choice but to master French and German at an early age.

By the time he was eleven years old, Vito was reading mathematics books on his own. The curriculum at the Scuola Tecnica Dante Alighieri, the junior high school in Florence he attended for three years, covered a broad spectrum of subjects: Italian and French; arithmetic, algebra, and elementary geometry; geography; history; drawing and penmanship; and course work in the natural and physical sciences. His favorite authors included the contemporary French mathematician Joseph L. F. Bertrand, who wrote a popular secondary-school textbook on arithmetic that Vito read from cover to cover, and the eighteenth-century French mathematician A. M. Legendre, whose lucid exposition of Euclid's elements of geometry was still a classroom favorite in the 1870s.

Using mathematics to understand how the world works fueled Vito's imagination. In 1873, after he finished reading Jules Verne's *Autour de la lune*, the budding mathematician decided to calculate the trajectory of the rocket launched from the earth to the moon in the story, thus unwittingly confronting the notorious three-body problem that had confounded mathematics since Newton's time. He divided the time traveled by the rocket into small intervals. After taking into account the combined gravitational force of the earth and the moon on each interval, he found that the trajectory could be pictured as a series of very small parabolic arcs. In effect, he had found an approximate solution to the infernal problem in this special case. He was then thirteen years old. Thirty-nine years later, in 1912, Volterra would demonstrate this very solution in a lecture on "the evolution of fundamental ideas about the infinitesimal calculus" at the Sorbonne.[3]

In 1874, at the age of fourteen, Vito Volterra enrolled in the Istituto Tecnico Galileo Galilei, the local technical high school, on Via San Gallo (today

---

*Dr. Giovanni Paoloni, e-mail to author, July 11, 2006.

Via Giusti), which specialized in agronomy, business, industrial training, and other technical subjects, including mathematics and physics. Unlike the *liceo* (classical high school), where Latin and Greek reigned, the Istituto Galilei and other technical secondary schools that had sprung up in Italy after 1859 offered German and English. After reading Michael Faraday's *Chemical History of a Candle*, Vito fell in love briefly with chemistry and physics. He organized a small laboratory at home and performed some scientific experiments. But "the fixed idea," as his uncle Alfonso put it, of becoming a mathematician could not be unfixed; indeed, as the pressure built within the family for Vito to quit high school after the first year and find a job, Vito became more determined than ever to finish the Istituto Tecnico's three-year program and get his diploma in mathematical-physical studies.

An uneasy truce had settled over Alfonso's household in the fall of 1875 when Cousin Edoardo Almagià, who had started his own railroad construction business in southern Italy in 1867, came to visit. It did not take him long to observe that the once-sparkling young boy was now a sulky, defiant young man, determined to resist his mother and uncle's insistence that he give up his studies and "devote himself to a job." When Angelica and Alfonso told Edoardo that they could not afford to keep Vito in school on Alfonso's tiny income, Edoardo assured them that he and his brothers would provide the financial help that Vito needed to complete his education. He also took Vito aside for a private chat and before heading south a few days later, he left him an envelope and a note: "Here is 60 *lire* [roughly the cost of student housing for a couple of months in Tuscany at that time] which might be of help to you in registration at the 'Istituto.' Use it as you see fit; but don't forget my advice, whatever you decide to do."[4] While no record of their earlier private conversation exists, it appears from some of the comments Edoardo later made in his letters to Alfonso that he urged Vito to curb his temper and pay more attention to the sort of mathematical calculations that civil engineers, such as himself, used in the real world. In the escalating domestic war over Vito's future, Alfonso and Angelica had lost the first battle but won a moral victory.

It was around this time, in 1875, that Vito on his own began reading Bertrand's two-volume treatise on calculus. For fun, and working entirely independently, he prepared a report (now lost) on a geometric transformation of his own invention, which his mathematics teacher—one Giacomo Bellacchi, trained at the Scuola Normale Superiore at Pisa and a respected textbook author—praised profusely. As his studies progressed, Vito also calculated the centers of gravity and moments of inertia of different bodies. Applying what he had learned from Bertrand's book, he divided each body into very small parts and then applied the method of limits. In autobiographical notes prepared many years later, Volterra remarked that as he worked through the problems, he realized that the process of integration was the inverse of differentiation, a discovery first appreciated by Newton and

Leibniz.[5] Armed with this insight, he proceeded to solve a great many problems, until his progress was stopped by an encounter with a more advanced form, an elliptic integral.

In July 1876, Vito came down with rheumatic fever, but he bounced back by the fall and breezed through his makeup examinations. That November, with the registration deadline for the third and last year at the Istituto Tecnico fast approaching, Alfonso appealed to Edoardo to help find a suitable job for Vito. "I don't want to annoy you," he wrote to Edoardo, "by airing things which you know in part, and in part I wouldn't be able to explain to you clearly how things stand except in person." Instead of his financial condition improving over the past year, Alfonso continued, "it has become worse, since there have been extraordinary expenses of no small consequence for someone like me who has a small income and very limited resources." Under the circumstances, it was imperative that Vito find employment quickly, but he continued to look down on going to work. What Alfonso had to say next, however, makes it clear that whatever his limitations as an intellectual mentor, this conscientious, harried, overworked uncle possessed genuine insight into his nephew's temperament and aspirations.

> [T]he kind of job I would be able to find for him here is too different from his inclinations for him to adjust to, especially in the beginning.... You would need somebody who knew him well and would be willing to overlook initially the kind of repugnance he would show, and it would be necessary to find the kind of job that would not be too far from his beloved studies and in which his knowledge could find, even at a distance, some application, and he could take satisfaction in putting to use in some way the knowledge he's gained.[6]

Alfonso then added that Vito himself had made a proposal: If his Uncle Alfonso would allow him to finish up at school, he would then ask Edoardo to take him into the construction business. Alfonso seems to have taken this with a grain of salt. "I don't know if he says this in order to be sure to be allowed to carry on his studies," he wrote to Edoardo, "but certainly a job with you would please him much more than a job with a shopkeeper or a banker." Angelica ("who has no other assets or hope except in her son,") had reconciled herself to the prospect of living far away from her only child. And if Edoardo would be willing "to do us this favor" and take Vito under his wing, what would prevent him from joining Edoardo immediately? Did Vito really need another year of school? In any event, Alfonso had decided not to allow his nephew to enroll for the third year at the Istituto Tecnico. "I don't have time left even to ask your advice. [Just] help me out with [Vito's proposal]," Alfonso pleaded.[7]

Edoardo answered Alfonso's letter a week later, on November 20. He had been traveling in the Campania region, visiting Salerno and finishing up a project in nearby Contursi—which accounted for the delay—before rushing back to Ancona for the first birthday of his daughter Virginia (who would grow up to marry Vito in 1900). "I have really been a wandering Jew—as between Rome, Chieti, Potenza," he wrote, "because I go here and there looking for work, but there is so much competition on the one hand and perhaps too much prudence on my part on the other that up to now I have nothing settled." Under the circumstances, he continued, choosing his words carefully so as not to antagonize Alfonso, "I would not disapprove of Vito's completing the 3rd year of the Istituto Tecnico, because that way he will obtain a diploma, which could be very advantageous to him in starting a technical career." Edoardo explained the virtues of his plan:

> It is true that a year of further study and therefore of expense rather than earnings is the difference, as our businessmen succinctly call it, between profit and loss; and I well understand how in your condition it is impossible to accept this solution. And it is for this reason and also to satisfy a desire of Vito's, which from one point of view we cannot blame him for having, that I would return to the old proposal; that is, I would like Vito to finish at least the technical course, without your having to bear the full cost, and if you do not oppose it, I will ask Daniele [a cousin] and Vito [Edoardo's brother] to help as well, because as close relatives of the fatherless Angelica we should do everything for her that she deserves.[8]

Edoardo promised Alfonso that after Vito's graduation from the Istituto Tecnico, he would find a job for Vito on one of his construction projects, or failing that, he would work something else out.

As it happened—perhaps because he had anticipated what Edoardo's response would be, or perhaps because of the rueful understanding of his nephew reflected in his earlier letter—Alfonso had not followed through on his threat. At the last moment he had agreed to let Vito begin his third year of studies. But he extracted from him a "sacred promise" that at the end of the academic year, or sooner if the opportunity arose, he would follow the wishes of his family. As for accepting money from Edoardo and the other cousins, Alfonso seemed mortally offended by the suggestion. "So long as he is with me and so long as I personally agree to his continuing his schooling, I could never accept your assistance," he wrote Edoardo.[9]

In 1877, during Vito's last year in the physics-mathematics program at the Istituto Tecnico, he took a class in geometry from Cesare Arzelà. Like many freshly minted graduates of the Scuola Normale, Arzelà would teach high school mathematics for several years before moving up the Italian academic ladder, from the chair of algebra at the University of Palermo in 1878 to the chair of calculus at Bologna two years later. As Alessandro Faedo, the rector of the University of Pisa, recalled in a speech in Rome on the hundredth anniversary of Volterra's birth, Arzelà was still feeling his way as a secondary-school teacher when an inspector from the Ministry of Public Instruction walked unannounced into his calculus classroom at the Istituto Tecnico, stepped to the front of the class, and wrote down a problem in geometry on the blackboard. "Now show me how you solve it," the inspector said, turning to the instructor. Intimidated by the inspector's presence, and not wanting to make a bad impression in front of his pupils, Arzelà stumbled around and was on the brink of making a complete mess of things when suddenly a gangly teen-ager raised his hand. "I know how to solve the problem," he said, and proceeded to do so. This was Vito, flexing his powers and not in a modest mood.[10]

That final year in high school, Vito also took classes in analytic and descriptive geometry, mechanics, and physics. As part of the school's hands-on approach to the engineering arts, he spent more than fourteen hours a week practicing linear drawing, free-hand drawing of machinery, and drawing in descriptive geometry. Edoardo Almagià had advised Vito both to improve his handwriting ("Nowadays it is the key to the front door that gets us into the business world") and to round out his program of studies. Having dutifully followed these suggestions, Vito completed the academic year, passed all his examinations in the summer of 1877, turned in a sterling performance in mathematics ("with distinction" the examiners noted), and received his diploma in the natural sciences. All the while, the family waited impatiently for their masterful, prosperous, globe-trotting cousin to summon Vito to his side.

The summons never came. For every letter in which Alfonso inquired about when to dispatch Vito south to begin his career as a railroad engineer's assistant, Edoardo responded with a multitude of reasons for delaying Vito's departure. "A thousand times I had meant to write you," he began one letter to Alfonso from Romagnano, in June of 1877,

> but there was no time, the occasion, etc., and in short I still hadn't written and I am very sorry about this. Your letter arrived with an extraordinary delay because it went to Ancona while I was here; then it was forwarded here while I was in Rome and arrived there when I was terribly busy. Then what happened is I misplaced your letter; finally tonight I have a moment of quiet and here I am with you.

Edoardo had secured a government contract to do the construction on the Contursi-Romagnano trunk of the major railroad line between Salerno and Taranto. Romagnano, his base of operations, was a remote medieval town of around 1,000 inhabitants, built on the slope of a hill in the Campania region, inland and south of Salerno, on the Tyrrhenian side of the Italian peninsula. Earthquakes, bandits, and disease were the stuff of local life. Edoardo had nothing kind to say about his temporary quarters.

> Can you imagine that I live in a house above the Romagnano train station which has three rooms below and three above. The three below: kitchen, small dining room, and office. The

three above: two holes used as a dormitory for three of my clerks and one room that I share with Cesaroni [Edoardo's business partner] and Manni [Edoardo's accountant]. I had to add a shed as an annex for another office room. Now that summer is approaching, the heat is unbearable and it is not wise to remain at such a low altitude on account of the malaria and so we are looking for a house in Balvano, a little town that is higher up and promises good air. But since the government clerks, the pieceworkers, and others who depend on our construction projects have already occupied all the lodging available, it is difficult for me to find a house.

If Alfonso and Angelica still fancied Vito entering "this life of work"—and here Edoardo emphasized the word "work"—Edoardo would figure out a way to give Vito, "our able computer," a shot at the job. "You can be sure that we will alternate his office work with that in the open country, but only," he warned, "when the air will permit it." Then a curious thing happened: Edoardo neglected to mail his letter, dated June 13, 1877, until a good two months after he wrote it.[11]

The most charitable explanation is that Edoardo, himself a graduate of the University of Pisa, wanted to see Vito continue his education. In late August, when another flurry of letters from his cousins in Florence convinced him that they had no intention of abandoning their plan to apprentice Vito to him, he sat down and composed a fresh letter. He started it on the reverse of the June 13 letter, and when he had finished he sent both letters on their way. "You must believe," he wrote to Alfonso, "that these were two months in which I suffered so much that my head was no longer on my shoulders, but I don't know where it was." His problems had begun with the death of his business partner's daughter, followed by the sudden illness of two accountants, coupled with the usual ups and downs of work, leaving Edoardo to do everything alone. He complained bitterly about his railroad contract—the need to hurry the project along, the bad air, the substandard accommodations for his office workers—and he cursed his bad luck and the work in the same breath. "Damn this world to hell," he wrote Alfonso. "[T]he fact is, I curse the moment when I took this project. I would have preferred to break a leg the first time I came to see [Contursi]. By this time it would have healed."

Convinced that Alfonso (and perhaps Vito, too) nursed a romantic and glorified picture of a railroad engineer's life in rural southern Italy, Edoardo described the primitive conditions daily facing his engineers. "Now I am at Balvano—what a town! What a horror! We have been able to find only one lodging, of some 4 or 5 rooms in a shed that they call a *palazzo* and we are always obliged to sleep two in a room." His office workers slept under worse conditions ("The five clerks are at the top in a convent and I can assure you that it is not a place where one could let a young man live who is leaving home for the first time"). Getting to work and back also seemed straight out of an earlier century: "These are cursed places, there really aren't any roads, those few being paths for mules through the mountains and very bad. One travels on a mule and then descends on foot to the construction site." Malaria, he reminded Alfonso, remained a public health problem; the town's supply of water had almost dried up and what little there was contained worms. "We drink wine," wrote Edoardo, "but what about Vito?"[12]

What else could Edoardo tell Alfonso, his cousin with the steady, white-collar job in a bank, in a city full of genuine palazzos, about his unpredictable, chaotic, and dangerous life working on the railroad in the impoverished backwaters of southern Italy? Edoardo knew the towns and villages in these provinces well, having gone from one to the next on horseback or in a carriage, looking for work ten years earlier. In one town, he had seen a handsome bridge built in the foundries of Naples, when the Bourbons ruled the land without any assistance from the outside world. "This means," he wrote in his diary, "that in these provinces there is the possibility and the ability of a renaissance, with no need for others to try and make them civilized. The problem of these provinces is the decayed bourbon despotism. Once without that, time will cancel the bad effects and the 'mezzogiorno' will be the best part of Italy."[13]

Edoardo had written these optimistic words in 1867, the same year he started his own construction business. But by 1877, after sixteen years as a builder of railroads, he had grown weary of the endless quest for new construction projects, the uncertainty of the work, and especially of living for long periods of time in desperately poor, drought-stricken towns far from his wife and children in Ancona. "Look, I am now so tired," he told Alfonso in one letter that summer, "that I say I don't want to know anything anymore and want to give up business, but having said it many times and not having done it, *perhaps* I will not do it even this time, but will I ever get the opportunities?"

Given Edoardo's state of mind and the harsh conditions under which he labored, it is not surprising that he once again encouraged Alfonso to allow Vito to continue his schooling. "Practice and experience alone will never make a complete engineer," he assured Alfonso. Vito, in his opinion, would benefit from one or two more years of advanced technical training at a specialized school for engineers.

Engineering schools of the type Edoardo had in mind for Vito had only recently sprung up in Italy, mainly in the north. Vito could choose from among several, including the Istituto Tecnico Superiore in Milan, the forerunner of the Milan Polytechnic; the Scuola di Applicazione per gli Ingegneri in Turin (Edoardo had heard good things about that school's French director, and the engineers trained there had no trouble getting jobs with an Italian firm that made steel beams for roofs and bridges); and the Istituto Tecnico at Fermo (Edoardo could vouch personally for the quality of the engineers its faculty turned out), a city situated in the Marches, south of Ancona.[14]

In the summer of 1877, while Edoardo and Alfonso were each committing to paper their own vision of Vito's future, Antonio Ròiti, a professor of physics at the Istituto Tecnico Galileo Galilei and Vito's physics teacher, learned that Vito was slated to become an engineer. He advised him to apply for a provincial scholarship that would allow him to continue his studies in Milan or Turin, or in Pisa where Ròiti himself had matriculated. "As you can easily imagine," Alfonso wrote to Edoardo on August 3, after Ròiti's offer to help Vito (and before he had received Edoardo's long letter written partly in June and partly in August), "this idea inflamed Vito's head a bit, to whom it would start to be appealing, as I think it would to any young person, the life of a student as Giusti [Giuseppe Giusti, a political satirist and well-known Risorgimento poet from Tuscany] describes it." In his uncle's opinion, Vito needed an infusion of real exercise and practical experience far more than his customary diet of contemplation and study. "Neither I nor his mother," Alfonso continued, expressing himself to Edoardo more bluntly than usual,

> liked this idea of letting him study for another 4 or 5 long years, away from home, with considerable expense to the family (with only a remote possibility of a subsidy) and with the uncertainty of the result, for in the end, even though everything went all right and he obtained a degree, we might see him at 23 years of age looking for a job, as I do see every day Doctors of Engineering, who often end up having to be satisfied with a miserable job at the Bank or with the railroad.

In fact, Alfonso, with his nephew in tow, had already called on Professor Ròiti. After reviewing the family's financial situation, and hearing at some length how Vito would benefit from learning the construction business under the supervision of an experienced and well-intentioned relative, Ròiti changed his tune and said that he himself would not hesitate to recommend this course of action to Vito. He thereupon advised Vito to continue his mathematical studies on his own and wished him all the best in his new endeavor. They shook hands, and Vito, according to Alfonso, left the meeting seemingly reconciled to apprenticing himself to Edoardo. "So I am waiting now with great desire for a letter from you," Alfonso wrote at the conclusion of his August 3 missive to Edoardo, whose silence was giving Alfonso fits, "in order to know when Vito can join you and where. I propose to accompany him."

Had Vito really readjusted his thinking, as his uncle maintained? The brief postscript by Vito in his uncle's letter ("Many greetings from your very affectionate cousin Vito who wishes to receive a letter from you soon regarding what has been written above and is always disposed to follow your advice"), said just enough, presumably, to satisfy Alfonso, but left Vito room to maneuver.[15]

In late August, when Edoardo's two-part tale of woe from the backwoods of Campania finally arrived in Florence, Alfonso took the news hard. He offered to visit Edoardo either in Ancona or Rome or Naples—wherever and whenever Edoardo's schedule permitted. Alfonso was counting on doing some work for the Bank of Italy on the side during the trip, which would keep the cost down. In subsequent letters there are passing references to such a meeting between the two cousins. In the meantime, he reluctantly got in touch again with the physicist Antonio Ròiti. The plans for a pragmatic engineering career were on indefinite hold, he said. Could Ròiti help him find the necessary funds for Vito's continuing education? In response, Ròiti swung into action, lobbying local politicians on Alfonso's behalf and trying every which way ("a thousand efforts," Alfonso told Edoardo), although he did not succeed in prying loose any money from the province for Vito. But he told Alfonso in early September that it would please him greatly if "Mr. Volterra could continue his appropriate studies," because he knew that Vito had a natural bent for science. On September 8, Alfonso wrote the following terse note to Edoardo: "As to Vito I will keep you informed, but we are truly tormented. I hope that your business will get better."[16]

Alfonso and Vito followed up every lead provided by Ròiti. Upon learning that several University of Pisa scholarships existed—named for their donor, Giovanni Maria Lavagna, a native of Livorno (Leghorn) and a professor of mathematics at the University of Pisa—Vito turned in an application complete with supporting documents, including a financial statement from the local tax agent testifying to his family's poverty. Alfonso meanwhile asked a family friend, Cesare Finzi, a professor of mathematics at the University of Pisa, whether Vito could be a candidate for a Lavagna stipend. Finzi told them they were wasting their time applying—the chances of winning the scholarship were no better than winning the jackpot in a lottery.

In early October, Alfonso wrote to Edoardo about the unlikelihood of a Lavagna stipend ("I think that there's nothing more to say about the continuation of the studies"), and suggested it was all for the best:

> What do you want? The idea of seeing my nephew working hard and wearing out his brain for another 5 or 6 years and we ourselves, especially his mother, waiting for such a long time only to see him later go in search of a job to be obtained who knows where and when, does not appeal at all to us. What kind of life would he be leading during this whole time? When one is obliged to subsist in the midst of want, a little collected here and there, always with uncertainty about continuing, it is better to give up studying and the possibility of a more or less famous career and apply oneself to a more modest but more secure job. Only an irresistible urge, an

iron will, would be able to deal with the problems presented by such a situation.[17]

Vito, according to Alfonso, no longer displayed "that tenacity in his disposition" which had so worried them a few years earlier, and now appeared amenable to the idea of putting the family's needs first. Alfonso begged Edoardo one final time to give Vito a chance to prove himself in the field. Angelica also tried to change Edoardo's mind. "I know your excellent heart, and the affection you have for me, and for these reasons I am sure about the way in which you would take care of my son." Edoardo's response to both letters was the same as before: "If you wish to follow my advice," he wrote Alfonso without delay, "let Vito carry on with his studies for two more years, and I am always ready to provide my share as I told you."[18]

The immediate response from Alfonso was an angry letter, dated October 14, 1877. "Why should I rack my brain with such advice: 'Follow my advice, let Vito carry on with his studies,' just at the very moment when I'm telling you that as soon as possible he has to find a way to earn money." Alfonso declared that he was going to do "what we should have done last year"—namely, find a job—any job—for Vito immediately.

Almost immediately, matters took an unexpected turn. Alfonso did find a job for his nephew at his bank, Vito duly prepared to present his papers to the bank's director, and on the very day he was to do so, he heard from Ròiti. Ròiti wanted to know if Vito and Alfonso would consider accepting a private fellowship for the first year at the University of Pisa. While the fellowship (between 400 and 500 *lire*) was a lot for a private party to put up, it was "little, or nothing," as Alfonso put it, in a never-finished rough draft of a letter he wrote to Edoardo, probably later that month. "I'm writing to you again about this business of Vito because I don't want to have any guilt hanging over my head."

Determined to rescue Vito from life as a bank clerk, Ròiti then offered him a job as a laboratory assistant in his physics laboratory at the Istituto di Studi Superiori Pratici e di Perfezionamento in Florence, where he held the rank of lecturer in physics. In the end, Vito's iron will prevailed. Indeed, the fire in Vito's belly that Alfonso longed to extinguish had the opposite effect on Edoardo, who said it reminded him of how he had felt about mathematics during his own university years. At Pisa, Edoardo had been encouraged to study mathematics by Enrico Betti, founder of the school of mathematics there—a course of study that "I ungratefully abandoned," Edoardo recalled in a letter to Vito many years later, "although it was worth a great deal in obtaining, as one used to say, a position. But I had the excuse of not having enough time to devote to these studies—for which I still hold a thankful devotion—so much so that I have inspired even my nephews who to a greater or lesser extent have applied themselves to mathematics."[19] Edoardo would pledge 100 *lire* a year for the next five years toward Vito's education.

Believing (along with his family) that he would get university credit the following year for the courses he took at the Istituto in Florence, Vito became a full-time student in the Istituto's faculty of natural sciences. As there was no department of mathematics, he attended lectures in mineralogy, geology, physics, and general chemistry. In his spare time, he audited a course in calculus taught by his former high school mathematics professor Cesare Arzelà, tutored a pre-med student in physics, went to the opera, and hiked in the nearby hills. Most important, he had found a scientific mentor and kindred spirit in Professor Ròiti, who would champion his protégé's aptitude for mathematics and physics long after Volterra had left his employ. A deep friendship developed from the improbable scientific collaboration between the brash, impatient eighteen-year-old laboratory assistant and the plain-spoken, research-oriented thirty-five-year-old experimental physicist. Vito helped Ròiti with his experimental research (which involved measuring the viscosity of various liquids), worked on the numerical calculations and plotted the data on a graph, and was appropriately thanked in print when Ròiti's scientific paper appeared later that year in the *Proceedings* of the Reale Accademia dei Lincei. Vito came faithfully every day as needed to the makeshift laboratory in a villa on the outskirts of Florence, but was heard on at least one occasion to grumble that the editorial work involved in publishing a scientific article never seemed to end. The academic year closed on a high note: Vito passed the preliminary examinations for a teaching certificate in the natural sciences—that is, physics and chemistry—with high praise from the examiners, earning a 9 in both subjects, the highest possible score.

By this time, Vito had made plans to enroll in the University of Pisa. He was certain that the year he had spent at the Istituto Tecnico would allow him to enter the university as a second-year student, and he intended to submit a petition to the mathematics, physics, and natural sciences faculty at Pisa requesting permission to do so. "The crossing," as the family referred to Vito's petition, was deemed crucial to the family's finances, and everyone pitched their own recipe for a successful outcome. Vito also had designs on Pisa's Scuola Normale Superiore, Italy's elite teacher-training school. The school, which did not offer a degree but was connected to the university, functioned like a college. Only a handful of science and humanities students were admitted to it each year, and competition for the available slots was intense. It was not uncommon for alumni of the Scuola Normale to start their teaching career in a high-school classroom, often a poor one, in order to hone their teaching skills; the move up the academic ladder to a university position usually came later (if it came at all), and only if the aspiring professor's publications record supported it. Vito, who had set his sights on a university career, wanted to take the school's rigorous admissions test right away, despite the daunting requirement of proficiency in Latin and Greek. That fall he packed his bags and rode the train forty-nine miles west to begin a new life in Pisa.

CHAPTER 4

"Long Live the Republic," 1878-1882

Even in a country rich in religious architecture, the Piazza del Duomo is a showstopper, a monumental legacy of Pisa's wealth and military prowess during the Middle Ages. Of his own first glimpse of the city's storied square, Vito Volterra retained a vivid memory of being nearly knocked off his feet by a blast of wind that sent him running for cover. Once he settled into the routine of student life, Volterra would find many things to gripe about—the weather, the townspeople, the professors. Even the local opera company would fail to come up to his exacting standards—but when he saw the soaring medieval marble cathedral, the gleaming baptistery, and the famous leaning tower bathed in moonlight, he reacted more like a proud Pisano ("truly an astonishing sight") than like a sardonic, self-assured eighteen-year-old university freshman eager to take on the town's academic establishment.

Guidebooks of that era describe Pisa as a small, sleepy Tuscan town of about 40,000 inhabitants, and to this day it has retained what one writer calls "the character of a living museum." The river Arno slices through the city, twisting and turning its way to the Tyrrhenian Sea, seven miles downstream. What Pisa may have lacked in Volterra's time in size and sophistication, it made up for in its old and illustrious maritime history, striking architecture and broad promenades, and venerated university dating back to the fourteenth century. The University of Pisa boasted one of the country's finest schools of mathematics as well as the prestigious Scuola Normale Superiore, in which Volterra eventually hoped to enroll.

Like other Italian universities, Pisa inaugurated the start of the academic year with great fanfare; the faculty robed up according to their rank, and, escorted by a pair of ushers, filed into the main hall of the university, where they were greeted by the rector. Vito's experience as a freshman on this solemn occasion—he fled as soon as the principal speaker rose to address the audience—produced a literary outburst worthy of the brash, intemperate teenager. "I did not have the patience to listen to Professor Bonanici's speech," he wrote to Angelica. "Can you imagine, it dealt with nothing less than Roman Law."[1]

Soon after Volterra's arrival in Pisa, his mentor in Florence, Antonio Ròiti resigned his teaching position there for a more prestigious university appointment in Sicily. "I learned with joy that Ròiti, as he deserves, has earned for himself a beautiful and enviable position in Palermo," Volterra wrote to his mother. "[Cesare] Arzelà will go to the University also, as

Professor of Algebra and Analytic Geometry; I think that it was Ròiti who arranged for him to get such a good position." While pleased for his former teachers, Volterra nevertheless painted an unsparing picture of the academic life he had chosen for himself:

> Seeing the rapid advancements made by these professors does not persuade me that the career of a teacher is a beautiful one: there are very few of them who progress, while one counts by the hundreds those who remain to rot in a miserable little Liceo. Not everybody has the capacity and the *know-how* of Ròiti, who knows how to help himself and to help also his friends.

Keenly aware of his family's expectations with regard to his own career, Volterra cautioned his mother—although she probably didn't need cautioning on this point, since she, like Alfonso, thought it was a wildly impractical idea from the start—"not to build too many castles in the air for the future or to believe that there are only roses ahead."[2]

Volterra was homesick only briefly, while Angelica, who had never been separated from her son before, felt as though a limb had been torn from her body. His absence consumed her. Overwhelmed by loneliness and this sudden void in her daily life, she poured out her feelings. "My dearest Vito," she wrote, "Today I have remained deprived of a letter from you, but I still have some hope of receiving it this evening...if, as you told your uncle, you wrote in time this morning....Your departure has been so painful, that it seems to me almost impossible to have been able to divide myself from you....Write to me often, my dearest Vito, tell me what you do; think that this is my only pleasure and think often of your Mama."[3]

Before leaving Florence, Volterra had rashly promised to write every day. He could hardly blame the Italian mail system for any procrastination, for a letter posted in the morning at the train station in Pisa typically arrived at its destination in Florence by dinnertime. If his eagerly anticipated letters did not come, which happened with greater frequency once he became busy with classes and homework, Angelica fretted. Little reminders filled her letters to him: Avoid troublemakers; keep a record of your purchases; take good care of your clothes; be sure to go to sleep early; don't forget to behave correctly; refrain from talking back to your professors; be sure to write to Cousin Edoardo.

Angelica also begged her son to write her longer letters. "I understand that at the moment you don't have many topics to bring up, but I who miss you so much and feel deeply the emptiness brought on by your absence, would like to spend more time with you, at least by letter." When her son was living at home in Florence, Angelica had been accustomed to waiting up for him at night. Breaking the habit proved difficult for her. "At all the times you usually came home," she confessed in one letter, "it always seems to me that I see you arriving; especially in the evening, it seems I need to hear the loud thud of the front door that signaled your return, always anxiously awaited by me."[4]

Volterra found lodging near the university, two rooms on Via S. Elisabetta 12, for which he paid 30 *lire* a month. The location suited him perfectly. Mornings he rose at 7:00 A.M., dressed, made coffee for himself while listening to the ringing of *il Campanone* (the students' nickname for the university bell), and was in his seat at the Palazzo La Sapienza by eight o'clock for his calculus class, taught by the brilliant Ulisse Dini. He took his meals at a neighborhood restaurant called Il Duilio, where for 65 *lire* a month he dined on soup and a main course at lunch; dinner, served up in the late afternoon, consisted of more soup, boiled meat, another main dish, and fruit. Service was not included, much to Alfonso's regret, who advised his nephew to watch his diet: "In any case I urge you to choose always simple and healthy foods, and not too much wine, but of good [quality]."[5] After paying rent and board, Volterra found that of his monthly allowance of 127 *lire*—money borrowed from relatives on both sides of the family, which had to be repaid after graduation—only 32 *lire* were left to pay for books (Dini's alone costing 15 *lire*), laundry, breakfast, and other incidentals.

His professor Ulisse Dini was a star of nineteenth-century Italian mathematics. Born and raised in Pisa, he had graduated from its university in 1864 and, with the exception of one year of postgraduate study in Paris, spent his entire career in his hometown. By 1878, he held two professorial chairs, which allowed him to teach analysis and higher geometry as well as infinitesimal calculus. Besides this exposure to one of Italy's finest mathematicians, Volterra took algebra from Cesare Finzi, a gifted teacher, and a friend of the family. Chemistry, physics, and descriptive geometry rounded out his first-year course schedule. His two mathematics professors could not have been more different in the classroom:

Uncle asks how I like the lectures by Finzi. They are without doubt very beautiful lectures; he is a teacher who explains things with order and clarity, he never gets confused, he is *sufficiently* precise, speaks well and is also good and kind; one really could not ask for a better person to teach algebra. But I always prefer Dini, who gives lectures that are a bit confused, who some days explains and explains without reaching any conclusions and at other times is a bit obscure. I prefer him because he puts his whole soul into his lectures, because the things he explains are almost always his own discoveries—at least the method is really all his own. He speaks simply and gets a little confused but in the end his concepts are infinitely more clear, more concise, and, what is more important, more exact, than those of Finzi. Dini's lectures are always interesting.[6]

The same could not be said for chemistry, whose professor he found dull and obscure, and he complained bitterly about his descriptive geometry class, noting that "it is a miracle if one succeeds in keeping the eyes open, because for one-and-a-half hours continuously and always in the same monotonous voice Professor Nardi Dei chatters idly about very boring things which never seem to end."[7] Physics, which was closer to his interests, he tolerated better, mainly because the experimental physicist in charge of the class, Riccardo Felici (a friend of Ròiti's from college days in Pisa), could explain the concepts clearly. In the laboratory, however, Felici had the disconcerting habit of telling the students that the experiment to be performed was not going to work even before they tried it.

> Now we are doing some experiments in acoustics which the professor says are very important. Up to now, however, the experiments are limited to making holes in cardboard; it remains to be seen when we get to appreciate the tuning, in view of the great ear which everyone in the laboratory has except Felici. I am afraid that we will fail, since neither the professor nor his assistants have any will to do anything.[8]

Dini was deeply involved in pure research, teaching, and public life (he served often on Pisa's city council), and therefore his opinion of Volterra's petition to be granted the status of a second-year student during his first year—which Vito had presented soon after enrolling at Pisa—carried great weight. But he quickly discovered that Dini, like everyone else he spoke to, opposed the idea. Ròiti came to Rome to plead Volterra's case with Dini in person, but had no luck either. Vexed by this turn of events, Vito's Uncle Alfonso quizzed various school officials in Florence. Hadn't Ròiti been assured by none other than Enrico Betti, the head of the Scuola Normale Superiore, that Vito would be given credit for the year he spent—at Ròiti's behest—in Florence? "It's not true that Ròiti last year wasn't sure that I would have been able not to lose a year," Volterra eventually wrote to his uncle, "rather he was absolutely sure about it because of what Betti had indicated to him." Betti, according to Volterra's sources in Pisa, had the unfortunate "habit of saying yes, but then changing his opinion at the first opportunity."[9] Certainly Volterra had no desire to cross swords with Betti; the whole point of his going to Pisa was to gain eventual admission to the Scuola Normale. And before he could do that, he had to pass language exams in Latin and Greek, two subjects he neither knew nor expected to like. ("I am persuaded," Ròiti wrote to Alfonso, "that when the irritating classical language exams are behind him, he will go ahead without any other obstacles, and he will have a brilliant career.")[10] But soon Vito was pleading with his mother and his uncle not to lobby Arzelà or Cesare Finzi, on his behalf. As he wrote in one letter home, "it could do me more harm than good, with the ideas they have here [in Pisa]."[11]

His family's need to hear from Vito regularly and Vito's inclination to keep the obligatory correspondence to a minimum came to a head after the Christmas break. Alfonso went on the offensive first:

> Is this the way to begin? You arrived there the evening of [January] 7th and we did not get a letter from you until the morning of the 9th. And what kind of a letter! Two lines written down in a hurry and nothing more. You always write in the last fifteen minutes before the deadline for mailing. The following day, the 10th, we were expecting a longer letter, but nothing. This morning we were more than sure to find a letter of yours, but nothing again. We sent somebody to the post office three or four times and we couldn't find any letter. Is there a problem with the delivery, or did you forget to mail it? Naturally we were a little bit worried. So I sent a wire to you this morning, and today at *three* I get your answer by wire saying that you are well, that you wrote last night and that you will write tomorrow. Why didn't you write today, as I asked you in the wire? It was easy to understand from my message today that we didn't get your letter. And if it hasn't arrived today, it won't arrive tomorrow either, because today you are not going to write.[12]

Vito rushed to defend himself; it is the impassioned *cri de cœur* of a young man on his own for the first time in his life and desperate to enjoy his independence:

"Dearest Mama,"

> ...I don't understand why I should have written to you on the 9th when I wrote to you on the morning of the 8th. I don't understand why you complain about my short letter of the 8th when, having left the day before from Florence, I had nothing at all to say to you. I don't understand why I should have written to you yesterday when I had already wired you to say that I was well and had written the night before. I don't understand why you worried. I don't understand why you get angry. I can't understand what kind of idea you get into your head after having decided one thing to want another one. I understand one thing only, and that is that you did well to send a wire with a *pre-paid reply*. If every time one of my letters arrives one hour late, whatever the cause, you are alarmed in such an extraordinary way, I really don't know what to

do. I think it is an exaggeration of a new kind,
really incomprehensible and stupid. If you are
longing to wire, wire, wire until the end of the
centuries, I couldn't care less, so long as my
answer is pre-paid. I wouldn't write in this
way if I hadn't read my uncle's letter, *conceived* in such a strange and *unthinkable* way:
that is, to get angry with me if a letter is delayed a little bit....[13]

The rancor passed quickly, and Vito continued to write often to the
family, if not every day. "When you have a little time, write to Edoardo
Almagià," Alfonso reminded his nephew. "Considering that he more than
anyone else promoted and supported warmly the idea of continuing your
studies at the University, it seems to me highly appropriate, now that everything has fallen into place, for you to get in touch with him to thank him
once again for his kindness. You can give him some sense of your new life,
far from the family...." But, warned Alfonso, "Don't spell out in particular the number of years it will take you to complete your studies. Limit
yourself to telling him that you are hoping that the year spent at the Istituto di Studi Superiori in Florence will count as saving one at Pisa, but
instead etc. etc... (in case this chance is gone by the time you write to him).
If, instead, Dini's reply is favorable, you can allude to it, without however
directly mentioning either three or four years."[14]

By the time Vito heard officially about his petition to transfer into the
second year of the University of Pisa—"it has been rejected unanimously (I
don't know if by *acclamation*) by the whole Faculty," he told his mother—
he had resigned himself to the fact that it wasn't going to happen, and he
urged her to spread the word to his friends still in Florence: "[Y]ou will
need to say that there is still time to come to Pisa if they don't want to
lose a year as I have done."[15] Volterra also broke the news to Ròiti, who
gently rebuked him: "As a matter of fact I had advised [you] not to ask the
Rector....Instead you exposed yourself to a rejection and now there isn't
anything to be done about it." The rest of Ròiti's letter was friendlier,
thanking Volterra for his kind memories of Ròiti's tutelage and deploring
the rundown state of his new physics laboratory in Palermo.[16]

The University of Pisa drew students (in those days, all male) from
all over Italy, and Volterra made new friends quickly. Often boisterous
and eager for adventure, the students pumped money and energy into the
otherwise quiet town. In the evenings, they played dominos, talked, and
drank quarts of blistering *poncino nero* in the many coffeehouses catering
to the university crowd. Volterra spent a couple of hours almost every night
talking or playing dominoes at Burchi, the coffeehouse on the north side of
the Lungarno Reale favored by his own group of friends—mostly *Normalisti*,
as students who had been admitted to the Scuola Normale were called. The

Normalisti were divided into two sections, those in letters and philosophy and those in science. While professing to like them ("they are all good fellows, cheerful"), Volterra found his new friends a study in contrasts. The literature students wore him out ("they never stop talking, but there are only a few of them"); the lawyers-to-be ("there are an infinite number of them [and they] are also great chatterboxes") rushed to embrace every cause but produced more confusion than anything else; the future doctors failed to make any impression at all ("they have no special characteristics"); while the mathematicians and the naturalists, perhaps not surprisingly, earned Volterra's approval for "being a little more serious and sedate."[17]

He went to the opera, reporting to his mother on a performance of *La Traviata* at the Ernesto Rossi, a small theatre in the Piazza Carrara. The audience consisted mainly of fellow students:

> Democratic students in the pit, aristocratic students in re-
> served seats and in boxes, very few ladies, almost no Pisani.
> It looked like the university—only the gowns of the professors
> and the hats of the janitors were missing. The audience—
> as you can imagine after having made an accurate analysis
> of the composition of the cast—could not be said to have
> been the most quiet in the world; every once in a while one
> heard yelling, laughing; now and then someone jumped over
> the seats to go from one place to another; again, two peo-
> ple called to each other and started talking from one end
> to the other of the theatre, wits repeated in a loud voice
> what was being passed around the pit, and this produced a
> prolonged and general cheerfulness; altogether a noise and
> enough screaming to ruin the delicate ear.

The orchestra was bad and the singers, Volterra assured his mother, "sang like dogs." Nothing about the performance measured up to the students' operatic standards:

> [T]he Prima Donna appears, thin and ugly with arms that
> look like toothpicks [and] a very weak voice, which makes her
> seem pitiful. Then the tenor hastens on the scene, fair hair,
> fair beard, using the French "r," mouth open like a volcano
> spewing wrong notes instead of fire...; the toast and the
> waltzes, the love duets follow, and the curtain closes amid
> boos and cheers and general laughter. In the second act,
> the baritone father enters dressed appropriately for a force-
> ful part like his, i.e., *red trousers, jack-boots, gray face, rough
> voice.* Amidst boos, he tells of his family's misfortunes, the
> Prima Donna cries out due to the wrong notes she sang;
> she repents for her dissonance, and exits. The tenor enters,
> despair shown by great opening of the mouth and arm move-
> ments; he listens with a singular coldness to the lament sung

by his father but, taken by a sudden fury, runs out, afraid
that the public will throw potatoes at him.[18]

At a performance of *Rigoletto* some months later, Volterra gave a thumbs-up to the soprano; the other singers, he announced, were "dogs who deserve to die of rabies."[19]

Besides the opera, Volterra's social calendar in Pisa that year included parties and dinner invitations. He often spoke of these evenings in his letters to Angelica, sometimes unkindly:

> Yesterday, I got an invitation to Mr. Perugia's. He hoped
> that I would accept, because his wife and his mother wanted
> *very much* to get to know me (but I did not have the same de-
> sire toward them). I had to go: and at 8:30, with a trembling
> hand inside my black glove, I rang the bell of the Perugias'
> house. I enter and I find many unattractive and rather old
> men and women. Mrs. Perugia has yellow hair, a big nose,
> and a lot of spirit. The other Mrs. Perugia...is horribly
> fat, has no teeth, but has a great quality: She will stuff any-
> body who is beside her (unfortunately it was not me) with
> cakes. I'd rather not talk about the other women—two are
> above their forties—and I'll mention a newly married bride,
> not ugly, not unpleasant, who suffers from the cold, with
> ugly hands; I think so because I've never seen them. She
> was with her husband, big beard, enough spirit, big glances
> at his wife....I didn't know anybody's name, even if I had
> been introduced to everybody. I spent three horribly annoy-
> ing hours devouring ice creams, punch, and pastries. I hope I
> don't have to go anymore: I am not made for etiquette....[20]

* * * * * *

A few months earlier, in January of 1878, Victor Emmanuel II had died, and scores of university students had rushed to Rome for the funeral. Some made the trip out of grief and the desire to honor the unifier of Italy; others used it as an excuse to visit Rome. Among them was the future Dante scholar, Fedele Romani, then a student at the Scuola Normale. He had expected to see a city in mourning, but as he later recalled, "people were going about their business with the usual facial expressions," as if nothing had happened.[21]

The indifference that Romani sensed reflects the fact that the struggle for national unification had involved many diverging points of view among the Risorgimento leadership. Followers of the patriot Giuseppe Mazzini, who favored a democratic republic elected by universal suffrage once Italy became unified and independent, refused to have anything to do with the parliamentary form of government finally established in 1861. There was no room for a king in Mazzini's vision of a united Italy. Garibaldi, on the other hand, would have settled for a dictatorship. Cavour, who cringed at the possibility of a democratic and republican revolution sweeping across the Italian peninsula, used his considerable diplomatic skills to unify Italy with a king already on the throne in Piedmont, one of the country's oldest and most conservative regions. Indeed, monarchist sentiment had never been strong in Italy except in Piedmont and Naples. In the end, the creation of a constitutional monarchy satisfied very few Italians and left many nationalists deeply disappointed; some critics of the new nation state continued to wage war against it, with a small number of unrepentant reformers flirting with anarchism and armed rebellion.[22]

Romani, who had never seen the king when he was alive, walked past the body as it lay in state at the royal palace. He stared at it. Was it real or only a wax replica, as many people around him insisted? On the day of the funeral, he waited from ten in the morning until midafternoon for the procession to arrive at the Piazza del Popolo. There he watched as ordinary Italians and members of Europe's ruling families followed on foot behind the horse-drawn coffin as it circled the towering obelisk in the center of the square. From the Pincio above the square came the repeated sound of cannon fire, followed by the muffled reply from Castel Sant'Angelo's cannon in the distance. After the funeral, Romani returned to Pisa, but not before catching a glimpse of the new sovereign as he entered Parliament to swear allegiance to the country's constitution.

In November of 1878, the new King Umberto, and his Queen Margherita (a religious fanatic who would outlive her husband and survive long enough to embrace Mussolini) visited various cities in Italy, including Pisa. Classes were canceled in honor of the visit, and Volterra watched the royal procession from the window of a friend's room before joining the milling crowd gathered on the Lungarno, that series of broad promenades lining both banks of the river for the entire length of the city. The apathy of the locals surprised him. To his mother, he wrote:

> I have just returned from the Lungarno, where I saw the King and Queen go by. The King was received with great coldness....[I]t was raining, there was a lot of mud on the ground and few people in the street. The royal coaches arrived with the police in disorder. Someone took his hat off, but there was neither a hurrah nor even a greeting of welcome to be heard. By the time the King reached the palace, the weather had become beautiful [and] a large number of people had gathered on the Lungarno. For a half an hour I was so crushed by the crowd that I was unable to breathe, but the King did not show himself because there was not even one hurrah to be heard from the throng.

"These Pisani," he added, "are cold and unpleasant."[23]

The Italian—and particularly the Tuscan—coldness toward the monarchy is described by the Englishman Oscar Browning in his memoirs. While visiting Florence in 1871, Browning went to the racetrack. As he later reported, "We saw the races in the Cascine and were surprised that so little enthusiasm was shown for King Victor Emmanuel as he drove down to them. I asked my driver why this was so, and he only replied, '*Se fosse Garibaldi;*' 'If it were Garibaldi, the feeling would be different.'"[24]

The ability of Italians to participate in the electoral process was severely restricted. In the newly unified country of 22 million people, only 2 percent of the population had the right to vote, and of those 440,000 privileged citizens only 300,000 did so in the first parliamentary elections. Broadly speaking, only the propertied classes were for the king; opposition to the monarchy ranged from the peasants and the small craftsmen and artisans to the emerging class of factory workers and underemployed university graduates. Jews and Protestants welcomed the fact that the new Italy was a secular state, while Pius IX railed against its legitimacy and forbade Catholics from voting or holding office. As a member of a religious minority, Volterra, like many other Italian Jews, felt a special kinship to the king and the constitutional guarantees his position embodied, which may help to explain his ill-concealed contempt for the average Pisano.

Later that November, while visiting Naples, King Umberto had his first brush with an assassin. (In 1900 the anarchist Gaetano Bresci would shoot and kill Umberto at Monza, in retaliation for the monarch's role in repressing street riots in Milan two years earlier.) Giovanni Passanante, a cook, carrying a small red flag proclaiming "Long live the Republic" and a sharp knife, lunged at the king in his open carriage. Umberto escaped unharmed thanks to the swift intervention of Benedetto Cairoli, the prime minister and a leader of the Left, who was wounded by the attacker.

News of the attempted assassination triggered demonstrations across Italy. In Florence a bomb exploded during a monarchist rally, killing four people; the violence spread to other cities in Tuscany. In Pisa, Volterra joined a noisy demonstration of students cheering and booing outside the headquarters of the prefect, the state official who supervised the new government of Italy's administrative and judicial system in Tuscany. From Palermo to Pisa, all university professors now worked for a centralized state and had to take an oath of allegiance to the king. Those demonstrators in favor of a monarchical form of government cheered the prefect; those who didn't booed loudly. A homemade bomb sent the crowd scattering, but it exploded harmlessly, and it seemed that no one would be hurt until one of the students took off after the bomb-thrower, who stabbed his pursuer in the back and fled while the wounded student fired a revolver wildly into the air. After this incident, Volterra immediately sent a telegram to his mother telling her not to worry, but the operator at the telegraph office censored the news about the bombing, leaving his mother in Florence puzzled and undoubtedly frantic about her son's safety. "The city is reasonably quiet,"

Volterra assured her by letter the following day, "and there is nothing to fear, but now it will be necessary for the authorities to keep their eyes wide open and to punish this infamous pack of internationalists"—a reference to the followers of the Russian Mikhail Bakunin, the leading advocate of nineteenth-century anarchism, whose revolutionary message resonated with many urban artisans in Italy, especially in Tuscany.[25] Over the next several days other incidents rocked the city; the police and soldiers stepped in and arrested several hundred people. People described by Volterra as "known for their socialist ideas," even those from good families, found themselves behind bars.

Pisa had long been an urban battlefield. Despite its outwardly sleepy appearance, the city had harbored reformers of various political stripes since the days of the Risorgimento, and the subversive undercurrent lingered. Alessandro D'Ancona, who was appointed professor of Italian literature at Pisa in 1860, recalls stepping into "one of the most active hotbeds of new ideas" in Tuscany.[26] Ten years later, a socialist organization known as the Società Democratica Internazionale sprang up in Pisa, through the initiative of a group of university students. The followers of Bakunin's doctrine of violent insurrection were also pressing their case in the city. In 1873, the first regional congress of the international anarchist movement was held in Pisa. Riots and strikes broke out, there and throughout Tuscany, the following year. A liberal underground press thrived in Pisa; newspapers, broadsides, and other tracts were published there clandestinely and circulated. By the time Volterra arrived, in 1878, *Il Radicale*, a republican journal, could be found on the newsstands. That same year, Pisa hosted (in secret) another congress of the international anarchist movement, this one attended by groups from all over Italy.

Town-gown relations were tense: The students felt physically threatened by the local agitators and the city officials deferred to the prefect to maintain law and order. The prefect, in turn, took his orders from Rome, where Prime Minister Cairoli, trying to mollify the competing factions on the extreme Left, was slow to arrest and crack down on the various socialist and international groups. Like many Italians of his generation, Volterra was quick to criticize Cairoli for not dealing firmly with the extremists. A powerful voice despite their small number, these radical republicans and socialists kept pressing the government to honor its original commitment to social reform. On December 19, 1878, when the cabinet fell and Cairoli resigned, Volterra cheered the news. "The students will certainly be pleased by the fall of this ministry which was unable to do anything good," he wrote to his mother, "and instead, with all its best intentions, did so much harm and placed the country on such a very ugly and dangerous road."[27]

Local officials groped in the first few weeks of January 1879 to restore order in the city, but the nightly muggings increased. In his memories of undergraduate life, Fedele Romani remembers a climate of hostility, that became more and more ruthless, between students and the riff-raff [*plebaglia*]. The students were harassed in a thousand ways; and it was dangerous to go around alone in the evening on secluded streets. The students often reacted; but the louts were more ruthless and more industrious naturally; and then, little by little... the initial reasons for the hatred were forgotten: the bomb-thrower and the arrest didn't figure in it anymore, and the struggle no longer had any social or political complexion but was simply a struggle between oafs and students. The city, which owed its vitality and its livelihood to the students, became worried.[28]

As rumors spread that the students might abandon Pisa and depart en masse to Bologna or Siena, the students themselves called for a boycott of their university lectures in an attempt to force the city to curb the hoodlums preying on them. The plan succeeded on the first try. "Today, not one student attended classes," Volterra boasted in a letter to his mother. "The grinds who entered the buildings were booed, and the professors who started to teach had to stop as everybody cried out, 'Enough!' 'Enough!'" Several hours later, the students assembled for a meeting at the university's school of medicine. Vito had just begun his letter to Angelica by describing the personality of the chancellor—"[T]his is a man who is unsuited and doesn't have the ..."—when he abruptly broke off in midsentence to participate in the meeting.[29] When the motion to leave or stay in Pisa was put to a vote, the students elected to postpone a final decision until they saw whether the Pisan police would increase their patrols. In the meantime, many of them decided to go home, including Volterra, whose opinion of the chancellor we can only guess at, as he never got around to finishing his letter. When he returned to Pisa a week later, he found the city slowly recovering from its wounds.

> Our prank seems to have shaken up the town a bit. City Hall is waking up a little from its usual sleep and they are circulating the names of people to be rounded up on the street. The Pisa socialists... have tacked up an official statement at street corners (without a tax stamp) to tell us that they never dreamed of exterminating us. The patrols as usual stroll about Pisa in the evening at a slow pace and half asleep, and the Prefect shakes for fear of being transferred or pensioned off. The students carry on the same life, the professors increase the number of lectures. This is Pisa after the famous events.[30]

Classes resumed with a vengeance. Ulisse Dini's calculus class now started at 7:30 in the morning and lasted until 9:45 A.M., five days a week. That month Volterra and Corrado Padoa, another student who also wanted to gain admission to the Scuola Normale, arranged for a tutor—a *Normalista* in classics—and started lessons in Latin and Greek. Padoa was a nephew of Professor Alessandro D'Ancona, one of the Scuola Normale examiners. "[H]e is almost sure not to be flunked," Vito wrote his mother, "and concerning Latin he knows perhaps, if this is possible, even less than I."[31] The exams were scheduled for the Easter holidays, and, as the date for taking them approached, Volterra's work pace increased. When he wasn't studying Latin and Greek in the evening, he worked on geometry and calculus with friends; and his attendance in the physics laboratory plummeted. Meanwhile, Ulisse Dini had pushed the starting time of his lecture back another half-hour. "If it continues in this way," Volterra complained to his mother, "in a few months he will start lecturing at 5 A.M."[32] But in spite of the brutal schedule, he rarely passed up any opportunity to venture beyond the city's walls ("Today... some of the professors were sick—would that God and their doctors keep them in such a condition....We went to the 'Bocca d'Arno' by boat and came back on foot. A very delightful excursion.")[33]

By the end of March, he had taken and passed his language exams. He sent the family a celebratory telegram, rendering an Italian idiom in fractured and distinctly unclassical Latin: "Passatus essi rotum cuffiae, suppungo" ("I think I passed by the skin of my teeth"), the meaning of which was nevertheless grasped immediately.[34] "Laudamus Deus," ("Thanks be to God"), replied his uncle, whose Latin was not much better.

The following November, Volterra registered as a second-year student in mathematical physics at the University of Pisa. Earlier that month he had also taken the entrance examination for the Scuola Normale Superiore. The results of his oral exams, preserved in the archives of the school, indicate that he passed with a 30/30, the highest score possible. Admission meant free room and board and a cash stipend, which Volterra would enjoy for the next three years until his graduation. Ten students were admitted to the school that year, four in the letters and philosophy section and six in the science section, including Volterra and Carlo Somigliana, another promising mathematician. "When I knew him in Pisa," Somigliana would later recall, Volterra

was a young man not yet twenty, with thoughtful and clear eyes, with a sweet smile. His extraordinary mental vigor appeared sometimes, but seldom. He was usually calm, facetious, witty, and ready to make friends [in] fraternal collegiality. He never showed off his superiority; everyone acknowledged it without discussion. These characteristics of the young twenty-year-old remained, with slight variations, in the mature man, even after many honors, even in the universal esteem that surrounded him. To me he was always my fellow student and friend from the University of Pisa.[35]

The Scuola Normale Superiore di Pisa, which started out as a branch of the French École Normale in Paris, traces its roots to 1810, following the annexation of Tuscany to Napoleon's empire. Closed in 1815 after Napoleon's defeat at Waterloo, the school reopened in 1847 under a new patron, Leopold II of Lorena, in the Palazzo Carovana, Giorgio Vasari's Renaissance masterpiece, which once housed the knights of San Stefano. The most dramatic change in the curriculum before and after unification had to do with the role of religious instruction: In the new Italian Scuola Normale, no religious exercises and requirements were required or expected, and the teaching of theology was suppressed. As Tina Tomasi and Nella Sistoli Paoli point out in their history of the Normale, "From this moment the presence in the school of Jews among the students and the teachers became a not precisely definable—and besides never studied—but unambiguous constant."[36] Early beneficiaries of the new nation's anticlerical stand included Alessandro D'Ancona, the Scuola Normale's first Jewish professor, who joined the faculty in 1860; Cesare Finzi, Volterra's algebra professor, who came to Pisa eleven years later; and of course Volterra, who settled into his lodgings in the Carovana in November 1879.

These changes aside, the mission of the school remained fairly constant—to prepare and qualify exceptional university students for teaching in secondary and teachers colleges. Students had to follow the science curriculum or sign up for courses in philosophy, history, and philology. The instructors (*professori interni*) were drawn from the university's faculty. When he entered the school's program, Volterra selected the compulsory two-year course of study in the mathematical and physical sciences; he moved to physics in 1881. In his senior year, he would take a number of electives—mineralogy, higher analysis, mathematical physics, and a course described simply as *prova pratica di Fisica* (roughly, "competency in a physics laboratory").

Several decades later, another matriculating mathematics student, Giovanni Sansone, took up residence in the Palazzo Carovana. While unpacking his things and putting them away, he saw Volterra's signature emblazoned inside the desk drawer. In 1918, while the Great War was still on, the two mathematicians met in person, when Volterra, dressed in his military uniform, came to monitor the early sonar apparatus installed at Sansone's command post at Pero sul Piave, on the Austrian front.

CHAPTER 5

"Professor by Deed," 1880-1883

Students admitted to the Scuola Normale were expected to participate in research and teaching. Volterra's apprenticeship as a mathematician started and quickly accelerated under the watchful eyes of Ulisse Dini and the school's director, Enrico Betti, both of whom had guided the Pisa school of mathematics to the pinnacle of academic distinction after Italy's unification. Passionate about teaching and devoted to his students, Dini petitioned (and was granted) permission in the fall of 1879 to lodge in the Palazzo Carovana. Living under the same roof as Dini and taking his courses in calculus and higher analysis seems to have sparked Volterra's interest in this field.

Dini introduced Volterra to pure analysis. Enrico Betti, whose courses on higher mechanics and mathematical physics Volterra took in his last two years ("We followed the lessons with passionate fervor," a classmate later recalled[1]), broadened his pupil's taste in mathematical problems. Over the years, Betti had worked in many different fields of mathematics, including algebra, analysis (in particular the theory of elliptic functions), and topology. Early in his career, while still a high school teacher in Pistoia and Florence, Betti specialized in algebra, which paved the way for his appointment in 1857 as professor of higher algebra at the University of Pisa. Not long afterward, he migrated from algebra to the field of analysis, and in 1859 he was appointed professor of analysis and higher geometry. In 1864, Betti succeeded Ottaviano Mossotti, his former teacher, as professor of mathematical physics. From that time on, as Ernesto Padova, one of his students and later a professor at Pisa himself, wrote, "all his scientific papers directly or indirectly concerned questions of mathematical physics and mechanics." In 1870, Betti formally relinquished his chair of analysis and geometry in favor of the chair of celestial mechanics at Pisa. As a teacher, he had few peers. Recalling his approach, Padova described Betti as follows: "Scrupulous in carrying out his duties as a teacher, he read everything that related to the topics that he developed while lecturing, and very frequently a doubt, or even only an objection raised by a student, drove him to think about the theories he presented, to complete them and to solve new problems."[2] Betti's interest in the applications of mathematics exerted a lasting influence on Volterra, who would not only spend the rest of his life moving back and forth between purely analytical work and the application of mathematics to the real world but would also become, almost despite himself, an inspiring teacher.

In February 1880, only three months after becoming a *Normalista*, Volterra sent to Giuseppe Battaglini's *Giornale di Matematica* in Naples some observations on point-wise discontinuous functions, the first of two papers dealing with mathematical analysis, a subject close to Ulisse Dini's heart.[3] Indeed, Dini's place in the pantheon of nineteenth-century Italian mathematics rests in part on his role in the early 1870s in placing the field of analysis on a solid foundation. His *Fondamenti per la teorica delle funzioni di variabili reali* of 1878, one of the first textbooks that Vito the student had purchased in Pisa, remained a classic in its field for many years. It offered the exceptional student both an introduction to "the principal concepts of analysis, beginning with that of a real number...set down with maximum rigor and with the greatest generality" and a starting point for further research work.[4] Volterra spent his first summer after university at home in Florence, reading Dini's book and making calculations.

In the autumn of 1880, Volterra returned to Pisa after three months of vacation, expecting to continue the work he had done on his own over the summer. Instead, he found himself pushed and pulled in different directions by the Scuola Normale's rules and regulations. He sulked at having to teach mathematics to the younger students. His own teachers led busy lives, leaving little time for him. Betti lectured every day, Dini whenever he could. Elected to Parliament in 1880, Dini was now spending much of his time shuttling back and forth between Pisa and Rome. Even when Dini was in Pisa, Volterra found it almost impossible to talk to him. As he wrote in November of that year to his mother, Dini "is always very busy and he doesn't have one minute to spare. But he has already advised me to read some papers (in German) that refer to the subject which I studied during these holidays." Mostly, Volterra railed against the academic regimen that separated him from his first love, mathematics. By the time he'd finished fuming, his letter to Angelica had turned into a litany of complaints about the school's core curriculum:

> Now it becomes very difficult for me to be able to study because lectures and annoyances take away a lot of my time. They want to make me attend the chemistry laboratory, every day for four hours a day, although I did chemistry regularly in Florence. I have not given up hope of being able to get rid of this annoyance; moreover, when I am there [in Florence] I will speak to Ròiti about this. In order to finish enumerating the *antipatiche* [disagreeable] occupations of this year I will tell you that they are making us *Normalisti* of the last year teach the students of the first year. You, who know how annoying it is for me to teach, can imagine how I welcome this novelty.
>
> It seems impossible that these gentlemen who overload us with teaching cannot convince themselves that we would be

able to do much more if they would leave us a little freedom to study at our will instead of imposing upon us what to them may seem more useful.[5]

"On the principles of the integral calculus," the paper Volterra had begun to work on in Florence that summer and completed in Pisa the following April, marked him as a deft analyst.[6] As his Scuola companion Carlo Somigliana recalled sixty-five years later, this second paper already had "the character of an organic work, rich in interesting results, which could have bestowed credit on a skilled mathematician."[7]

Volterra's paper concerned the mathematical operations of differentiation and integration. Differentiation involves measuring the slope of a function by taking the limit of its finite differences; indefinite integration is the inverse operation. Several decades earlier, Bernhard Riemann had been able to prove that under certain conditions, definite integrals can be computed using indefinite integration. Dini suspected that Riemann's conditions were not strict enough to enforce his theorem in all cases, and in his *Fondamenti* he came close to saying as much. Dini was right—there are cases in which definite integration according to Riemann is not the inverse operation of differentiation—and in his paper Volterra provided the counterexample, one that is still used in many calculus textbooks. Volterra showed explicitly a class of pathological functions (functions that oscillate an infinite number of times in a finite interval) that are covered by Riemann's theorem but for which definite integration is not the inverse of differentiation.[8]

That Volterra discovered this important counterexample while a student set him apart from his peers. It also earned him a personal introduction to the distinguished Swedish mathematician Gösta Mittag-Leffler, who came to visit Dini and Betti in Pisa in 1880. He was sitting in on one of Dini's lectures when either by chance or by design, Dini summoned Volterra to the blackboard. Despite the fourteen-year difference in their age, a warm friendship immediately sprang up between Dini's protégé and Mittag-Leffler that lasted until the latter's death in 1927. Mittag-Leffler, who founded the international mathematical journal *Acta Mathematica* two years after his visit to Pisa, did much to promote Volterra's work abroad.

Having reached the age of twenty, Vito now faced the prospect of required military service in the Italian Army. Although there was seemingly no danger of his being called up (Italian law exempted the only male child of a widow), both his mother and his uncle took pains to remind him that he could not afford to miss the appointment with the local draft board in Florence, scheduled for the 16th of November, 1880, a Tuesday, at eight-thirty in the morning. When Angelica pressed him to spend a few days with her beforehand, Vito refused to commit himself, saying only that he would try to arrange such a visit. The record is silent on whether he in fact did so. Still the affectionate son, he had clearly developed an equally strong attachment to his new-found independence.

The visit to the draft board proved inconclusive, except that Volterra was apparently not excused on the basis of the "sole male child" exemption. He was ordered to come back the following year, and several months after his twenty-first birthday he did so, reporting for a physical examination. Among the certificates Volterra's family preserved for posterity is the official *dichiarazione di riforma*, issued one month later, declaring him medically unfit for military service. After recording Volterra's height at 1.64 meters (roughly 65 inches), the examiner ruled his chest too small, an "imperfection" spelled out in Article 1, List B of the army's rules and regulations. Volterra would be able to complete his schooling without the threat of military service hanging over his head.

His friend Carlo Somigliana was not so lucky. Somigliana had transferred from the University of Pavia (having studied there under the versatile mathematician Eugenio Beltrami) to the Scuola Normale in 1879, where he and Vito met and became instant friends. He graduated a year ahead of Volterra, in mathematics, in 1881. When Somigliana got his orders to report for military service, he rushed to take his final exams early, only to discover that he was missing some lecture notes. He needed his friend's help.

"Dearest Volterra," wrote Somigliana, "Confident of your kindness I beg a favor of you, even at the cost of breaking your 'b__.' Could you send me the two celestial mechanics lectures after the famous one in which the very clear Prof. Padova has such a shining role?"[9] Once in uniform, Somigliana started a new round of exams to become a corporal—"a great many exams," he wrote to Volterra in late spring 1882, from Pavia, where he was stationed, adding, "[H]aving seen my great application and diligence they did not reject me, by some miracle. It helps me shout at the top of my lungs and [put] on a great display of military energy when I am in charge of 20 or 30 soldiers in the presence of our colonel." After bemoaning the lack of time for his postgraduate mathematical studies, Somigliana outlined how he spent his days in the Italian Army: "Now I have target practice, now guard duty, now fortification work, and a thousand other nuisances ... that have stopped me" from attending Beltrami's lectures at Pavia.[10]

Mustered out of the army in 1883, Somigliana returned to Como, his hometown, and the daunting task of finding a teaching position. "Military life has been very good for my physique, which was already in excellent condition," he told Volterra, "but... not at all for the rest. One must really admit that it is a totally wasted year.... [O]ne can console oneself thinking that it is for the fatherland and the indivisible good."[11] In his quest for a teaching position, he enlisted Ulisse Dini's help, only to be told bluntly that at such an early point in his career he stood no chance of finding a *liceo* opening and that his only hope for employment lay in a technical school or a junior high school—and even these positions were scarce. Having spent two years canvassing numerous universities and secondary schools around the country, Somigliana reluctantly settled for a job teaching arithmetic in a *ginnasio* [junior high school] in Teramo, a village in the Abruzzi, in central

Italy, noting to Vito that "the position is rather modest; I don't know, actually, if anything less than this is possible. But of course I didn't refuse it, because it is always better than doing nothing."[12] Teramo had few frills, and he had to rustle up math books and journals on his own if he wanted to work on the theory of elasticity, a topic he had been interested in for some time.[13]

Academic success came slowly. As an assistant for calculus at the University of Pavia in 1887, Somigliana began by teaching a course on the functions of complex variables to a dozen or so students in a brand-new Scuola Normale there, in addition to assisting with the recitation sections for first-year students. Meanwhile, he continued to look for better positions elsewhere; in 1891, he was put in charge of the mathematical physics course at Pavia, and in 1892, more than a decade after graduating from Pisa, he won the *concorso* (national competition) to become professor of mathematical physics at Pavia, replacing Beltrami, who had transferred to the University of Rome. Somigliana's circuitous path to a university chair is typical of his time; Volterra's relatively rapid ascent was the exception, not the rule.

Long before his own graduation in June 1882, Volterra had become set in his ways. When he wasn't procrastinating, he found things to worry about. Sometimes Angelica couldn't decide whether she should chide her son for being lazy (he wrote even fewer letters after becoming a *Normalista*) or compliment him for persevering with his mathematical studies. "The pen certainly doesn't weigh you down when you are dealing with problems to solve, it is your strong point, and you are not lacking in ideas, nor is it difficult for you to express them," she wrote.[14] Upon learning that his nephew's breakthrough paper on differential integration had received Dini's stamp of approval prior to publication, Alfonso remarked, "So it wasn't completely wrong, as you kept telling yourself."[15] When, after much complaining, Volterra passed his language examination in German with a nearly perfect score, his uncle gently teased him: "And you were afraid!"[16] Volterra also had to demonstrate proficiency in English. Perhaps weary of hearing his brilliant nephew make excuses for himself, Alfonso wrote to him, in exasperation: "So, English frightens you! It was to be expected. Why do you leave it to the last minute to study a subject in which you know you have to take exams? Dealing with a language, it is even more difficult, if not impossible, to cover everything from the beginning in a few days. In any case let's hope for your lucky star. At the worst, you can repeat the exam in November."[17] As it turned out, this wasn't necessary—Volterra passed easily, much to his uncle's professed amazement: "[T]he examiners must have had a whole lot of cotton in their ears that day, not to hear the blunders you certainly made, if not in something else, in the pronunciation."[18]

The first mention in Vito's correspondence of Enrico Betti's role in charting the future course of his academic career is hesitant, almost as if Volterra feared not being able to make it as a mathematician. In the spring of 1881 he wrote to Angelica, "Of the work which I was doing with Betti, I did some more; we will see if it is all right."[19] Meanwhile, he was also taking Betti's course in celestial mechanics. Betti must have deemed the work "all right," for Volterra's efforts, "On the potential of a heterogeneous ellipsoid on itself," appeared soon after in *Nuovo Cimento*, Italy's sole physics journal, under Volterra's name alone.[20] He sent a reprint of the paper to Alfonso, who, not surprisingly, found it incomprehensible. "Just the title gives me goose bumps," he informed his nephew.[21] Describing her own reactions to his growing list of publications, Angelica wrote: "You cannot believe, my dearest Vito, how much pleasure I derive from seeing your efforts in print, even though for me these writings are truly indecipherable *hieroglyphics*."[22]

For reasons he never disclosed, Vito forbade his mother and uncle to send any of his publications to Edoardo Almagià, the engineer who had taken such an active interest in Volterra's education. But Volterra's activities didn't stay a secret very long, and when news of his success trickled back to Ancona in the summer of 1881, the Almagià relatives started to clamor for a visit from the budding mathematician. Edoardo wrote to Angelica: "Let us speak now of your Vito....Could we have the pleasure of having him among us? On my part I wish it very much, as I think all the relatives here do who are pleased by his rapid progress and the excellent renown he has already acquired. From now on we will call him the "Professor." If he is not yet a Professor by name, he is by deed."[23]

Volterra chose instead to spend the summer at Soffiano, a suburb of Florence in the hills above the church of S. Maria Novella. There his uncle had rented for the summer a small two-story villa set among untilled fields and patches of woodland. Volterra had a room of his own on the second floor, where he could do mathematics to his heart's content and come and go as he pleased. He did not take up the invitation to visit his relatives in Ancona until 1889, when he was twenty-nine years old and a full professor at the University of Pisa.

By the time he had satisfied the requirements for graduation from the university in 1882, "Professor" Volterra had written and published five papers in three Italian mathematics and physics journals; each of them carried a byline that identified him as a "Student of the Royal Superior Normal School of Pisa," a rarity even in those days. He received a doctor's degree in physics at the end of June 1882 (unlike the pomp and circumstance marking the start of Pisa's academic year, no ceremony marked the occasion). Two days after his final examinations, Volterra, now entitled to be addressed as Dr. Volterra, although not yet as *Professore*, packed up his belongings and returned to Florence to stay with his family for the summer. By fall, he was back in Pisa, where Betti had invited him to continue his education as his assistant.

Only recently established in Italy, the position of university assistant was highly coveted by young graduates as a way to begin a postdoctoral career within the university. As Betti's influence on Volterra increased,

Dini's diminished, and once back in Pisa, Volterra would no longer follow Dini's star. Consciously or not, Volterra very likely modeled himself after Betti. Both men had lost their fathers in early childhood and remained close to their mothers; Betti, in fact, never married. A fervent patriot, Betti had fought at Curtatone and Montanara during Italy's 1848 War of Independence. During World War I, Volterra would enlist as a lieutenant in the Italian Army Corps of Engineers and see action on the Austrian front. His mentor served for a time in Parliament; Volterra would become a senator. Betti had spent years fostering mathematical physics in Italy; Volterra did likewise. Much later, in recalling Betti's role as an educator, Volterra wrote that his abilities could "be fully understood only by someone who was his student."[24] Betti tended to remain on warm terms with his students after they graduated; his letters to Volterra are addressed "To the Very Distinguished Doctor" and routinely signed "Your very affectionate friend, Enrico Betti."

Inspired by the interplay of analytical methods with applications to physical phenomena in Betti's courses, Volterra had written his dissertation on a topic in hydrodynamics—the motion of spherical solids in incompressible fluids—under Betti's direction. In tackling this classical nineteenth-century physics problem, Volterra emulated George Gabriel Stokes, J. P. G. Lejeune Dirichlet, and Gustav Kirchhoff, well-known nineteenth-century mathematicians with a strong interest in physics. The neat mathematical technique he used in his thesis (the "image charge" method) had been employed by William Thomson (later Lord Kelvin) to solve various problems in electrostatics. Volterra submitted his work to *Nuovo Cimento*; while correcting the galleys, he learned that W. M. Hicks, an English mathematician, had just published a paper on hydrodynamics using the same technique. Now his disappointment at being scooped was tempered by Betti's invitation to become his assistant.

Ernesto Padova, one of Volterra's professors at Pisa, had moved to the University of Padua, and his chair of rational mechanics remained vacant until a new professor could be chosen by *concorso*. The faculty had petitioned the Ministry of Public Instruction to open the search immediately, and Betti had agreed to take over Padova's teaching load, on the condition that the state would cover the expenses for an assistant. He expected the Ministry to accept both faculty proposals, writing to Volterra, "As soon as the approval arrives, I will need to propose the [assistantship] and I would like to propose you. Write to me if you are inclined to accept...."[25] Meanwhile Volterra had already arrived in Pisa hoping to talk to Betti about applying for a postgraduate scholarship, the *posti di studio Lavagna*, not realizing that his professor had gone to the country for the week. It was Dini who told him about the vacant chair and Betti's scheme and who also advised Volterra to submit his scholarship application anyway, just in case he ended up working as Betti's assistant without any stipend.

"I cannot hesitate for an instant over accepting your very honorable offer," Volterra immediately wrote to Betti. "If I were to be awarded the Lavagna scholarship, I could give up the compensation for the assistantship in favor of whomever you wish, but keep the assistantship itself, for which I would be deeply grateful. Now I ask you to let me know what day it is necessary for me to be in Pisa and to be at your disposal. Accept my best thanks and allow me to call myself your most devoted and obligated, Vito Volterra."[26]

The appointment as Betti's assistant officially took effect in December 1882; by then Volterra had moved back to Pisa and once again into the Palazzo Carovana, to begin work at the Scuola Normale on his *abilitazione*, a certification process (involving another year of study, another thesis, and a public lecture) that would qualify him to teach and hold the rank of professor at a university. He was also competing for the *libera docenza*, the next step in the Italian academic system and a prerequisite to being put in charge of a course. While not a permanent professorial position, even an interim assignment as *libero docente* would boost his academic career. And because Volterra knew all too well how many aspiring university professors fell by the wayside, he made plans to take the examination for a magistrate's certificate, which would allow him to teach in a *liceo*.

That December, shortly after he had begun preparing topics to discuss with Betti's students, Volterra learned from Dini that he could not continue with the Betti assistantship if he won the Lavagna scholarship. When Antonio Ròiti in Florence learned that Volterra was on the verge of withdrawing his name from the Lavagna competition, he contacted Alfonso and told him that his nephew was making a big mistake. Alfonso lost no time writing to Vito that Ròiti felt he ought to "insist to Dini and Betti that you receive the Lavagna scholarship, giving up the assistant's salary; that there are previous cases of young people who have had both positions, but that at the worst, by renouncing the assistant's salary, it wouldn't be too [egregious to accept] the Lavagna scholarship."

The Lavagna scholarship—whose donor, Giovanni Lavagna, taught mathematics at the University of Pisa for many years—paid 2,000 *lire* a year and was renewable for a second year, whereas the salary of an assistant (equivalent to that of a university lecturer) amounted to 1,250 *lire*. Given the disparity in pay, it is easy to understand why Ròiti felt so strongly about Volterra's proposed withdrawal from the Lavagna competition. But Vito should do his best, Alfonso counseled in the letter's closing lines, to "make Betti understand you would be sorry not to receive the salary" from the assistantship as well.[27]

In the end Volterra opted to keep the assistantship, and by the late winter of 1882-1883, Ròiti had persuaded him to apply for Pisa's vacant chair of rational mechanics. Or so he thought. When he learned that Volterra was reluctant to submit the official paperwork for the *concorso*, he urged Alfonso to intercede. From Florence on March 5, Alfonso wrote a long letter to his nephew that contained the admonition: "[Ròiti] told me that if you don't follow [his advice] this time, he will never again give you any suggestions."[28]

For reasons that Volterra never makes explicit in his letters home, his mentor Betti apparently did not encourage him to apply for the vacant chair. Before Betti himself had become a professor at Pisa, he had spent eight years as a *liceo* teacher, and Volterra may have sensed that Betti wanted him to make his way up the academic ladder the same way. It is also possible that Volterra still thought of himself as Betti's student rather than as his colleague. Then, too, Betti may have conveyed to Volterra that he would benefit from being his assistant for another year or so. Perhaps Betti, who was one of the five judges on the selection committee for the vacant chair, wasn't sure that the other four would choose Volterra over the other candidates. Volterra knew that the other applicants would have had more experience. Candidates were required to submit their scientific papers and little else; there were no written examinations, no interviews, no oral presentations. Under the circumstances, who could blame him for feeling uncomfortable about putting his own name forward, particularly without Betti's encouragement—or worse yet, doing it behind Betti's back?

Predictably, on this point, Volterra and his uncle did not see quite eye-to-eye. "As regards your scruples toward Betti and the others," Alfonso wrote in the same letter in which he had warned that Ròiti's patience might be wearing thin, "you should blame it on him [Ròiti], saying that he wanted you to do this, and you have obligations toward him that do not permit you to decline his advice." Having himself participated in such university competitions, Ròiti suspected that a deal might be worked out among the judges behind the scenes—one that would allow the runner-up to be appointed at another university. As Alfonso recounted their conversation, "Professor Ròiti believes that you will have great probability of being successful, if as he believes an event will happen which he did not explain to me." Alfonso's own advice to his nephew was short and to the point: "Don't pass up this opportunity. Act first, and then speak." Fifteen days remained before the deadline for filing but Alfonso, leaving little doubt as to who was really in charge, peremptorily summoned his nephew home: "It has been decided that you come [to Florence] next Sunday, bringing with you all the documents that you can. When you are here, you will settle with Ròiti how to prepare the application [for the chair] and give him the documents."

In what seems an excess of caution, Alfonso suggested that Vito tell Betti that some family matter had come up and that he needed to go home. "If necessary, and you would like me to write it, I could even send you an ordinary telegram before Saturday saying, for example, that there is a relative here who would like to see you, or some other excuse."[29]

As instructed, Volterra took the March 11 train to Florence where he filled out the application for Padova's erstwhile chair and assembled the dossier of his papers, certificates, and other affidavits to present to the Ministry of Public Instruction. He then returned to Pisa, and on March 13, he defended his *abilitazione* thesis—which dealt with several problems related to potential theory—before an audience of professors and *Normalisti*.

The next day Ròiti left for Rome, taking with him Volterra's dossier, including an affidavit signed by the mayor of Florence attesting to Volterra's good conduct. Afraid that his protégé would get cold feet at the very end and not go through with the plan he had meticulously orchestrated, Ròiti went himself to the Ministry of Public Instruction, where he personally submitted Volterra's application. Writing to Volterra from Rome on March 17 ("I hope you do not regret giving in to my insistence"), Ròiti reminded him that he also had two papers in press and to be sure to send the galleys of these papers to the Ministry "*subito*" (immediately).[30] By then, Volterra had almost certainly told Betti about his application, perhaps even along the lines suggested by Alfonso ("Tell him Ròiti *wanted it this way* and Ròiti will write to you from Rome that...he *has acted arrogantly* in handing over your application.")[31] Showing Ròiti's letter to Betti would support any claim that he had coerced Volterra into taking the initiative.

Shortly after his application was submitted, Volterra contracted typhus aggravated by a heat rash (not malaria, but something called miliaria). Angelica hastened to Pisa to nurse him. He ran a high fever, and the quinine he took to reduce it made him agitated; he brooded about his health and his future. Friends who visited noticed his uncharacteristic distress. It was almost three weeks before his physicians felt confident that their patient would recover. His former classmate Corrado Padoa confided to a friend, "Vito retains his usual clarity of mind and if only his temper were better he could speak as usual of mathematics. However, his spirits have changed; he is restless, annoyed, and very worried."[32]

Meanwhile the selection process for the University of Pisa chair had begun. The committee—composed of Betti, Eugenio Beltrami, Cesare Razzaboni, Francesco Siacci, and Ernesto Padova, the chair's previous tenant—singled out Volterra as possessing the qualifications for a full professorship, judged him superior to the six other competitors, and gave him a nearly perfect score: 47/50. Dino Padelletti, a professor of rational mechanics at the University of Naples, came in second. Volterra was still sick in bed, dealing now with the fulminating and infuriating heat rash, when he learned that the chair was his. ("I believe that this news truly revived him because what was tormenting him for many weeks was the fear of having cut a poor figure," Padoa noted.)

In the fall of 1883, Vito Volterra, joined the ranks of the faculty at the University of Pisa. At the age of twenty-three "*il professore*" had both feet firmly planted on the academic ladder.

CHAPTER 6

"Our Professor of Small Intervals," 1883-1893

In the summer of 1885, shortly after his twenty-fifth birthday, Volterra found himself in a classroom in the Sicilian city of Catania, staring down the barrel of a gun. Just ten years after that perilous moment, in 1895, he would receive the Accademia dei Lincei's mathematics prize, for "various memoirs on pure and applied analysis." The co-winner that year was the University of Turin geometer Corrado Segre, the heralded leader of the Italian school of geometry and, like Volterra, from a Jewish background. From 1879 until 1944 (when the Academy of Italy, the creation of Benito Mussolini that annexed the Lincei in 1939, collapsed) seven of the sixteen prizewinners in mathematics were Jewish.* They were prime examples of a pattern in Italy's newly emancipated Jewish population discerned by the social historian H. Stuart Hughes, who observes: "In countless cases a single generation sufficed to bring individual Jews to the forefront of whatever calling they had chosen."[1] With the exception of Guido Fubini, the 1919 prizewinner and a 1900 graduate of the Scuola Normale, Volterra's golden circle of mathematicians belonged to that privileged generation. They were the first wave of Italian Jews to participate fully in all aspects of Italian life, from science to politics, and they had the medals and membership in Parliament to prove it.

Between his eventful visit to Sicily in 1885 and the receipt of the national academy's mathematics prize ten years later, Volterra matured from an alternately cocksure and insecure youth, by turns soliciting and disdaining guidance (and nagging) from his mother and other relatives, to a seasoned, cosmopolitan world-class mathematician.

Volterra's voyage of self-discovery began in Sicily, an island of "people that know only the extremes," as the Sicilian chronicler Leonardo Sciascia has observed.[2] He had gone there at the behest of the Ministry of Public Instruction in Rome, as an external examiner for the local technical high school. He had never yet visited southern Italy, let alone set foot on foreign soil, and he welcomed this new assignment as a break from a relentlessly "boring life" in Pisa—if we are to believe the picture he painted for Angelica in his infrequent letters home.

Her son's grumbling puzzled Angelica. "It's true that you have always said that Pisa didn't offer any distractions," she had written to him in

---

*The others were Salvatore Pincherle (1889), Enrico Castelnuovo (1905), Tullio Levi-Civita (1907), Federigo Enriques (1907), and Guido Fubini (1919).

the late winter. "But you always added that you found the necessary distractions in your jobs, especially in your work, such that you didn't feel deprived and the time passed quickly for you."[3] Still, after seven years in the same small university town in Tuscany, Volterra was ready for a change of scenery. Over the course of the next decade—in fact, down to the end of the nineteenth century—he would take part in many diploma candidacy examinations across Italy. But for sheer excitement, nothing came close to matching what would happen to him as a neophyte examiner in Catania.

Traveling on a reduced second-class ticket, Volterra took the train to Naples and boarded a Società Florio steamship to Palermo, that cosmopolitan city of Phoenician origin on the island's northern coast. Departing at five in the evening, the passengers had their first glimpse of Capri one-and-a-half hours into the voyage and, soon after, their last view of Vesuvius before the steamer, rounding the point into the open sea, headed for Sicily. Early the next morning, the Lipari Islands and the Stromboli volcano could be seen to the south, sending plumes of smoke aloft. After a while, the volcanic island of Ustica beckoned to the west, and then the soaring mountains of Sicily rose on the horizon, signaling the steamer's approach to the Bay of Palermo and the city itself. The sea voyage of some sixteen hours agreed with him—not even a whiff of seasickness, he informed Angelica. From Palermo, he made his way to Catania, which sprawled between a wide bay and the lava-strewn slopes of Mount Etna. In 1885, Catania boasted a new astrophysical observatory (Italy's first), a concentration of odorous sulphur-treatment plants near the rail terminal (Catania was the only city in southern Italy with an industrial base), and a population as volatile as the volcano at its back door. An English visitor to Catania that year recalled being "struck with the peculiarity of its stuccoed [lava material, most likely] buildings, with Mount Etna standing boldly out as a good background, sending forth its volumes of smoke and steam, yet capped with snow and wreathed in clouds." The drive through Catania in an open carriage offered vivid impressions of the city and its inhabitants at every turn:

> The harnesses of the horses are very gay, being one mass of colored and gold or brass brocade, with a very peculiar collar covered completely with polished brass and bells. The carts are mostly painted yellow, the panels decorated with brilliant landscapes of the locality. You see scores of these carts coming down to the harbor from Mount Etna laden with the yellow sulphur of commerce.... We drove through the principal streets to look at the people and the place, and were stared and jabbered at, as we supposed, as though we were barbarians.[4]

Sicily was exotic, its ways and customs inbred for generations and often impenetrable to northerners from the mainland, like Volterra. As Angelica and Alfonso hastened to point out to him, more than a body of water separated north from south. "Please don't be too rigorous," Angelica implored, imagining the fierce young candidates for the high school diploma at Catania's technical institute being grilled by her son. "Many people tell me that in these southern places, it helps to treat people with great respect, as one says, with kid gloves, since they are by nature resentful. Behave yourself." Alfonso's advice echoed Angelica's. Aware of his nephew's tendency to be severe and uncompromising in his opinions, he counseled Volterra to be "somewhat tolerant in your judgments with respect to the candidates and [their] professors. The students especially are touchy and hot-tempered; you need to go overboard in praise, restrain yourself in censuring, taking into account the local character....[I]t's always better to sin on the side of flattery rather than rigor."[5]

Catania had recently experienced a wave of secondary school expansion that culminated in the creation of a *liceo*, several junior high schools and technical schools, and a quartet of specialized technical institutes, including the Istituto Tecnico Gemmellaro, the school to which Volterra was assigned. Students at Catania's Technical Institute were expected to work hard. The Sicilian-born economist Epicarmo Corbino, who graduated from the institute with a degree in accounting in 1908, recalled

> very few outlets for diversion, and therefore scant occasions to spend [money]. There weren't any movie-houses yet; one went to the theatre only on a few Saturdays for drama performances, for which one paid very little (50 *centesimi*).... Certainly with these limited means one could not live with abundance; but then we were so busy studying that, for many other things, there would not have been time to realize the impossibility of having them. We hardly ever went out at night, except towards spring; we went to school on foot....In that era, really studying, one was easily promoted without exams.[6]

In this test-free environment, the arrival of the strict perfectionist from Pisa must have come to some as an unwelcome surprise. Clearly, a new spirit was abroad in the classroom—and, as Volterra's relatives had anticipated, it was not always appreciated. While the local newspaper, the *Gazzetta di Catania*, praised Volterra's work in connection with the final exams, other newspapers faulted him for harshness. Volterra mailed the *Gazzetta di Catania* to his family; he wrote to them about the negative reviews too, although he didn't bother sending any.

"Yesterday we received the *Gazzetta di Catania*, in which you are favorably mentioned, Angelica wrote back. "I am sorry to hear from you that in other newspapers...it is not so. I understand that the task you have assumed is anything but easy. It is a serious one and of great responsibility."

"You can't please everyone," Alfonso added, fearful that Vito might be distraught by what the newspapers had said about him.[7] He might have done better to worry about the volatile students: one of them, a senior, outraged by his failure to pass the Volterra-proctored exams, pulled a revolver out of his trousers and fired at the insufferable visitor. The bullet missed its mark.

At this stage of his life and career, Volterra does seem to have been "difficult." Long before the Catania incident, Angelica had felt it necessary to remind him not to be too hard on his own students in Pisa: "Remember, my Vito, to be benevolent toward those you have to examine. Consider that they are young people to whom a humiliating defeat could be perhaps very damaging both materially and morally." In the same letter, she took aim at another perceived shortcoming, admonishing him "to be less distracted in order not to fall into forgetfulness, which can cause grief and sometimes hurt you."[8] And on another occasion: "You are so absent-minded that you often forget the things you must do, and this is why I am reminding you about them."[9]

Ever the vigilant mother, Angelica laced her letters with instructions that ran the gamut from whom Vito ought to visit to how he should find someone to mend his clothes. When Volterra let slip that he had skipped the university's convocation procession and lecture marking the beginning of the 1884-5 academic year, his parent scolded him roundly: "From the tone of your letter it seems to me that you are in a foul mood, and I regret that very much." She couldn't believe that he hadn't marched up the staircase leading to the *Aula Magna* (great hall) like the other professors, dressed in his academic finery. Angelica was offended by her son's behavior: He had disgraced the family. "It displeases me to hear that you made *una brutta figura* by not taking part in the opening ceremony," she snapped. "I hope you will do what's necessary [referring perhaps to buying or renting the proper regalia] to participate next year, so this won't happen again."[10]

Clearly, Volterra had his share of rough edges, and just as clearly, his family undertook a dogged campaign of improvement to smooth them out. "We are waiting for your letters," Alfonso wrote him on one occasion, "but if you speak about politics, please be more circumspect, especially if you write postcards." Volterra avidly followed the local elections in Pisa, which pitted radical and left-wing critics of the monarchy against factions loyal to the crown and parliamentary government. As in his student days, Volterra sided with a moderate group on the Right that supported the monarchy. On this question, he and his uncle were in agreement, and he had undoubtedly complained to Alfonso about the antics of the Left in local and provincial politics.

Along with discretion, Alfonso counseled tolerance. "You are not wrong, but it is necessary to allow everybody to express their opinions freely," he replied when Volterra griped to the family about city and university officials who planned to commemorate Giuseppe Mazzini, the revolutionary Risorgimento leader who had championed a united and republican Italy. "To avoid troubles of any kind," Alfonso continued, never argue politics with others (especially with young people). This argument, particularly among young people, can often degenerate and cause personal disputes. Besides, if we, the followers or admirers of Cavour, Massimo D'Azeglio [a writer of historical romances and polemical tracts and prime minister of Piedmont, who championed the emancipation of Italy's Jews], etc., do not organize commemorations to honor them, it doesn't mean that Mazzinians can't commemorate their leader's death. As long as the demonstrations are not excessive and stay within limits, everyone has the right to them.[11]

Despite the solid good sense contained in this advice, Alfonso was no admirer of Mazzini, who symbolized to him subversion and violent uprisings in the cities. Cavour's moderate nationalist program, in contrast, had helped establish a constitutional monarchy in Piedmont, along with a body of law that, until it was struck down by Mussolini in 1925, guaranteed Italian Jews the same rights enjoyed by other Italians in the Kingdom of Italy. D'Azeglio, a constitutional monarchist, had repudiated the secret societies and abortive revolts favored by Mazzini. Consciously or unconsciously, the family had begun to groom Volterra to play a larger role in Italian life. Many of Alfonso's political heroes, the great statesmen of the Right in the early years of Italy's formation as a nation, would later become Volterra's role models as well.

Here the question naturally arises: Were those qualities that Angelica and Alfonso considered shortcomings—Volterra's seeming inability to compromise his own gold standard of excellence, his inflexibility regarding student performance, and his opinionated attitudes—strengths or weaknesses in his character? The unsociable tendencies they deplored may well have contributed to the strength and fortitude with which he would, late in his life, stand up to Mussolini's Fascist regime. Volterra himself has provided no obvious clues about the development of his personality. Although he wrote pages and pages of mathematical calculations, public discourses, and monographs, he left behind no memoirs, no autobiography, no diaries—nothing that would help us to better understand the fastidiously private man who was catapulted into the ranks of the international mathematics community two years after his close call in Catania.

Between 1883 and 1886, Volterra's rank as an associate (*straordinario*) professor at the University of Pisa came with an annual salary of 3,000 *lire*. But after he had sent money home each month—repayment for the money borrowed for his college education, as well as funds to cover Angelica's current expenses—very few *lire* remained in his pockets. Having depended for many years on the generosity of others, Angelica treated money respectfully and admired those who had it; Vito barely noticed its absence. Give him a sum of money and he was frequently at a loss how to spend it. "I believe

you're the only person who could say, 'I don't know what to do with it,'"
Angelica wrote him in June 1884, upon learning that he had come into 200
lire unexpectedly. She advised him to pay off his debt to the bookseller
and keep the rest for himself, but also hinted that she could always use
more money. "If you like, you can send me the rest, for one never lacks
for wanting things, even taking in account what's necessary only." Alfonso,
being the banker in the family, recommended paying the bookseller on the
installment plan; he too volunteered to relieve Vito of any remaining funds,
referring in passing to "large expenses."[12]

That December, Volterra attended a festive faculty banquet and enjoyed
himself tremendously ("You certainly must have been the youngest among
all the professors and the invited guests," Angelica exulted. "It is a great
satisfaction, truly, to find oneself in the company of such distinguished and
celebrated persons.") only to discover that he was short of funds with which
to pay his share of the bill. Once again, it was an opportunity for his mother
to crow: "I really can't reproach you for having some extra expenses, but
you see how wrong you are in assuming that you can take care of everything
with just a few *lire*. I tell you always that it is necessary to think about
the unexpected expenses: whatever they say about *vil metallo* ["the vile
metal"], it is unfortunately indispensable, and one must appreciate it."[13]

Above all, Volterra loved books. A habitué of the Scuola Normale's library, he started helping out there and gradually took on more and more
responsibility for its operations. In 1886, with Enrico Betti's blessings and
the approval of the Ministry of Public Instruction he was put in charge of
the library on a temporary basis, until his promotion to full (*ordinario*)
professor the following year, when his annual salary would rise to a lordly
5,000 *lire*. The library position carried no stipend, but it came with room
and board in the Palazzo Carovana and, of course, the luxury of using the
library at any time. Betti had assured him that his promotion to *professore
ordinario* was coming, and sure enough, in April 1887 Volterra became a full
professor. That year, he also received the first of many honors and prizes,
the gold medal for mathematics of the Italian Society of Sciences (known
as the Forty). He now found himself serving on many national committees
to advance the academic careers of other mathematicians, and he soon became a member of Italy's most prestigious scientific organizations, starting
with the Accademia dei Lincei. As he gained seniority, his administrative
responsibilities would increase: In his last year at Pisa, Volterra served as
dean of the science faculty.

Perhaps most important, with his salary increased to 5,000 *lire*, the
family no longer needed to find ways for him to save money. Angelica
radiated maternal pride: "The moral satisfactions that children give their
mothers are always amply rewarded, and I hope to be able to see you, as
much as possible, always happy," she wrote. Toward the end of her letter,
Angelica worried yet again about Vito's health. He led too sedentary a
life; with the arrival of spring he would benefit from more fresh air and

less book study. "I hope that now with this beautiful weather you will get some physical exercise, otherwise you will become fat," she pointed out to the son who at the age of nine had found literary, or at least epistolary, inspiration in his first piece of strudel. Then you won't be able to take long walks without tiring yourself, which is really a bad thing."[14] Volterra did indeed put on weight as he grew older; his once slender frame filled out, his shoulders became broader, the waistline thickened.

Early in 1887, Volterra's maternal grandmother, Fortunata, died, and later that year, Angelica left Alfonso's household in Florence to live with her son in Pisa, in a spacious apartment he had taken at Via San Martino 30. Her last surviving letter to him from Florence (they did not have to correspond by letter again until 1896, when Angelica made a visit to Edoardo Almagià and his family in Rome) makes it clear that she was as vigilant, protective, and fully engaged in her son's life as ever. He had earlier written to her about his plans to spend a day sightseeing in Milan with his former geometry teacher Cesare Arzelà, now a professor of mathematics at Bologna.

"Amuse yourself," Angelica wrote to her son, but she reminded him that if he hurried "as always to see everything, you get hot and you get sick." Domestic news followed. She and her brother's family were staying in the country, but the accommodations no longer measured up to Angelica's rising expectations. "As you know, the country house leaves something to be desired, there are no amenities at all, and some comforts [are lacking]. So what can you do, you have to adapt." Along with her "many loving hugs and blessings," Angelica reiterated an earlier piece of advice: "Be indulgent in your examinations," she urged Vito, referring to the oral tests in mathematics he would be administering to his students at the university.[15]

In Florence, Angelica, along with her mother and sister-in-law, had kept house for her son and her brother's growing family. In Pisa, she resumed this role for Vito, but now she had only the two of them to please. Her presence on the home front (including the maintenance of a kosher kitchen) left Volterra free to work, to go abroad, to attend conferences, to serve on commissions without any domestic distractions or responsibilities. He, in turn, introduced Angelica to his colleagues in Pisa and elsewhere, and in the summer mother and son vacationed together, often at Engelberg, in the Swiss Alps.

At the university, Volterra taught rational mechanics, a course designed to lay the groundwork for more specialized work in mathematical physics. As Eugenio Beltrami at the University of Pavia wrote to Volterra shortly after his professorial appointment, "the course work [in rational mechanics] has remained unchanged for many years [and] will take away less time from your own personal work."[16] Volterra would also give courses in mathematical physics and graphical statics (*statica grafica*), the latter a problem-solving course involving graphical construction without mathematical calculations— and a course still taught in Italian engineering and architectural departments today.

As predicted, Volterra's teaching load did not interfere with his "personal work." From the outset of his professorial career, Volterra had devoted himself to three subjects: the equilibrium of flexible, inextendable surfaces; linear differential equations; and an extension of Riemann's theory on the functions of complex variables in three-dimensional space. In autumn 1887, his mentor, and now his colleague, Betti, forwarded the first of Volterra's three papers dealing with differential equations to the Accademia dei Lincei's mathematics prize committee. He only told Volterra about it after the fact. "Leaving aside the bigger or smaller probability of winning the prize," he told Volterra, "it seems to me it would be very useful to solicit an authoritative opinion on the value of one's work." Then he appealed to Volterra's national pride. "On this occasion also I would like our country to show off all its good mathematical output."[17] However, Volterra was still feeling his way as a mathematician and seems to have wanted no part of Betti's well-meaning schemes. He thanked Betti for thinking of him, but "knowing the names of other candidates," he wrote back, he couldn't possibly imagine how he might win. The fragmentary draft of his hastily scribbled reply, a mishmash of barely legible jumbled phrases, unfinished thoughts and inked-out words, reveals Volterra's efforts to put the best face on his insecurity:

> I would compete only in order to get an opinion on my works...[yet] it seemed too early to me to ask for a judgment, because none of the problems I studied in the last four years has been completed in a way [the sentence breaks off here]....On the three linear differential equations, I just published the first part, and although I edited the second one, another difficulty to overcome is still keeping me from publishing it and from going on with the editing of the following part; and I don't believe that the problems studied in the first [paper], on which I realize I have spent too much time, could explain the first one in the study I did. Finally, in this last work, I am still far from getting over many difficulties and finishing it; I did this for itself and love of this, the idea that led me....I wanted to explain these reasons first; and [the draft breaks off abruptly here, in mid-sentence.][18]

Did Betti have his way, or did Volterra stand his ground? The written record reveals only that the mathematics prize, when it was finally awarded two years later, was shared by Luigi Bianchi and Riccardo De Paolis, professors of analytic geometry and higher geometry, respectively, at the University of Pisa; and Salvatore Pincherle, professor of mathematics at the University of Bologna (and one of the few notable mathematicians of Volterra's generation to publicly endorse the Fascist regime in the 1920s). Volterra's turn would come in 1895, by which time he had moved to the University of Turin, the unfinished studies on differential equations had been published, and he had inaugurated a series of ground-breaking papers in the *Rendiconti dell'Accademia dei Lincei* on the theory of functions.

In the field of analysis, Volterra is one of the inventors of a new branch of mathematics: functional analysis. Starting with the German mathematician Johann Peter Gustav Lejeune Dirichlet's definition of a function (a number dependent on another number) and Bernhard Riemann's generalized definition of an integral, Volterra methodically and systematically set about creating a new calculus, complete with definitions and examples, that was, in his words, "analogous to the ordinary calculus used for operations on functions."[19] In marked contrast to the hesitation he had showed over promoting his paper on differential equations, Volterra's first paper in 1887 on this topic, "On functions that depend on other functions," opens with this sweeping and self-assured assertion: "Permit me to point out in this Note several considerations that help to clarify some concepts that I believe necessary to introduce for an extension of Riemann's theory of functions of complex variables, and that I think can be useful in various other researches too." Several sentences later, Volterra illustrates what a functional is by giving examples of how it is used:

> In fact in many questions of physics and mechanics, and in the integration of partial differential equations, there sometimes arises the need to consider quantities that depend on *all the values* that one or more functions of a variable take in a given interval, or one or more functions of more variables take in a given field. Thus, for example, the temperature at a point on a conducting surface depends on all the values of the temperatures of surroundings points; and the infinitesimal displacement of a point on a flexible and inextendable surface depends on all the components of the displacements of surrounding points, parallel to a certain direction.[20]

Between 1887 and 1890, Volterra produced a steady stream of publications relating to functionals. They ranged from five-page notes in the proceedings of the Lincei to a comprehensive fifty-three-page summary, in French, in Gösta Mittag-Leffler's *Acta Mathematica*, a publication that hastened the spread of his ideas abroad. Volterra soon abandoned the designation "functions that depend on other functions" in favor of the expression "functions of lines." In 1903, the French mathematician Jacques Hadamard, who was among the first of Volterra's European colleagues to grasp the scope and magnitude of Volterra's research program and had steered his brightest students in that direction, coined the term "functionals." That name stuck and is still in use today.

Shortly before the end of World War II, Hadamard wrote a now-classic essay entitled "The Psychology of Invention in the Mathematical Field," in which he tackled the problem of where these ideas in mathematics come from and what role semantics, mathematical reasoning, and mental pictures play in the process. The last chapter deals with the question "How do mathematicians go about selecting problems to solve?" Turning to Volterra's work on functionals, he writes:

> Why was the great Italian geometer led to operate on functions as infinitesimal calculus had operated on numbers, that is to consider a function as a continuously variable element? Only because he realized that this was a harmonious way of completing the architecture of the mathematical building, just as the architect sees that the building will be better poised by the addition of a new wing. One could already imagine that... such a harmonious creation could be of help for solving problems concerning functions considered in the previous fashion; but that "functionals," as we called the new conception, could be in direct relation with reality could not be thought of otherwise than as mere absurdity. Functionals seemed to be an essentially and completely abstract creation of mathematicians.
>
> Now, precisely the absurd has happened. Hardly intelligible and conceivable as it seems, in the ideas of contemporary physicists (in the recent theory of "wave mechanics"), the new notion, the treatment of which is accessible only to students already familiar with very advanced calculus, is absolutely necessary for the mathematical representation of any physical phenomenon. Any observable element, such as a pressure, a speed, etc., which one used to define by a number, can no longer be considered as such, but is mathematically represented by a functional![21]

Long before he had made his name in mathematics, Volterra's fellow students at Pisa had enjoyed calling him "our professor of small intervals."[22] Little did they know then where his interest in classical analysis would lead.

CHAPTER 7

"The Life I Live," 1887-1895

By the late 1880s Volterra had begun to sample the other mathematics capitals of Europe, pushed in part by a fellow mathematician—one Giovanni Battista Guccia, an exuberant, energetic, and fiercely nationalistic Sicilian geometer, who knew and corresponded extensively with Europe's leading mathematics researchers. Independently wealthy, Guccia had not only founded and underwritten the *Circolo Matematico di Palermo*, Italy's oldest mathematical society, but also started, in 1885, a journal, *Rendiconti del Circolo Matematico di Palermo*, in which members of the society could publish. Within a short time, Guccia's *Circolo* had grown into an international organization and the *Rendiconti* into a respected mathematical journal.

Volterra was elected a nonresident member of the *Circolo* in December 1887. Convinced that Volterra's work would not necessarily get the attention it deserved elsewhere on the continent, Guccia counseled, "You have sent, I hope, some reprints to [Georges] Halphen, [Jean Gaston] Darboux, [Emile Picard], [Paul] Appel, [Georges] Humbert," while lamenting the shortcomings of his countrymen: "We Italians, you will pardon me, have the great defect of not knowing how to value our scientific production: we keep it under lock and key, hidden, in ways and means that foreigners never succeed in recognizing and appreciating."[1] Only a few years earlier, Alfonso Almagià had pleaded with his nephew to send a reprint to cousin Edoardo, the engineer in the family, and had been soundly rebuffed. Guccia's message, coming from a fellow mathematician, in a professional context, now struck a more self-confident Volterra as sensible advice.

In spring 1888, Volterra journeyed to Paris—the city he would come to prefer above all others—armed with a stack of his reprints and selected volumes of the Scuola Normale's *Annali*, intended for various French mathematicians, including Gaston Darboux, professor at the Sorbonne and dean of French geometers. Celebrated for "approaching geometrical problems with full mastery of analysis and differential equations, and working on problems of mechanics with a lively space intuition," Darboux was a mathematician after Volterra's heart, and upon arriving in Paris Volterra went straight to the University of Paris campus, in the Latin Quarter.[2] But he had not taken into account the local customs of his colleagues. "Unfortunately, the geometers all obey a common law; they all travel at the same time. At Easter, it is very rare for me to remain in Paris," Darboux wrote to Volterra several weeks later, promising a warm welcome on his next visit. Meanwhile, without having met any of his French counterparts, Volterra was voted a member of the Mathematical Society of France—an unmistakable sign that he had made the grade as far as the nation's mathematical establishment was concerned, even if he had picked the wrong week of the year in which to introduce himself. Volterra's election that July to the venerable Accademia dei Lincei in Rome only confirmed his growing reputation in mathematics circles at home and abroad.

Volterra was then twenty-eight. He left Pisa that summer as soon as classes had ended, heading first for Soffiano, in the countryside near Florence, where the family had rented a house for the summer. He stayed for a week or so, then set off again for some rest and relaxation in the mountainous Engadine of Switzerland, to be followed by an excursion in the Alps with Gösta Mittag-Leffler, the Swedish mathematician he had met during his student days in Pisa. When they rendezvoused at Le Pre, a village near Lake Geneva in the foothills of the Alps, Mittag-Leffler had a better idea.

"He persuaded me to go with him to the Harz [Mountains] in Germany, where [Karl Theodor] Weierstrass and Mme. [Sofia] Kovalevskaya are staying, telling me that along the way I would have the opportunity to see most of the German mathematicians," Volterra told Betti, in a letter written from Heidelberg at the beginning of August, explaining why he had not yet returned to Italy. "The trip is a very long one, but a more favorable occasion than this to make the acquaintance [of mathematicians] I could not have arranged as easily, and I accepted very willingly Mittag-Leffler's kind invitation."[3]

The urbane and cosmopolitan forty-two-year-old Swedish analyst and the rather shy and socially awkward Volterra, fourteen years his junior, became good friends as they sprinted from one German university town to the next before finally heading for Wernigerode, a picturesque little municipality in the heart of the Harz Mountains. In Göttingen, Volterra met Felix Klein and Hermann Schwarz; in Halle, he chatted with Georg Cantor; in Marburg with Heinrich Weber; and in Heidelberg with Leo Königsberger. In the space of three weeks, he got to know most of Germany's elite mathematicians.

In Wernigerode, Volterra spent several days with the venerable Berlin University mathematical analyst Karl Weierstrass and the celebrated Russian mathematician Sofia Kovalevskaya, Weierstrass's former pupil and a professor at the University of Stockholm. Mittag-Leffler, also a former student of Weierstrass, had met Sofia for the first time in St. Petersburg a dozen years earlier and had been swept off his feet the moment he laid eyes on her. "More than anything else in Petersburg what I found most interesting was getting to know Kovalevskaia," he gushed to a Swedish colleague.

> ....As a woman, she is fascinating. She is beautiful and
> when she speaks, her face lights up with such an expression

of feminine kindness and highest intelligence, that it is simply dazzling....[S]he is in all respects a complete "woman of the world." As a scholar she is characterized by her unusual clarity and precision of expression....I understand fully why Weierstrass considers her the most gifted of his students.[4]

Kovalevskaya—the only woman to hold a university chair in mathematics in Europe before 1900—had come from London to Wernigerode to consult with her former teacher about an old problem in classical mechanics that had consumed her for five years. Her research involved solving the differential equations for the general case of the motion of a rigid body rotating around a fixed point: Leonhard Euler and Joseph-Louis Lagrange, the two greatest mathematicians of the eighteenth century, had considered two special cases of rotating rigid, symmetric bodies, such as spinning tops. "I would hate to die without discovering what I am looking for," Kovalevskaya told a friend shortly after she began her own investigations. "If I succeed in solving the problem on which I am now working, my name will be listed among those of the most prominent mathematicians."[5]

What she was looking for—and had found by the time Volterra and Mittag-Leffler caught up with her in the Harz Mountains in the summer of 1888—was "a new special case where the equations could be completely integrated." As the historian of mathematics Roger Cooke points out in his study of her mathematical work, "Kovalevskaya's new case represented physically a *nonsymmetric* body, and the rotation of such a body under the influence of gravity is extremely complicated."[6] Then on the verge of submitting a paper about her discovery to the French Academy, which had dangled a prize on this very topic before her eyes, Sofia grumbled about having to put the finishing touches on it while the other mathematicians and their families who had joined Weierstrass and his sisters in the mountains spent their time talking mathematics. Still, Kovalevskaya found time to tell Volterra about her work on the rotation problem. "I learned from Signora Kowaleski [sic] what her discovery is relative to the motion of a solid body subject to gravity and it truly is a very important discovery," he wrote to Betti from Florence in the late summer of 1888.[7] That December, Kovalevskaya's memoir won the French Academy's Prix Bordin; in its report, the prize committee cited her use of the freshly minted and highly complicated theory of Abelian functions and hyperelliptic integrals—one of Weierstrass's principal research topics—to solve a physics problem.[8]

In the meantime, Volterra had gotten a taste of Karl Weierstrass's newly developed theory of Abelian functions from his traveling companion, Mittag-Leffler. Regarded as "the mathematical conscience par excellence, methodological and logical"[9] among his contemporaries, the seventy-three-year-old Weierstrass impressed Volterra as someone who "carries his years well for his age."[10] Even before setting foot in Wernigerode, Volterra knew, from browsing through Mittag-Leffler's copy of Weierstrass's two large volumes of lectures on Abelian integrals, that the eminent German analyst, legendary for his polished and well-organized lectures at the University of Berlin, had made great strides in revising and expanding his theory. From Heidelberg, Volterra wrote to Betti: "I was able to look at it a little; they are much more interesting than the lectures that were copied for the Scuola [Normale] several years ago, because they contain a part of Weierstrass's theory that hasn't yet been published or cited."[11] In fact Weierstrass published very little during his lifetime; his lectures were the main vehicle for spreading his ideas, which is why Volterra went on to recommend that Betti pay someone in Germany to copy the newest version of Weierstrass's lecture notes.

An aura of mystery surrounded Weierstrass's lectures. In general, the Berlin mathematician forbade his students to "reproduce his propositions," as Eugenio Beltrami once remarked in a letter to a colleague in France, Jules Houël. He was baffled by the exceptions that Weierstrass apparently permitted, or at least ignored. A case in point was Salvatore Pincherle, who, after returning to Italy from a year of postgraduate work at the University of Berlin with Weierstrass in 1877-78, had published a comprehensive exposition of Weierstrass's theory of analytic functions in the *Giornale di Matematiche*, which became the main vehicle for introducing Weierstrass's new theory to Italian mathematicians.[12] "I never wanted to ask him if he was doing it with or without Mr. Weierstrass's permission, for fear of awakening his scruples, if by any chance he did not have [permission]; because, truthfully, I think that the German geometer's veto is so extraordinary, that [Pincherle] has all the reason to disregard it, with the help of his youthful candor," Beltrami told Houël.[13] Several years later, Beltrami mentioned to Carlo Somigliana at Pavia that the Scuola Normale had the course notes—in what form Beltrami couldn't say—for Weierstrass's lectures on the general theory of functions. Somigliana dashed off a short note to Volterra. After remarking that Pavia possessed a manuscript copy of the course on elliptical functions, he came straight to the point: How could Pavia get its hands on Pisa's copy of Weierstrass's function theory notes? "I beg you to tell me in what manner your library has been able to procure these lectures," he wrote, and dispatched his old friend *Mille affettuosi saluti* ("A thousand affectionate greetings").[14] Volterra's answer has not been found. But if he did write back, it is likely he would have suggested solving the problem the same way Betti did, by spending money. It often works.

The Volterra / Kovalevskaya / Mittag-Leffler story does not end in the Harz Mountains. Among Volterra's personal papers is an envelope in Kovalevskaya's handwriting addressed to him at the University of Pisa, postmarked March 27, 1890. There is no record of what the envelope contained, nor do Volterra's papers at the Lincei contain any correspondence between the two mathematicians. Eleven months later, on February 10, 1891, Sofia Kovalevskaya died of influenza in Stockholm; she was only forty-one and at the height of her powers.

Volterra wrote to Mittag-Leffler that June:

Monsieur et cher ami,

> I cannot begin my letter without expressing to
> you my deepest regret for the loss which the
> *Acta*, the University of Stockholm, and the
> mathematical sciences have suffered through
> the death of Mme. Kowalevski [sic]. This
> very cruel loss unfortunately precedes only by
> a few days a very sad communication which
> I consider it my duty to make to you be-
> fore anyone else, and which I think will quite
> amaze you.[15]

Kovalevskaya, it seems, had made a mistake in an earlier paper on the
refraction of light in crystals, published in 1885. (In fact, her teacher, Weierstrass, had suggested both the problem and the method for attacking it,
although he had never published the method himself.) Starting with a set
of differential equations derived by the French mathematician Gabriel Lamé
twenty years earlier to explain the propagation of light as a disturbance in an
elastic medium,[16] Kovalevskaya had worked diligently on the light refraction
problem for two years and believed, as did Weierstrass, that she had found
the general solution of Lamé's equations. Her paper had been proofread by
Carl Runge, a colleague in Berlin, given an approving nod by Weierstrass,
and forwarded to *Acta Mathematica*, Mittag-Leffler's journal, for publication.[17] Six years later, in 1891, Volterra stumbled on Sofia's error while
preparing his own class lectures on elasticity theory and optics. "I verified
that the functions she gave as integrals of Lamé's on light waves in doubly-
refracting media...do not satisfy the differential equations," Volterra wrote
his Swedish colleague. Anticipating Mittag-Leffler's response to this unexpected news, he added: "To convince you of this very strange result, I ask
you to examine the formulas given on page 297 [of Kovalevskaya's article]."[18]
Just as surprising, continued Volterra, Lamé's solution was not right, either.

In June 1891, before leaving Pisa for a return visit to Göttingen, Volterra
sent Mittag-Leffler a sixty-two-page paper on the integrals of double refraction, another example of his mastery of analysis and differential equations.[19]
In reviewing where Kovalevskaya had gone astray, Roger Cooke has argued
that "[h]er mistake was a purely technical one to which she was probably
led by the firm conviction of Weierstrass that his method must work on
these equations. (It doesn't)."[20] As for Volterra, adds Cooke, he "built on
the foundation Kovalevskaya had laid, avoiding her mistake—she claimed
that it is possible to differentiate a certain integral under the sign of integration when it is in fact not possible—and [Volterra] produced correct
solutions." Volterra, one might add, also showed himself to be a careful
reader. More important, his name had begun to resonate in the halls of
Europe's mathematical capitals, including Paris. When Volterra sent Emile

Picard, professor of differential calculus at the Sorbonne, a copy of the light refraction paper, he received the following reply:

> *Cher et éminent collègue,*
>
> I have long wished to make your acquaintance....Your beautiful works of analysis and mathematical physics have interested me extremely, and, just now, your memoir relative to Mme. Kowalewski's [sic] studies have been for me the object of a very deep examination. You have also been much occupied with generalizing the functions of a complex variable and the memoir that you have published on this subject in *Acta Mathematica* contains many profound and original views; they touch as well some of the questions that occupy me. I hope that some day you will go on with this memoir, and I think I will be able to profit from your work in the continuation of my research on the algebraic functions of two variables, a matter that preoccupied me greatly some years ago and that I am considering taking up again soon....
>
> I do not wish to abuse today, dear sir, the liberty that I am taking in starting a correspondence with you. If you would pay me the honor of a personal correspondence, I would have a great deal to learn from you.[21]

In effect, Picard's letter offered Volterra the assurance he needed as a young scientist that one of France's leading mathematicians, and by extension his colleagues at the Sorbonne, welcomed him as an equal, a connection that would matter greatly to him later on.

Volterra arrived in Göttingen, a red-roofed town of 18,000, in early July 1891 for a month's stay. A destination for mathematicians since the time of Gauss—the mathematical virtuoso who studied there at the end of the eighteenth century and later served as professor of astronomy and director of its observatory—Göttingen hummed with activity, from scientific societies and student taverns to companies churning out precision scientific instruments. A medieval town hall, a monument to Göttingen's storied past as a trading center, stood in the center of the old town, its open arcade, Great Hall, and Gothic heating system still intact. Ancient half-timbered stucco houses, many of them intricately decorated, and shops flanked the narrow, crooked streets. An ancient wall, its ramparts lined with lime trees, enclosed the older part of the town. To this day, after Sunday lunch, as David Hilbert's biographer Constance Reid reports, "the townspeople 'walk the wall'—it is an hour's walk."[22] Outside the wall, in the town's newer sections, stood the university named for its founder, Elector George Augustus (George II of England). Its buildings included the Auditorienhaus, a graceful three-story, yellow-brick structure on the Wilhelmsplatz, erected in 1865 and home to Göttingen's mathematicians.

Like Pisa, the university had gone through periods of student and faculty unrest. The most famous incident occurred in 1837, when Ernest Augustus, the new king of Hanover, revoked the liberal constitution of the kingdom and demanded that all public servants swear allegiance to him instead. Gauss, then sixty years old, protested privately but did not resign. But seven other Göttingen professors (among them the comparative-philologist brothers Jacob and Wilhelm Grimm and the physicist Wilhelm Weber) refused and were summarily dismissed—a story that eerily foreshadowed Volterra's refusal to swear allegiance to Mussolini almost a century later. Göttingen's prestige and enrollment plummeted; after Hanover became a Prussian province in 1871, the university's reputation gradually revived, receiving a significant boost in 1885 with the arrival of the accomplished mathematician Felix Klein.

Klein's reputation rests on his work on geometry, function theory, and the theory of groups. In his inaugural address at Erlangen University in 1872, Klein had used the group conception as a way to classify much of the mathematics of his time. Known as the Erlangen program, the address "declared every geometry to be the theory of invariants of a particular transformation group. By extending or narrowing the group we can pass from one type of geometry to another," mathematics historian Dirk Struik writes.[23] Klein's program not only made space in the mathematical world for non-Euclidean geometries and topology, it also pointed the way to a synthesis of the geometrical and algebraic work of earlier generations of mathematicians, from Gaspard Monge to Gauss to Riemann.

Besides his talents as a writer and lecturer on the theory, history, and teaching of mathematics, Klein, as the historian of mathematics David Rowe writes, had an ambitious "long-range plan: to rehabilitate the Göttingen tradition of Gauss and Riemann, with its strong emphasis on the interplay between mathematics and physical reality" and to unseat Berlin, the leading school for German mathematics for decades.[24] By the time of Volterra's second visit to Göttingen, Klein had become a supremely successful academic empire builder (he would bring Hilbert there in 1895) and had set in motion his plans to make Göttingen the finest research center in Germany for the exact sciences—which may explain why Volterra chose to spend a full month there rather than in Berlin.

Among the professors he sought out was Hermann Schwarz, a gifted mathematician who combined great intuition with a flair for geometry. Volterra rented a room in town and settled into a comfortable routine. As usual, he kept in touch with his mother:

Here's the life I live. I get up at 7, have breakfast and then around 9 I go to the university where I usually remain until one o'clock and then I go to lunch. Here everybody eats at a round table and the meal costs very little. I have a contract at the Union, one of the best hotels and I spend one mark and 10 pfennig, which is to say, 1.40 lire, for lunch... soup and two generous courses of meat with all sorts of vegetables, potatoes, salad, cooked fruit, etc. and afterwards butter and cheese. No one drinks wine usually, which costs at the rate of at least 10 lire per flask, but in compensation there is excellent beer. At night I have supper at home alone. Supper consists of tea. Here, by tea one means: Tea, butter, bread, sausages, ham, roast beef, etc. But I often have dinner at the hotel and then instead of drinking tea with the meal, I drink beer.

Göttingen is like Pisa both in size and way of life; but I see that people have much more fun here than in Pisa. Every night, in fact, invitations arrive, from professors naturally, because unlike Pisa, here [I know] only professors and their families. Last night [Ernst] Schering [a mathematician] invited me to a dinner at the Union and afterwards there was dancing. Tonight there is a party to celebrate the anniversary of the foundation of the Matematische Verein [German Mathematical Society] but this kind of party is rather tedious. All it consists of is drinking a lot of beer; then everyone is given a book and has to sing together the whole evening.[25]

Despite his frequent claims that "I have nothing to say," Volterra's letters brought home to Angelica her son's keen interest in observing how other people lived and his growing passion for travel. His lodgings in Göttingen may have lacked amenities he took for granted in Italy—night tables, candles, sheets on the bed—but whatever the locals used instead, he pointed out, served just as well. The house he lived in could have been cleaner: "the cleanliness bears little relation to the labor expended to clean it....In fact, every day there is a woman with a big wash tub full of water who continuously scrubs floors, staircases, etc."[26]

Overall, however, his accommodations pleased him. The weather, which Volterra remembered from his brief visit three years earlier as foggy and rainy, had improved—"it is neither hot nor cool"—and the thick forests and lush rolling hills beyond the town's wall now looked to his eyes "very beautiful." He explored the countryside, hiking in the nearby hills (Klein had christened them the "Viale di Colli," Göttingen's street of hills) and attending fancy evening balls at a luxurious resort ("the most remarkable place you can imagine"), where the cream of Göttingen society mingled and danced. He also worked on improving his language skills. "I speak German very badly," he wrote Angelica. "For ordinary things it is very easy to get by; the difficulty comes in the conversation, especially understanding it." (He traded Italian lessons for German lessons with a visiting humanities scholar.) The month in Göttingen flew by. "I would like to have worked more," he wrote Angelica toward the end of his stay, "but here there are too many distractions."[27]

Shortly before he left Göttingen, Volterra asked his mother to send him his white tie and tails, explaining that in Berlin, his next destination, where he knew many mathematicians, people dressed up more "than here in the provinces." Gone for good were the days when Volterra, in borrowed or ill-fitting dress clothes, attended formal events in Pisa under protest. Public appearances now demanded clothing befitting the distinguished professor he had become. He instructed Angelica to pack his vest, shirt, and "the decorations"—a reference to the gold medal of the Italian Society of the Sciences, the insignia of the Lincei, and his latest acquisition, the Ordine Cavaliere della Corona d'Italia—carefully, "in such a way that nothing will be damaged."[28]

By the time he reached Berlin, only Leopold Kronecker, Weierstrass's eminent colleague, was there to greet him; the other mathematics professors had all left on vacation. Unlike the weather in Göttingen, Berlin's climate did not cooperate—fog and rain prevailed—but the capital's vast imperial palaces, towering monuments, opulent government buildings, broad boulevards, and leafy urban parklands impressed him. After spending a few days sightseeing ("Berlin is a large city and there is a lot of traffic," he told Angelica), he headed for Leipzig, then Nürnberg (where the mathematician Carl Runge had invited him to his country house), then to Munich for a couple of days. In mid-August of 1891, having temporarily satiated his wanderlust, he returned to Italy.

The rapidity with which Volterra had vaulted into a position of prominence in the international scientific community helps explain why he became the spokesman for the new school of Italian mathematicians. He also became an impassioned champion of the young nation's scientific heritage. Whether his audience was an august group of mathematicians or the readers of a science journal for the nonspecialist, Volterra addressed the same themes. The teachers of his youth had played a significant intellectual role in working out the theory of functions in the nineteenth century, just as they had played a significant cultural and political role as midwives at the birth of Italy's parliamentary democracy. It's not an accident that when asked to prepare a general plenary lecture for the second International Congress of Mathematicians in Paris in 1900, Volterra chose to talk about Enrico Betti, Francesco Brioschi, and Felice Casorati—scientists who equated their own scientific ideals with the liberal, democratic state that Italy had become.

CHAPTER 8

"Demonstrations of Their Resentment," 1893-1900

When Enrico Betti died unexpectedly in 1892 at the age of sixty-nine at his country house outside Pisa, Vito Volterra remarked that only someone who had been Betti's student could truly appreciate what Betti had accomplished "as *maestro* and teacher"—which helps explain why Volterra had remained at Pisa for ten years, loyal to Betti to the very end.[1]

During the same decade, the university had been losing ground in mathematics to comparable schools, partly because Betti and Ulisse Dini had been unable to obtain positions there for many of their own students. In this, as in nearly every area of Italian higher education, universities found themselves essentially at the mercy of the central government, which through its Ministry of Public Instruction had the final decision-making authority over such academic matters as faculty rank, salaries, titles, promotions, and the number of professors in a given department. The result at Pisa was that the cadre of gifted and aspiring mathematicians who began their scientific careers there had been obliged to leave the university to advance their careers and, in fact, became important professors in their own right elsewhere. Cesare Arzelà, once Volterra's high school teacher, now a valued colleague and personal friend, taught at Bologna, as did Salvatore Pincherle. After earning his degree at Pisa in 1891, Federigo Enriques joined the Bologna faculty in 1894 as a temporary instructor for the projective and descriptive geometry courses, thanks in part to Volterra's intervention on his behalf. Two years later, Enriques took first place in a national competition for the vacant chair at Bologna, which remained his home until 1922, when he joined the faculty at the University of Rome. Meanwhile, Dini's one-time assistant, Gregorio Ricci-Curbastro, had become professor of mathematical physics at Padua; he is best remembered for the invention, in collaboration with his student Tullio Levi-Civita, of the absolute differential calculus, which later became the mathematical scaffolding for Einstein's general theory of relativity. Volterra's oldest *Normalista* friend, Carlo Somigliana, had a chair at Pavia but fancied one at Turin, where a potent group of mathematicians, including Enrico D'Ovidio and Corrado Segre, the founders of an important school of algebraic geometry, and Giuseppe Peano, a pioneer in the development of symbolic logic, had built up a department of mathematics that rivaled Pisa's.

Then the unthinkable happened. In spring 1893, almost a year after Betti's death, Antonio Ròiti passed the word to Enrico D'Ovidio, then serving as dean of Turin's science faculty, that Vito Volterra would not be opposed to joining Turin's math department. Volterra had heard rumors that Francesco Siacci, recently appointed a senator of the Kingdom of Italy, wanted to move closer to Rome. The official announcement in July that Siacci had been appointed professor at the University of Naples would pave the way for Turin's science faculty to invite Volterra to take Siacci's chair of rational mechanics at Turin. "It will be a great acquisition for Turin and a great loss for Pisa, not easily remedied," Enriques (who had worked at Turin) remarked to Guido Castelnuovo, his scientific collaborator and close friend, as news of the impending transfer began to filter through Italy's mathematical community.[2] When Volterra told his colleagues in Pisa, who had nurtured his growth as a scholar and actively furthered his academic career, of his hopes to transfer to Turin, he assumed, perhaps naively, that his longtime colleague and early role model, Ulisse Dini, whose voice carried great weight in the Ministry of Public Instruction in Rome, would not object. He was wrong.

Volterra would wear himself out in the negotiations that finally brought him to the University of Turin. In Volterra's eyes, the villain in this contest of wills was, sadly, Dini, who worked feverishly behind the scenes to prevent Volterra from leaving Pisa and who seemed equally determined to punish him if he did. But some might also question Volterra's tactics in trying to get the best deal possible from the authorities at Turin; his machinations reveal a less appealing side of his personality.

At the start of the 1892-93 academic year, Pisa, despite some erosion in the overall quality of its mathematics program, had still seemed the ideal place for Volterra. Betti's death had forced the mathematics faculty to shuffle its teaching duties, and Volterra found himself for the first time teaching mathematical physics, which had been Betti's signature course for more than thirty years. Teaching a generic course like Betti's suited Volterra's breadth of interests, for it meant that he could pick and choose among any number of topics in mechanics or related subjects. Relations between Volterra and Dini appeared to be appropriately cordial. Dini had become a senator of the Kingdom of Italy; Volterra had become a leader of Pisa's science faculty and, in a move that reflected his growing interest in promoting the cause of science within and outside the country, had agreed to take Betti's place on the editorial board of *Nuovo Cimento*, founded in 1855 in Pisa. Nevertheless, by spring 1893, Volterra had begun to picture a new life for himself (and his mother) in Turin, the capital of Piedmont, a city and a province steeped in Risorgimento history, the birthplace of Italian parliamentary government.

Sparks began to fly several months later, after Volterra learned from Ròiti that the science faculty at Turin had voted unanimously to offer him the vacated chair of rational mechanics—the same chair Volterra held in Pisa—and an *incarico* (assignment) to teach the higher mechanics course as well. "I was not expecting this news, and I wrote to Turin saying that to deal with something so important required reflection on my part," Volterra told Dini in a letter in early July,[3] neglecting to mention that a while back he had indicated to Ròiti (who dutifully passed the information along) that he would not be averse to a call from Turin if Siacci should decide to leave. There was just one drawback: Ròiti, in informing Volterra of Turin's offer, noted that it entailed "*niente crescini*"—slang for "no extra compensation"—for the *incarico*.[4]

When the rumors began to circulate that Volterra would abandon Pisa for Turin, Riccardo Felici, Volterra's old physics teacher and now a good friend and colleague, took the news hard. "It's been some time since you spoke to me about [Turin]...but I always believed that you wouldn't do anything about it, considering the difference in climate....I also began to think what we on the faculty could possibly do to repair such a loss, to keep you among us." He noted that Eugenio Bertini, professor of higher geometry at Pisa, had been lured back to Pisa from Pavia the year before with the promise of a raise. Stay at Pisa, Felici advised Volterra, and perhaps "we can do for you what was done for our good Bertini."[5]

Volterra took away a different message: that he might well be able to use money as a bargaining chip. As Pisa's professor of rational mechanics, Volterra was now earning 5,500 *lire* (roughly $85,000 today) annually—comparable to the salaries of other mathematics professors of his rank and length of service in Italy's state-supported university system. In addition to his rational mechanics course, Volterra typically taught a second course—in 1892-1893, it was mathematical physics—for which he received an additional 1,250 *lire*. On top of that he earned 500 *lire* for his lectures at the Scuola Normale; in short, his total income at Pisa exceeded 7,000 *lire*. Could Turin do better?

Ròiti, eager to see his protégé reach the top of the academic ladder, promised to pass that idea on to the three scientists spearheading the campaign to wrest Volterra away from Pisa: Enrico D'Ovidio, the physicist Andrea Naccari, and Alfonso Cossa, the director of Turin's Royal Technical School for Engineers. "I [intend to] tell them that...they could win over your hesitation if you could be assured of being treated like Siacci when he was teaching the assigned course, because then you would have a plausible motive to put to the gentlemen of Pisa," he informed Volterra on September 21, 1893.[6] "The gentlemen of Pisa" referred principally to Dini. Ròiti had learned that Dini, who knew the minister of public instruction, was working behind the scenes in Parliament to thwart Volterra's transfer. If Dini could not prevent the transfer itself, he was prepared to block Volterra from benefiting financially from the move to Turin. Always the strategist, Ròiti also counseled Volterra not to count on other well-placed colleagues to plead his case: the astronomer Pietro Tacchini was in America; Eugenio Beltrami, now permanently at Rome, did not have the tenacity; Luigi Cremona, another Rome mathematician, could not be relied on to lead the fight within the Ministry; and Valentino Cerruti, Rome's professor of rational mechanics, lacked sufficient zeal. Ignoring this advice, Volterra apparently did make several attempts to enlist Cremona, who, as vice-president of the Italian Senate, was skilled politically as well as mathematically and belonged to that small group of Italian mathematicians in the latter half of the nineteenth century who put Italian mathematics on the world map.

In wishing to be "treated like Siacci," Volterra assumed that his counterpart at Turin had been generously compensated for teaching the extra course in higher mechanics. Since his salary for the rational mechanics chair would remain the same, Volterra reasoned that the University of Turin, by persuading him to take over Siacci's teaching load (its offer had not stipulated any compensation for the *incarico*), was attempting to save money by hiring one professor to do the work of two professorial chairs.

D'Ovidio urged Volterra to accept the offer, arguing that it would give him the ammunition he needed to press Volterra's case at the Ministry of Public Instruction in Rome, which determined the size of teaching stipends. The Turin faculty, D'Ovidio wrote, had had no say in how much Siacci received for teaching the higher mechanics course, although he thought that amount was 2,200 *lire*. Given Volterra's stature and the fact that the course would be taught at a higher level than usual, D'Ovidio could all but promise Volterra that he would receive a similar stipend: "But in the meantime I wouldn't delay starting the paperwork at the ministry"—a none-too-subtle hint to Volterra to make up his mind quickly.[7] Several days later, Volterra replied, explaining his qualms:

> If I could obey my leanings and my feeling of gratitude for the honor bestowed on me, I would at this moment accept without hesitation. But on moral grounds, my concerns must be for the university where I took my first steps and also for my family, for several reasons; on material grounds, the very large expenses I would encounter with the transfer leaves my resolve dangling for now. I will overcome, however, any hesitancy about my definite acceptance once I am assured of obtaining a fee not less than that earned by Prof. Siacci—and that is principally to justify my decision to the University of Pisa, though it will not save me from demonstrations of their resentment. So I will be most grateful for any and all efforts that you propose to make to facilitate the matter.[8]

D'Ovidio went immediately to Rome and met with an assistant to the minister for public instruction, one Ferrando. To his surprise, he learned that Professor Siacci had received only 1,118.40 *lire* for teaching the higher mechanics course—a little less than Volterra was making for teaching the *incarico* at Pisa. Moreover, it seemed that Rome wasn't all that happy with the arrangement that had lured Professor Bertini back to Pisa. "Ferrando also told me," D'Ovidio wrote to Volterra on October 5, 1893, from Rome, "that given the tightness of the budget, the minister does not intend to increase the payment for courses taught; that, on the contrary, he does not want to continue Bertini's higher salary...."

He further noted that Volterra's salary for teaching the extra course would remain fixed; that is, if he stayed at Pisa he would continue to receive 1,250 *lire* for an additional course, and if he left Pisa and went to Turin he would receive exactly the same amount for the *incarico*—no more, no less. "I believe that this is a promise made to Dini," he wrote.

To soften the blow, D'Ovidio suggested that Volterra could supplement his income by lecturing in the teachers' college in Turin. He also held out the promise of an invitation to join Turin's Royal Academy of Sciences, a learned society with an illustrious past—Joseph-Louis Lagrange, a founding member, had contributed many papers to its scientific journal, including his memoir on the calculus of variations—and still a vibrant presence in the cultural life of Turin. The privileges of membership included an annual pension of 600 *lire*—bringing the promised total to 7,350 *lire*, 100 *lire* more than Volterra was making in Pisa. Changing the thinking of the people in the Ministry, D'Ovidio added, seemed remote.[9]

Unsatisfied, Volterra went to Rome in search of Luigi Cremona, but Cremona was vacationing in the hills south of the city. But D'Ovidio was still in Rome, and he met with Volterra to bring him personally up to date on what he considered Dini's underhand activities. Upon his return to Florence, Volterra wrote to Cremona explaining what Dini had been up to:

> According to what D'Ovidio told me, the Ministry had been willing to grant this raise [i.e., for the *incarico* at Turin]. It seemed certain, but after some scheming on the part of the University of Pisa, and in particular by Prof. Dini, the promise was withdrawn. This is what Prof. D'Ovidio reports today from Rome. Such an attitude on the part of my colleagues at Pisa does not seem kind....Is it not an act of violence to deprive me of the possibility of benefiting from the University of Turin's proposal without suffering financial harm?
>
> It would seem that the University of Pisa and (through it) Prof. Dini have a monopoly on State franchises. Prof. Bertini's special treatment, when he was called [back] to Pisa from Pavia, is an example. The University of Pavia may have been disappointed by Bertini's departure, but it did not conspire with the Minister to damage him by objecting to the raise that was the condition of his acceptance. You kindly suggested that I accept the offer of the University of Turin, and you also saw fit to urge that faculty to promote a better salary in my favor: I would now like to ask you, if you think it appropriate, to mount your powerful maneuvers

against those of Pisa, so that the Minister may decide matters as they were originally communicated to Prof. D'Ovidio.[10]

As Ròiti had predicted, Cremona had no desire to continue fighting Volterra's battle; he wanted to put the backstabbing and finger-pointing to rest. "I think you should not hesitate to accept the call to Turin," he advised Volterra. "Later it will not be difficult to obtain what is denied now."[11]

From Florence, on October 22, 1893, Volterra wrote to Andrea Naccari, who was then Turin's acting rector, agreeing to come at the offered salary of 5,500 *lire* for the rational mechanics chair and 1,250 *lire* for the *incarico*.[12] During the seven years he spent in Turin, Volterra would teach rational mechanics and higher mechanics every year but one, and would receive 1,250 *lire* and not one *lira* more for teaching the higher mechanics course.

In explaining his decision to Ròiti, he wrote, "What will you think of me, who hasn't written to you in a long time, while my duty was to answer your last letter immediately? But if I must tell you the truth, I have two letters here for you, already written, which I never sent you. As you will realize, during this time I got such contradictory news that... the letters... no longer expressed the state of things and it wasn't worth sending them."

After recounting the convoluted saga of the negotiations on his behalf with Ministry officials over salary, he sounded a somber note. "As you see, things have become ugly, and I fear I have been harmed both here and there, and especially at Pisa, where people have clearly shown that they bear me a grudge, with no attempt to conceal it; on the contrary, they have tried in different ways to let me know that there are strong feelings of spite toward me, as if my leaving Pisa were a desertion or a betrayal, an intentional dereliction of duty." He had found himself "in an extremely embarrassing and painful situation." If he stayed in Pisa, he would become a *brutta figura* in the eyes of the Turin faculty, and if he went to Turin "I would find my friends turned into enemies....[P]eople let me know that in the case of my departure I would have no friends left in Pisa." He went on to praise the handful of colleagues who had rallied to his side: Felici, fellow mathematician Luigi Bianchi (whose name figures prominently in differential and non-euclidean geometries), and Cesare Finzi, the longtime family friend who had taught algebra at the university for many years. Still, Volterra could not shake the fear that he had become a victim of character assassination.[13]

Pisa had not bothered to nominate a successor for Volterra's chair, and as a result the chairs of mathematical physics and mechanics would remain vacant during the 1893-94 academic year. Perhaps Dini had thought he could bully Volterra into staying; if so, he had miscalculated badly. Nor is it easy to explain why Volterra insisted on tying his acceptance at Turin to a fatter paycheck; despite (or perhaps because of) his mother's lectures on the subject, he had never been particularly interested in money. Finally, why did he wait until the final hour to bring Ròiti up to date? Perhaps he was embarrassed. Who can blame Volterra for not wanting to admit that he could have spared himself a lot of pain simply by accepting Turin's offer as presented? He had tried to hold Turin up for as much money as he could get, without considering the consequences if he failed—the act of someone untutored in academic infighting and an amateur at academic politics.

* * * * * *

Turin was new territory for Volterra. A baroque city, its Risorgimento history could be read in the halls of the Palazzo Carignano, where the Italian Parliament met from 1861 to 1865; in the Piazza Carlo Emanuele II, where the towering forty-six-foot monument to Camillo Cavour still stands; and in coffee bars such as the Caffè Nazionale and the old Caffetteria Baratti and Milano, where conspiracies were hatched and rebellious proclamations issued. During its brief reign as the capital of Italy, Turin "seemed more alive than ever," the writer Edmondo De Amicis observed, recalling the horses and carriages clogging the streets; the bustle of government clerks in the ministries, the state's central bank, and the mint; and vast numbers of military officers and courtiers milling about.[14] The Jews of Turin, eager to put their stamp on the ascension of King Victor Emmanuel II to the throne in 1861, had hired the architect Alessandro Antonelli to design a temple (the future Mole Antonelliana). Intended to celebrate the emancipation of Italy's Jews, the Mole never became a synagogue. The staggering cost of building the soaring brick structure, some 500 feet high—"that enormous exclamation point right in the center of our city," Primo Levi called it—finally prompted the Jewish community's leaders to sell the uncompleted "money-devouring edifice" to the city.[15] The Mole may have been an extravagant gesture of local patriotism, but, as the Italian-Jewish writer Enzo Levi (no relation to Primo) recalls, "In my family, as in many bourgeois families, especially Jewish ones, there really existed a blind veneration for the House of Savoy"— a veneration that can be traced back to the generation that came of age following political unification.[16]

After the capital of Italy moved to Florence in 1865, Turin was in danger of losing not just its political primacy but its identity as well. The city suddenly found itself mired in the past, its cultural life turned inward, its citizens marginalized. For upward of twenty years, it would remain a city of soldiers and bureaucrats, while the region's landowning aristocracy continued to supply the state with diplomats, magistrates, and administrators. Unlike elected officials from other parts of Italy, Piedmont's senators continued to come from the ranks of the old rural aristocracy, whose interests remained narrowly focused on two goals—to block the departure of any remaining state institutions and to keep a sizable contingent of Turin's inhabitants in the army (a respectable haven for the city's unemployed). Then, in the 1880s, a severe financial crisis struck the city, once Italy's major banking center. Turin's financial institutions had speculated heavily in Rome's huge urban renewal program. When the bubble burst, real-estate brokers and builders alike went bankrupt, dragging the banks and the small investors of Turin down with them. The city's stock exchange lost its prestige, the middle class saw their savings evaporate, and fear and distrust of any kind of investment permeated the marketplace. Milan quickly replaced Turin as the country's financial center. In the meantime, an agricultural depression, touched off by massive imports of cheap grain from the United States and elsewhere, swept across Europe, dealing a harsh blow to Piedmont's agrarian economy. By the end of the decade, Italy's banking system had collapsed, investment capital from abroad had plummeted, and the government had launched an ill-conceived and costly tariff war with France.

In the midst of these wrenching economic conditions, Turin slowly turned into a modern industrial city. The Italian historian of science Roberto Maiocchi has pointed out that Turin, unlike other Italian cities, cultivated a small cadre of physicists who were interested in and "capable of treating technical arguments mathematically."[17] Growing up in that atmosphere, Galileo Ferraris, a physicist and electrical engineer by training, began doing experiments in the laboratory of Turin's Royal Industrial Museum, which would merge with the Royal Technical School of Engineers in 1906 to form the Royal Turin Polytechnic. Ferraris is remembered today for the development of the alternating-current electric induction motor. The electronics program and specialized laboratory he started at the Industrial Museum is widely credited with jumpstarting Turin's school of electrotechnics.

The agricultural crisis also contributed to the city's industrial birth. Unemployed farm workers began moving to the city, where they joined a burgeoning industrial workforce—the so-called industrial proletariat—whose ranks included large numbers of women and young children who gravitated to factory jobs in the cotton industry, often laboring under very harsh conditions. The expansion of textile manufacturing coincided with the rise of brand-new industries, from sheet metal to electric power companies, along with a dramatic increase in the city's population. From 204,000 inhabitants in 1865, the number rose to 250,000 by 1881, and by the time Volterra took up residence in 1893, Turin's population had swelled to 320,000, more than ten times the population of Pisa. As Turin's transformation into an industrial city intensified, the pool of factory workers began to demand better working conditions, higher wages, and shorter hours; the demands led to strikes and organizations of workers, including Turin's first Chambers of Labor, a prototype union established in 1891. The following year, at the Genoa Congress, the Italian Socialist Party would make its debut.* Membership in the Party ranged from anarchists to "evolutionary socialists," the latter being the chosen designation of its parliamentary leader, Filippo Turati. As one historian of the period has pointed out, the platform of Turati's moderate faction of the party "appealed to the belief in progress through science and education."[18]

---

*The name became official in 1895.

In the last decade of the nineteenth century, according to the historian of Italian mathematics Livia Giacardi, who has written extensively about the mathematics faculty at the University of Turin,

> the scientific scene in Turin was extraordinary. The University took on a key role in research which, with regard to certain sectors, gravitated at first around the Accademia delle Scienze and the Regie Scuole di Artiglieria e Fortificazione [Royal Schools of Artillery and Fortifications.] Schools of thought developed, scientific discussion flourished, new journals sprang up, University and industry worked together, and publishers turned to the popularization of science.[19]

Volterra arrived in Turin in November 1893, in time to experience a wave of anticlerical student demonstrations at the university, triggered by the sculptor Ettore Ferrari, who unveiled his statue of Giordano Bruno in the Campo de' Fiori in Rome on the spot where Bruno had been burned alive at the stake in 1600. He taught both his rational mechanics and higher mechanics classes three days a week in the palazzo dell'Università, a spacious rectangular eighteenth-century building at Via Po 15, with high sweeping porticos flanking a large courtyard in the center, well protected from the pedestrians and horse-drawn trams on Via Po. Some years later, Beppo Levi, an Italian mathematician and one of the hundreds of students Volterra taught, described what it was like to hear him lecture. Levi recalled "a day in the winter of 1893-94 in a dim classroom on the first floor of the University of Turin where... Volterra was teaching rational mechanics, the voice a little shrill in comparison with his powerful build, sparkling eyes that would often drift toward the ray of light that fell from a high window on his left."[20] That winter, Levi, aged eighteen, had started his second year of engineering and mathematical studies; he and Volterra, then thirty-three, became fast friends. Levi would see his former professor for the last time in 1939, after Mussolini's racial laws had robbed Levi of his teaching post at the University of Bologna and shortly before he left for South America, where he would become director of the Mathematics Institute of the Universidad del Litoral in Rosario, Argentina.

Compared to Pisa, Turin in the last decade of the 19th century was a hotbed of student activism. Early in 1894, most of the university's 2,300 students once more disrupted classes, destroying the massive wooden doors to the university's main hall and rampaging through the city, in the mistaken belief that school officials had canceled special make-up exams. In December the students rioted again, this time over an increase in university fees, and the university shut down. Determined to put a stop to these disturbances, the new rector of the university, Luigi Mattirolo, wrote an open letter to the student body on December 15. He prudently began by pointing out that the higher fees applied to new students only, and went on to say:

The disorders that have taken place in recent days must absolutely be censured. No matter what the complaint, if made with rioting and violence it is illegitimate and cannot be effective. Neither you, nor any other citizens, are lacking the means to legally present petitions and observations that you think opportune: It will always be my concern to immediately take them to the competent authorities and to support them when they appear to me to be just....Next Monday, the 17th of this month, the university will reopen. I have firm faith that order will no longer be disturbed and that I shall not be compelled to fulfill my burdensome duty of using all means that the law gives me to repress disorders and maintain complete freedom for study.[21]

As Volterra was about to learn, some of the members of the mathematics faculty at Turin could be just as obstreperous and combative as their students, among them a professor of calculus named Giuseppe Peano.

CHAPTER 9

"God Liberate Us from His Symbols"

In 1895, Volterra entered into a long polemical exchange with his Turin colleague Giuseppe Peano, a mathematician renowned for, among other things, the construction of the first space-filling curve, the modern definition of an abstract vector space, the axioms for the natural numbers (now universally known as Peano's axioms), and the geometrical calculus (which would turn out to be of considerably less significance).

Peano taught calculus both at the university and at Turin's Royal Military Academy. Despite his acknowledged contributions, he had, by the time Volterra arrived in Turin, been somewhat marginalized by the mainstream Italian mathematical community. He had become obsessed with using his own peculiar notations of mathematical logic in the classroom, much to the dismay of students and faculty alike. In 1891 Peano embarked on an ambitious program to recast mathematics in its entirety, using his symbolic notation. Between 1894 and 1908, he would publish five editions of his *Formulario Mathematico*, and when the calculus volume of the *Formulario* appeared in 1898, he used it almost to the exclusion of all other texts in his classes. One of his students recalled that Peano taught from it

> with the greatest love and much patience, the first pages [being] dedicated to the symbols of logic and then several lines of several other pages dedicated to accurate definitions of the concepts, to the various operations, and several passages of various parts of mathematics. Only in the last months of the academic year did Peano arrive at briefly treating, and always with his symbols, the calculus by the system of vectors, and of explaining some applications of curves, with calculations of lengths, areas....We disliked having to give time and effort to the "symbols" that in later years we might never use.[1]

The dispute between Volterra and Peano arose over a classical problem in rational mechanics: the wobbling of the earth's axis of rotation. In studying the rotational motion of solid bodies, the eighteenth-century Swiss mathematician Leonhard Euler had developed equations suggesting that the earth shimmies slightly as it spins on its axis—enough to cause the location of the poles to wander with a period of ten months. The confirmation of Euler's equations came more than a hundred years later, when astronomers succeeded in measuring the periodic variation in the latitude of the poles.

Volterra's interest in the rotational problem dates from this time; and in 1894, in his rational mechanics class, he had used it "to illustrate [Heinrich] Hertz's concept of substituting hidden movements for the consideration of forces in the investigation of natural phenomena."[2] At the suggestion of the astronomer Giovanni Schiaparelli, director of the Brera Observatory in Milan, Volterra submitted a nineteen-page paper, "On the theory of the movements of the earth's pole," to *Astronomische Nachrichten* in February 1895, showing that hidden motion of matter within the earth can cause the earth's poles to shift slightly. Two days later, having just been elected a resident member of the Royal Academy of Sciences of Turin, he presented a short note on the same argument, adding that other types of constant motion on the surface of the earth—ocean currents, rivers and streams, evaporation of water, and rain—might also explain the shifting of the earth's axis. In March, he presented three more notes on this subject.

Although it ostensibly focused on what today would be considered a problem in geophysics, Volterra's quarrel with Peano was really rooted in their divergent approaches to mathematics. A pure mathematician by trade, Peano had an uncanny gift for choosing topics that would turn out to be important in the development of mathematics in the twentieth century. Volterra, too, was a pure mathematician (some would argue that his most significant contributions lie in pure mathematics), but unlike Peano he did not shy away from the notion that his work might have real-world applications, and indeed actively embraced a range of mathematical problems pertinent to physics, celestial mechanics, astronomy, geology, and geophysics— and ultimately even biology and economics. Peano was fundamentally interested in the logic—and beauty—of mathematics' underlying structure. He also specialized in finding counterexamples to apparent theorems, and delighted in holding other mathematicians to his standards of rigor. Peano had taken his degree in mathematics, Volterra in physics, an important difference between them. Volterra's interest in the problem of the earth's wobble led him to formulate a mathematical theory that could be either verified or falsified by hard data—the gold standard of scientific investigation. Peano was less committed to finding a detailed solution of the problem than in illuminating its intricate mathematical architecture. Single-mindedly bent on promoting his geometrical calculus, Peano used the wobble problem to demonstrate his system's superiority over integration, which was Volterra's forte. The use of the geometrical calculus in Peano's very first paper on the wobble topic set the stage for their dispute. From that moment on, the battle between two giants of Italian mathematics was joined.

One Sunday in early May of 1895, seated in the elegant high-ceilinged meeting hall of Turin's Royal Academy of Sciences, Volterra listened in amazement as Peano launched into a talk about the displacement of the North Pole. "[A]lthough I saw him every day, without saying a word to me he set out to repeat my calculations," Volterra told a colleague afterward.[3] Peano also implied that an article he had published in the January 1895 issue of the *Rivista di Matematica* (a journal founded and edited by himself), about the rotation of a falling cat filmed by the French physiologist and pioneer cinematographer Étienne-Jules Marey, had piqued Volterra's interest in the problem. Drawing an analogy to the descending cat, Peano asked, "Can our Earth change its orientation in space by means of internal motion, just as every other living being? From the viewpoint of mechanics, the question is identical. But to Professor Volterra belongs the credit for having proposed it first. He made it the object of several notes presented to this Academy."[4]

The next step, Peano said, would be to consider how the geometrical calculus he had developed could be brought to bear on the problem, for which, he implied, no satisfactory treatment currently existed. Volterra jumped to his feet at the end of Peano's talk, his words laced with irritation. "[T]he numerical calculations of academy member Peano start from ideas already expounded in [my] various notes but are founded on rather unreliable numerical data," he told the assembled group, according to the minutes of the May 5 meeting.[5] Warming to his subject, Volterra insisted that calculations based on rigorous data did exist and that he had made such calculations himself a while ago, starting with important results obtained by the American astronomer Seth Carlo Chandler, who had found a period of about fourteen months in the polar motion. And while Volterra had delayed presenting his own work until he had developed a general theory, he had come to the May 5 meeting intending to present "A theorem on the rotation of bodies and its application to the motion of a system with internal stationary motions." After listening to Peano, however, he wished to present a second note, which would test his mathematical model against Chandler's observational data. He asked, and the Academy granted, permission to go home and get it.

Volterra once said of Enrico Betti that "when he talked mathematics he more often than not thought physics."[6] As his second note, "On the periodic motions of the earth's pole," presented later that same day, made plain, Volterra, too, excelled in using mathematics to solve physics problems.[7] Starting with the hypothesis that polar shift could be decomposed into a series of harmonic motions, he analyzed in detail the relationship between the rotation of the earth and its stationary internal motions. (By "stationary," Volterra meant the cyclical motions of mass inside the earth.) A comparison with observational data followed. Was it possible to reconcile the data with Euler's classic equations? Carlo Somigliana, who watched his friend's argument with Peano unfold, described how Volterra managed it:

> Volterra set about calculating the value of the angular momentum corresponding to the internal motions necessary to produce a variation in the Eulerian period that would make it equal to Chandler's period. He found that the component along the terrestrial axis of this angular momentum had to

be 1/1,053 of the angular momentum of motion of the Earth, assumed rigid. This result is not verifiable, but it is plausible.[8]

On May 19 at the Turin Academy, Peano presented another paper on the topic, but he pulled it before publication, citing an error. Then, on May 30, he sent Volterra a letter that opened on a conciliatory note ("After your comments, I recognize the influence of the earth's deflection on the polar motion") and included a promise to read Volterra's articles on the subject. The closing lines, however, struck a dissonant chord. "But you'll agree that your way to deal with the problem introduces three integrations, *with as many superfluous arbitrary constants* [emphasis added], which come from the differentiation of the first three equations."[9] Volterra sent back a terse reply: "The method of integration that I used doesn't present three superfluous arbitrary constants, as you say. The differentiation of the equations expresses the principle of the area, which includes three arbitrary constants, for the special choice of the x,y,z axis, [and] is necessary to establish the equations of rotation—that is to say, to remove the nine cosines from the equations." After rather unkindly mocking Peano for his ignorance of classical mechanics ("After all, the process of integration is the same followed for every problem of rotation from Euler to the present day; for this, therefore, you can consult any mechanics book"), Volterra asked him if he intended to withdraw his May 5 note as well or else modify his remarks at the next session of the Academy—in which case, Volterra would cordially refrain from any public remarks about the matter.[10]

Peano apparently declined to do either, reportedly telling Andrea Naccari, who served as the Academy's secretary for the class of physical, mathematical, and natural sciences, that he didn't repent of anything he had written. In a letter to Corrado Segre, Volterra passed on a remark made by Enrico D'Ovidio to the effect that all Peano was willing to concede "is that the earth is not a sphere," prompting Volterra to declare that Peano was "ignorant of mechanics from Euler up to the present."[11]

Ever since Peano's note on May 5, Volterra had been fuming about what he perceived as his lack of collegiality and his cursory treatment of Volterra's work. In the privacy of his own home, he took to pen and paper to express his rage and resentment. One can almost hear him lecturing the walls (or more likely, Angelica) about Peano's various affronts. Among Volterra's unnumbered and undated notes housed in the Accademia dei Lincei in Rome are several that clearly relate to this period, such as this.

> It was his duty to inform me [of his interest in the problem of the earth's wobble]: it is something that any educated person would do; [and] What is absolutely false is that, as Peano affirms, once the idea is set, the problem is solved. In fact: 1) it is not so easy to put the problem into an equation, since several calculations are needed, which he has to include

by copying my results; 2) from the point of view of developing any analytical solution, the path is very difficult....Professor Peano dares to recount a story of how I came to develop and apply the idea of cyclical motions [the story of the poor falling cat proffered at the May 5 meeting]. His affirmations on this matter are perfectly gratuitous; I will say more: they are completely false, as is always the case with those who want to read minds, with incredible irresponsibility and no basis in reality.

And there was this, which may have reached to the heart of the matter: "That the note of Professor Peano does not have any scientific purpose is shown clearly by this: That he uselessly repeats works and researches that I have already published."[12]

While overseeing final examinations as a visiting proctor at the University of Pisa at the beginning of June, Volterra asked Segre to find out, if possible, what Peano was planning to do about the offending May 5 note. Segre replied that he didn't know, but did say "I believe it will appear in the Academy's proceedings. I don't know if Peano intends to publish some words of correction or not....In any case, I believe it is better for you to wait and learn about Peano's intentions: the more so since if he does nothing at the June 9 session, there is still another session remaining before the holidays. I am very pleased to see things advance so as to be able to hope that you will not be annoyed and distracted by a polemic later on."[13]

Segre knew whereof he spoke, having exchanged acrimonious words in print with Peano just a few years before—and worse, discovering that Peano enjoyed playing the role of *provocatore*. "I hold the scientific polemic to be one of the forms under which ideas may sometimes usefully be expressed,"[14] he reportedly told Segre.

On June 9, at the Turin Academy of Sciences Volterra presented a theory that modeled the pole's motions, supposing the earth to be plastic. There was no recantation by Peano. On June 23, the last session before the summer recess, Volterra delivered yet another note, which he described as "observations" on his second paper of May 5; it brought the number of his papers on the subject to seven. No sooner had he finished his presentation than Peano rose to make his own on the motion of the North Pole, taking pains first to acknowledge recent work in the field by others—save one, Vito Volterra, whose name he conspicuously left off the list. To make matters worse, he insinuated that Volterra was a sloppy mathematician: "To decide such a question, one must make complete calculations, omitting nothing," Peano told his audience.[15] Once again his paper featured the notations and symbols that he had developed for use in his geometrical calculus, which most contemporary mathematicians, Volterra included, found less than edifying.

Determined to correct the record, Volterra now sought a larger arena. On July 2, 1895, he submitted his eighth paper on the subject to the *Rendiconti dell'Accademia dei Lincei*, the monthly published proceedings of the venerable national academy, and let fly at the outset:

> Prof. Peano, in a note presented to the Academy of Turin in the session of 23 June of this year, and which has just now been printed, shows that a system which is symmetric about an axis, and which constantly maintains its form and density distribution, may have variable internal movements that follow a law such that the rotational pole moves continually farther from the inertial pole. Seeing that this result can be obtained as an evident and immediate consequence of formulas considered by me and explained by me in several preceding memoirs, which Prof. Peano forgot to cite, although they were published this year in the same *Acts* of the Academy of Turin, I may be allowed to show this here, avoiding the employment made by said author of methods and notations not generally accepted and proceedings hardly suited to making clear the path taken and the result reached.[16]

If Volterra thought that scolding Peano in the Academy's proceedings would put paid to their quarrel, he was mistaken. In fact, Peano was delighted. Here was an opportunity not only to goad Volterra further, and on the latter's home ground, but also to showcase the superiority of his geometrical calculus in solving a classic problem in rational mechanics. Since he had not yet been elected a member of the Lincei (he became a corresponding member in 1905), he asked Eugenio Beltrami to submit the paper on his behalf. Volterra's brief note, "On the motion of a system in which there are variable internal movements," had appeared on September 15th; Beltrami presented Peano's note (to add insult to injury Peano had appropriated Volterra's title as his own) on December 1st. He began by revisiting the falling-cat experiment—("Discussing the question of the displacement of the earth's pole, produced by movements of parts of the earth, such as the ocean currents, I pointed out to several people the identity of the two questions, seeing that instead of a cat and its tail, one could talk about the earth and its ocean"). He derided Volterra's technique for solving a very difficult mathematical problem ("Now, if the question can be easily and completely solved by my way, the reason others run into great difficulties, I believe, depends on their habitual use of long formulas to indicate simple ideas"), and, not surprisingly, decided the question of who deserved credit for solving the problem of the earth's polar motion in his own favor ("I believe it useless to add anything else, seeing that finally Prof. Volterra agrees with my result that 'relative movements, however small, acting for a sufficient time, can displace the earth's pole, even supposing the continents rigid' "). For good measure, Peano "translated" the opening calculations of Volterra's maiden paper at Turin's Academy on February 3 into vectorial language, explaining that Volterra had begun "by writing three equations that, understood geometrically, say that a certain vector is constant. He differentiates them, transforms them, and in the general case that interests us here, arrives at a single integral (on the last page of the memoir), which signifies 'The length of this vector is constant.' "[17]

On January 1, 1896, Volterra in effect walked away from the dispute. As Segre had warned, engaging in a polemic with Peano could end up costing the other disputant far more than it was worth in time and energy. In a three-page letter to Francesco Brioschi, president of the Lincei and senior statesman of Italian science, Volterra made his case for the final time, spelling out his side of the story:

> Relative to what is said in the beginning of [Peano's] note, it seems to me that it is not worth the effort of spending any words, seeing that no one can doubt my priority, whether with respect to treating the question or with respect to the fundamental idea which forms its point of departure; nor can any doubt arise about the originality of my idea, as I explained in my lectures of last year... and it is not necessary for me to justify myself with the cat question, as Peano hints—a question, for that matter, about which he limited himself to writing in his journal a simple and brief review of the work of others. . . . Having thus shown to be empty and unfounded any of the points of criticism made of me by Peano, and that his assertions are neither original nor exact, he himself having recognized them as such, for my part I hold this polemic definitively closed.[18]

Brioschi transmitted Volterra's letter to the Lincei several days later, for publication as "Replica ad una nota del Prof. Peano."[19] A comment that Volterra later made in a letter to Tullio Levi-Civita, an exceptionally talented young mathematician at Padua, shows how hard he struggled to extricate himself from the controversy: "At the end of last year I was very busy preparing several translations I needed before the start of the New Year, and the holidays, which I was to hoping to have free, I was forced to spend on work excessively serious and in many ways unproductive, which I could not avoid."[20]

Two years later, in 1898, Volterra published his definitive work on the polar shift problem, some 156 pages of closely reasoned arguments embedded in a forest of calculations and diagrams, in Gösta Mittag-Leffler's journal, *Acta Mathematica*. True to his word, he made no further attacks on (and did not even mention) Giuseppe Peano.[21]

Meditating on Peano's behavior in a letter to Carlo Somigliana, Volterra concluded, "It is very strange to recognize one's error and then publish [it] again; but there is a lack of balance to such a degree in Peano's mind that we should not be terribly surprised." To which Somigliana replied, "God liberate us from his symbols, if these are the results to which they can lead."[22]

In time, Peano's refusal to teach a traditional calculus course cost him his teaching post at the Royal Military Academy, whose engineering faculty succeeded in finding an alternative calculus instructor. Peano remained at the University of Turin, teaching calculus until 1925. No action was taken there; he was an important mathematician, and if the students suffered, so be it. The mathematician Francesco Tricomi, who came to Turin in 1925, long after Volterra's departure, suggests in his autobiography a darker strain of discord in the university's mathematics department:

> Among the colleagues I found... were, besides Peano—and [A.] Terracini, who arrived almost together with me—C. Somigliana (1860-1955), G. Fano (1871-1952) and T. Boggio (1877-1963) as well as G. Fubini, who was appointed at the Polytechnic but taught higher analysis at the University, and C. Burali-Forti, who taught only at the Military Academy. But I must say that, as unfortunately often happens, these professors did not get on very well among themselves, and on one side [was] the "Jewish" group (headed until his death by C. Segre) with conservative tendencies, to which Fano and Fubini adhered, and on the other side the "vectorialists" group of Peano, Boggio, and Burali-Forti, who breathed the spirit of rebellion instead. As for the noble C. Somigliana, descending directly from A. Volta, who was then dean of the faculty, he oscillated between the one and the other but leaned toward the "Jewish" group, notwithstanding that he was not completely immune from a little anti-Semitism.[23]

* * * * * *

Eighteen ninety-six turned out to be a banner year for Volterra. Having divested himself of the Peano matter at its start, he proceeded to publish six papers between January 12 and April 26, dealing with the inversion of definite integrals—the solution of the type of integral equations that are now called Volterra integral equations of the first kind. He had been sitting on his results for some time. The impetus for what became in Volterra's hands a systematic treatment of integral equations owed something to his young colleague Levi-Civita, who had reported on the solution of several "interesting" cases of integral equations of this type at the November 17, 1895, session of the Turin Academy, taking the opportunity to bemoan the absence of a systematic study of the problem. Levi-Civita's lament may have inspired Volterra's first inversion paper, which he delivered at the Academy in the early winter of 1896 and in which he offered "a small contribution" to the problem, limiting himself, as he put it, "to the simplest case."[24] In a letter to Levi-Civita at the end of February, Volterra confessed that he had long despaired of publishing this work, because his method lacked elegance (*"poco elegante"*), but he now had found "a more general and direct expression" that he intended to communicate to the Lincei at its March 1st meeting; other results needed to be cast in an appropriate form before publishing, and after that he planned to move on to the case of constant limits. In closing, Volterra begged Levi-Civita's indulgence: "Excuse me if I have made you lose time with my concerns, but it is very pleasing to carry on a correspondence with you about an argument that has always interested me."[25]

Quick to appreciate the subtlety and penetrating intellect of his junior colleague, Volterra undertook to mentor Levi-Civita in much the way that Ulisse Dini and Enrico Betti had looked after him in his early years at Pisa. Like Volterra, Levi-Civita had studied under a noted mathematician (in his case, Gregorio Ricci Curbastro) and distinguished himself in mathematics at an early age, advancing rapidly up the academic ladder. He took the hand of friendship that Volterra extended. The two mathematicians would exchange many letters over the years. A man of enormous scientific versatility, Levi-Civita was appointed to the chair of mechanics at Padua in 1898 (again like Volterra, he was only twenty-three at the time), with Volterra's strong backing. Their personal friendship dated from that autumn, when Levi-Civita visited Turin; it would be cemented when Levi-Civita joined the faculty of the University of Rome in 1918.

Volterra's circle of friends in Turin included Giovanni Vailati, a mathematician turned logician and historian of mathematics, a champion of pragmatism and a habitué of cafes. A polymath with an irrepressible laugh and a passion for music and books, he was above all, a scholar with a vast breadth of interests ranging from Archimedes to school reform (his classmates at Turin had dubbed him "the Philosopher"). Several years after graduating from Turin in 1884 with degrees in mathematics and engineering, he had become a teaching assistant in Peano's calculus course. He published a number of papers on mathematical logic in Peano's *Rivista di Matematica* and later collaborated with Peano in the writing of the early chapters of the *Formulario*. Following a stint as assistant in Luigi Berzolari's projective geometry course, Vailati became Volterra's assistant in the fall of 1896, a position Volterra had personally arranged on a visit to the Ministry of Public Instruction. Vailati wanted to offer a course in the history of mechanics, and to do that he needed official standing. Volterra secured for him the title of "voluntary" (i.e., unpaid) assistant and arranged to have Vailati's name listed in the university calendar; as a result, Vailati was allowed to teach his own course for several years—though without a paycheck.

Early in 1897, Volterra, along with 2,000 other mathematicians, was invited to an upcoming mathematical congress in Zurich. According to its organizers, Switzerland had the advantage of being centrally located

and well versed in hosting international conferences. In selecting a neutral country, the organizing committee may also have felt the need to avoid
stirring up bitter memories of the Franco-Prussian war, still a sore point
with many French. The three-day meeting that August, billed as the first
International Congress of Mathematicians, attracted 242 participants from
16 countries, including 38 women. Although Vailati begged off, Volterra
decided at the last minute to attend the congress, accompanied by Angelica. He was promptly elected secretary for the meeting's Italian-language
papers—not an onerous assignment, as only two of his countrymen, Peano
and Francesco Gerbaldi, professor of analytic and projective geometry at
the University of Palermo, presented their papers in Italian. All the sessions
took place at the Swiss Federal Polytechnic school, whose lavishly decorated lecture-hall on the second floor provided ample seating for talks by
plenary speakers Henri Poincaré, Felix Klein, Adolf Hurwitz, and Peano.
Living up to their reputation as good hosts, the local arrangements committee organized a pre-Congress reception on Sunday evening, August 8, at the
Tonhalle, a stunning complex by the lake, and a banquet in the Tonhalle's
main hall at one o'clock on Monday, followed by a steamboat excursion on
the Lake of Zurich. The steamer returned to Zurich around nine o'clock in
the evening, where a parade of boats decorated with wreaths, flowers, and
flickering lanterns greeted the guests as they disembarked. Mathematics held
sway on Tuesday, with sessions devoted to arithmetic and algebra, analysis
and theory of functions, geometry, and history and bibliography. On the
last day of the meeting, it was announced that Paris had been selected as
the site of the second congress, to be held in 1900, with Germany poised to
host the third congress—perhaps in five years. Since that first meeting, the
International Congress of Mathematicians has become the most important
mathematical meeting in the world.

After Peano's address ("Mathematical logic") and Klein's ("On the question of instruction in higher mathematics,") the congress adjourned, and the
participants boarded special trains for one last social event, a banquet on
the summit of the Uetliberg, a ridge of hills overlooking the city. "The afternoon was warm, but not uncomfortable," the American mathematician
William Fogg Osgood, later recalled. "The panorama of the Alps was unusually distinct and many remained till a late hour in the evening, enjoying
the moonlight landscape that lay before and beneath them."[26] Writing to
Vailati, from Zurich, Volterra had only one complaint: "As I had predicted,
given the limited time, it was difficult to get to meet everyone who was
present."[27] He did speak with two who sang Vailati's praises—the Danish
historian of mathematics Hieronymus Georg Zeuthen and the Polish mathematician Samuel Dickstein, who would later commission the translation of
Vailati's inaugural lecture at Turin, "The deductive method as an instru-
ment of research," into Polish.

The rapport that existed between Volterra and Vailati shines through
their letters, which exhibit far more camaraderie than most of Volterra's

other correspondence with colleagues. Volterra writes to his junior colleague almost as if he were affectionately addressing a younger brother—he encourages and advises Vailati, praises his research, asks his opinion. In the Vailati correspondence, Volterra reveals another side of his personality: humorous, self-deprecating, gifted with a reporter's eye for detail. Flashes of a witty and unique cast of mind gleam from a letter he wrote to Vailati in 1898 from Morges, in the Swiss Alps:

> Morges is a microcosm... a beautiful panorama, a good hotel, clean streets with stores, many bicycles, intelligent people and... even a philosopher: a poor soul struggling between materialism and mysticism, with a strong tendency toward the latter, according to fashion of the last quarter of hour, which suggests long sleeves for the ladies, tight pants, and spiritualism in philosophy, with the related recollection of religious ideas. Indeed, it is for this poor soul that I have a request for you: to send me—if you have a copy available—your review of the history of spiritualism by [Count Cesar Baudi de] Vesme and the reference—if you are aware of it—of the journal that published (in French) the articles by [William] Crookes and [Oliver] Lodge about spiritualism. Every other article or review of yours would be very much welcome, but I won't say so, since I can't allow myself to ask for them.
>
> I think you would enjoy yourself very much if you were here.
>
> Unfortunately I am too weak, too far away, and too miserable an echo of your philosophical ideas, and I have more of a habit of contradicting than of supporting them, although I am persuaded that ideas in philosophy are... so vast and abused that there is the same tiny difference between supporting and contradicting an idea as there is between $1/\infty$ and $-1/\infty$.[28]

In 1899, when Vailati decamped for Sicily to become a schoolteacher in Syracuse, Volterra lost not only an assistant but a good companion. In those earlier years, as they strolled the streets of Turin, Vailati regaled Volterra with insights into books he was reading and together they discussed what contemporary philosophers were up to. "What you tell me about the most recent philosophical novelty fills me with wonder," Volterra wrote to him on one occasion. "To understand everything I need another animated discussion between Piazza Castello and Via della Cernaia."[29] After Vailati left Turin, he and Volterra would continue their "animated discussions" in the mails.

Like Vailati, Volterra was a bachelor in his Turin years. Many Italian mathematicians of Volterra's era remained, so to speak, single integers, all their days: Cesare Arzelà, Betti, and Somigliana were among those who stayed lifelong bachelors. As the nineteenth century drew to a close, Volterra found himself facing that most deceptively simple of mathematical questions: How do you move from one to two?

CHAPTER 10

"It Is the Greatest Desire of My Life," 1900

Angelica Volterra did not figure prominently in Professor Volterra's entanglements with Professor Peano, but it is not hard to imagine that the episode left the strong-minded Turin matron with the distinct impression that her son, now nearly forty, had entirely too much time on his hands. At the end of the nineteenth century, the idea that marriage was too important a decision to be left to chance remained an article of faith in many Italian families, Volterra's included. Like Lady Bracknell, who remarks in *The Importance of Being Earnest* that "an engagement should come on a young girl as a surprise, pleasant or unpleasant as the case may be,"[1] Angelica had decided that the moment had come for her not-so-young boy to marry and, moreover, had the perfect—and, it may be added, entirely unsuspecting—young woman in mind. In early November 1899, she left Vito to fend for himself in Turin with the help of their housekeeper, Giuseppina, and set off for Rome, carrying with her a question that she had at least briefly canvassed with her son before her departure: What would it take to arrange a marriage between Vito and his second cousin, Virginia Almagià, fifteen years younger than himself and Edoardo's daughter?

Although Angelica's motivations were primarily maternal ("Where are my grandchildren?" was as common a refrain among Jewish mothers in 19th century Italy as it is across the world today), she had also longed for years to live closer to her brother. Alfonso had moved with his family to Rome several years earlier, when his employer, the Bank of Italy, transferred him to the main office. A frequent visitor to Rome after that, Angelica always lodged with them in their apartment on Via Liguria 26, steps away from the Via Veneto and the extensive gardens of the Villa Borghese. On this visit, too, she stayed with them.

Volterra, had with great trepidation, acquiesced to his mother's plan—perhaps in part because Virginia was rich beyond his wildest imagination. His opinion of her other qualities has not been recorded, but he first laid eyes on Virginia in 1889, when he visited her family in Ancona. One wonders what a girl of fourteen made of a man twice her age who thought mostly about math and physics. Later on, after Virginia's family moved to Rome and Volterra started spending more time in that city on academic business, their paths probably crossed again—especially if Edoardo, who took a keen interest in Volterra's career, made it a point to invite him to their apartment from time to time. Named for the aunt who had extended such kindness and hospitality to the widowed Angelica and her young son many years earlier, Virginia was by then a slim and shapely young woman, with saucer-shaped hazel eyes and a tangle of thick, dark, curly hair that framed a warm, appealing face often etched with a coquettish smile. By the time she had reached her early twenties, her father had already turned down many suitors. Volterra may well have noticed that his little second cousin had blossomed into a lovely young woman; he may have heard her play the piano; they may even have exchanged pleasantries at the Almagià dining room table—and he would have shared his impressions of her with Angelica. But in that era, and within their respective families, that's as far as it would have gone had Angelica not intervened aggressively on her son's behalf.

By this time, Edoardo Almagià's once small railroad construction company had grown into a formidable international engineering firm, engaged in large-scale projects in Eastern Europe and the Near East. He had built railroads in Turkey and Romania and dredged harbors and constructed piers in Alexandria, Salerno, Bari, and Brindisi. In 1889 he had established his headquarters in Rome, in an office on Via Agostino Depretis 86. After relocating several times to increasingly upscale dwellings in the city, the family took up permanent residence in 1898 on the second floor of the Palazzo Fiano. Built in the 15th century atop ruins dating back to the time of Augustus Caesar, the handsome four-story building stood on the corner of the Piazza in Lucina and the Via del Corso, then as now the most important street in Rome. The Palazzo at Via in Lucina 17—a metallic marker there today reads "Palazzo Fiano-Almagià"—also housed Edoardo's business. Virginia herself figures prominently in the family's account of how the formidable palazzo came to be acquired by her father. The story goes that she was visiting a girlfriend one afternoon and overheard her friend's relatives talking about a palatial apartment house for sale on the Corso that they were thinking of buying. That evening she told her father about it. A consummate businessman always on the lookout for a good investment, Edoardo proceeded to buy the Palazzo Fiano himself. Later on, he was approached about buying a second palazzo on the street. No, he is reported to have said, "One Jew should not be the owner of *due palazzi* on the Corso."*

Arranged marriages were the norm on both sides of the Almagià-Volterra family.† Edoardo and his wife, Eleonora, seemed interested in Angelica's

---

*However he did go on to construct several houses in a compound on Via Scialoja, on the banks of the Tiber (he and his wife, Eleonora, would move there in 1905) and buy buildings in many of the new residential zones in Rome: a complex of new apartment houses on Via Porta Pinciana, including the luxurious Eden hotel around the corner on Via Ludovisi. Some of these properties would be included in Virginia's dowry. Over the years, Edoardo would diversify his real estate portfolio: Between 1866 and the outbreak of World War I, he acquired nearly a hundred working farms in the Marches and in Tuscany that yielded a bountiful array of cash crops, from unfermented grape juice to tobacco, corn, wheat, and silk.

†Ginetta Montecorboli, one of Edoardo's granddaughters, in an interview conducted by Carlotta Scaramuzzi for the California Institute of Technology Archives, recalled her

proposal, provided that Virginia, who appears to have been a family favorite, did not have to move very far away from them. No doubt remembering her own brokered union, Angelica played the matchmaker with gusto. "I am sorry I have to stay away from you," she wrote to Vito a few days after her arrival, "but if it is for your good, there is no sacrifice that can discourage me, believe it."[2] Her visits to the Palazzo Fiano only increased her eagerness to broker the marriage between the two cousins. "It is a consolation to see in that house, more than the awesome furnishings, the harmony and the great affection that those wonderful people have for each other," she told Vito. "I assure you that in this atmosphere one is transformed and acquires serenity."[3]

Born in Ancona in 1875, Virginia was the third of Eleonora and Edoardo's two sons and six daughters, and her vigilant and attentive parents had spared little expense in providing their children with the means to improve themselves, stretch their minds, and hone their artistic talents. Virginia learned French and German as well as croquet and badminton from a series of governesses, and became proficient on both harp and piano. (Her piano teachers, who included Giovanni Sgambati, a major figure in Italy's musical world, are reported to have said that it was a pity Virginia did not have to work, for her skills approached the professional.) In 1898, Edoardo commissioned the Italian artist Antonio Mancini to paint a portrait of his twenty-three-year-old daughter. Mancini posed his subject in the studio Edoardo had set aside in the palazzo for the family's budding painters, Virginia and her sisters Amelia and Sandra. In the full-length portrait, she is wearing a pink evening dress. Skintight ballroom gloves in a lighter shade of pink reach to her elbows; her hair is swept into a graceful topknot, and a pendant hangs from a ribbon clasped around her neck.

Almagià family tradition says that Virginia inherited her father's business acumen along with a strong sense of filial duty. By the time she had turned twenty, she could handle the company books, deal with the payroll, and run the household, servants, and siblings alike, in Edoardo's absence. Nevertheless, the protracted marriage negotiations remained a closely guarded secret. Virginia suspected nothing and would not learn of Angelica's intentions, her father's acquiescence, or that of her suitor for that matter, for many months.

---

grandmother, Eleonora Baruch, a Greek Jew from the island of Corfu, as "a very beautiful young woman who had lost her parents. And this Edoardo Almagià, perhaps for work, went to Corfu—or perhaps someone said to him, 'There is this beautiful young woman there.'" In another such interview, Luisa Almagià, the granddaughter of Edoardo's brother Roberto, reported that Eleonora was a cousin on her grandmother's side of the family: "My grandmother was the daughter of an Olivetti and a Baruch and Eleonora was the daughter of a Baruch and somebody else....It probably was an arranged marriage; almost all turned out to be arranged. After all, the families were closed [to non-Jews], the women did not circulate, and therefore there was always someone, even [outside the family], who would say, 'Here's someone you can marry, who would be a good match, an honest person,' and so forth."

Vito, of course, could not be similarly kept in the dark, although he might have preferred it that way. His mother trusted Antonio Ròiti, the family's longtime confidant and Volterra's sometime Svengali, to bolster his resolve with regard to her matchmaking. She felt sure that Ròiti, who had guided Volterra's career since Vito's high-school days, would be flattered to be involved in the proposed marriage of his former protégé, and that Vito would listen to his old mentor. After her arrival in Rome, she advised her son to contact him: "Write to Ròiti at once... so you may come to a conclusion."[4] Torn, as he would later explain, between responding affirmatively to "the famous business"—Angelica's pithy description of her efforts—and leaving himself open to criticism for marrying a wealthy woman, Volterra at first pointedly ignored Angelica's reminders from Rome to solicit Ròiti's advice. But further inducements from his mother soon followed, as she assured him by mail that he would get a warm welcome at Via in Lucina 17. "I mentioned to Eleonora that you will come to pick me up, and I said that in this visit you can stop only for a very short time, while for Christmas you could have stayed for the entire vacation; and she said very diplomatically, 'Since you are already here, stay here a little longer, so Vito will be able to spend at least a fortnight here. We'll be very glad of this.'" Still Vito dragged his feet.[5]

Volterra was now just past his fortieth birthday. Strongly built at just under five-and-a-half feet, he had a trim figure, pale blue eyes, fine-textured skin, and thick dark hair beginning to show streaks of silver around the temples. Thanks to Italy's indulgent policies toward her public servants, we have a good idea of how he would have looked to Virginia when they first met years earlier. An identification photo taken when he was in his mid-twenties for a discount train pass (a state-sponsored perk for teachers) shows an intense, diffident looking young man with a gentle, rather abstracted gaze, well-shaped mouth, neatly groomed beard, and hair that has clearly been close-cropped to constrain any unruly tendencies. By the time his mother launched him on the marriage market a dozen years later, his appearance had appreciably matured. He had not yet begun to resemble the nearly spherical scientist-sage of later years, but he had gained weight, acquired more gravitas in his face, and now accented his strong features and steady expression with a luxuriant, curly-tipped moustache and a gray-flecked goatee, again very elegantly trimmed. His girth would increase with age.

He had not breathed a word about his prospects to anyone, not even to his friend Corrado Segre and his other colleagues in Turin. In the second week of December, he finally wrote to Ròiti about the possible marriage and his potential fiancée's wealth. Volterra's letter is lost to us, but Ròiti's reply speaks to the heart of the younger man's dilemma. After telling Volterra that he took the news as good ("There is the probability that you might start a family") and valued the trust that Volterra placed in him ("I am really touched"), Ròiti characteristically got to the point:

I, too, am of your opinion that the man must raise the woman
to his level, and it is therefore necessary to think seriously
before taking a very rich wife. I must confess that I have
always felt a strong repugnance for this. To give you my
opinion without any hesitation, I need to know the education,
aspirations, and habits of this young lady and the kind of life
she will have to lead to be fully happy after she has been
married for many years.

   To have an idea of all this, it is necessary to talk to you.
For now, I can say at once with a clear conscience that although on one side there is great wealth, on the other side,
given your intelligence, your industriousness, and the universal esteem you enjoy, no one who has any sense will deny
that your social position is already much more brilliant and
admirable than that of a person who has millions. Thus,
an open-minded woman with noble feelings—however rich—
should feel honored and elevated by your choice.

But apparently Volterra had also needed to hear that people would not
think he was marrying for money. Rest assured, added Ròiti, "I am convinced that no one among those who know you well could think you would
take a wife for self-interest... and for your tranquility I am returning to you
the small sheet of paper [presumably Volterra's letter] that deals with this
delicate matter."[6]‡

Volterra seems to have lost little time relaying Ròiti's encouraging words
to his mother, who happily urged him to take Ròiti further into his confidence: "You can tell him about the affection that has bound our two families for so many years; that each family knows intimately the habits of the
other... and that the young woman's family knows that you would not want
to change ours; thus, if her family (as I suppose) agrees to this union, it's
only because they know the very humble spirit of the young woman." She
urged Vito to describe for Ròiti Virginia's quiet demeanor, her simple tastes,
her filial piety, and assured him that the Almagià girls shunned receptions,
balls, parties, and other insidious amusements. "They go to the theater,
sometimes, or to the Pincio in their carriage, but only if they are forced
to," she wrote Vito. Moreover, they spent their days immersed in study—
"but for their pleasure, not for ostentation; [and they have] great affection
for their family, an affection that (I'm sure) the young woman would feel
equally for you and for her husband's home."

---

‡Volterra's concerns were prescient. At a 1982 conference at Brandeis celebrating
Volterra's life and work, the American economist Paul Samuelson, who was not on the
program, asked to be allowed to say a few words. Striding across the stage, hands waving
in the air, he declared that it was not so terrible that Volterra had been forced to resign
his university chair in 1931 for refusing to sign the Fascist loyalty oath. And why not?
Because, Samuelson explained, Volterra had married a very rich woman and could afford
to make the gesture.

Despite this glowing and almost certainly misleading account of her prospective daughter-in-law's life of demure devotion, selfless denial, and relentless self-improvement, Angelica does seem to have recognized a certain ambivalence on Volterra's part, which she did her best to address: "My dear Vito, nobody should force you. You feel some affection for this woman, and it would be a shame if you missed an opportunity that you, too, consider auspicious, now that we have excluded what most pained you—that is, the idea [that the union would be] a little delicate."[7]

A dutiful son, Volterra spent the university's Christmas holidays in Rome at Alfonso's house and made several visits to the Palazzo Fiano. Relishing his role as patriarch of a large, extended family, Edoardo Almagià encouraged his relatives to gather under his roof, and Volterra would have had ample opportunity to meet and observe Virginia there frequently without being conspicuous. He brought with him from Turin Angelica's fur coat and a new wardrobe for himself, "of good quality," as Angelica had instructed him to do.

Returning to Turin in early January 1900, Volterra resumed teaching, and Angelica, too, returned and resumed looking after him. But she had not abandoned her plans. The matrimonial negotiations shifted into high gear in February, following the death of Eugenio Beltrami, who only months before had added the title of senator of the Kingdom of Italy to his portfolio of honors, from president of the Lincei to professor of mathematics at the University of Rome. Virginia's mother, Eleonora, promptly seized this opportunity to inform Angelica that Volterra's chances of wedding her daughter now depended on his being appointed to Beltrami's vacant university chair.

Signora Almagià's bombshell exploded just as some members of the Rome science faculty launched a vigorous campaign to recruit Ulisse Dini for the now-vacant position, an effort that the Pisan mathematician did nothing to discourage. If Beltrami's chair went to Dini, it appeared that Vito could kiss his chances to wed Virginia good-bye. By insisting that Volterra get a job in Rome, Eleonora Almagià obviously hoped to keep her daughter close to her family, but it is also possible that her ultimatum reflected a shrewder understanding of Volterra's character than his own mother had shown. Yoking the unfamiliar act of a marriage proposal to the intimately familiar one of proposing himself for a new academic post seems to have finally galvanized the reluctant suitor into action. He lost no time in enlisting the aid of both Ròiti and a younger mathematician at Rome, the thirty-five-year-old professor of geometry Guido Castelnuovo, in his quest to occupy the vacant chair.

By the beginning of March, Castelnuovo had signaled Volterra that it would be an uphill battle. Dini had a strong following among Beltrami's colleagues at Rome. "Notwithstanding the high esteem I have for Dini," Castelnuovo wrote, "for various reasons this does not seem to me to be the best choice for our faculty, which needs an active young person who can and wants to guide students and does not have political connections, already too brilliantly represented among us." Furthermore, if Dini came he would teach analysis, which was already being taught by Valentino Cerruti, who had once taught mathematical physics but had no desire to do so again. Unless the faculty insisted on designating the vacant chair for mathematical physics, Castelnuovo added, no one would teach that subject. How is it possible for a faculty like ours, he mused in a letter to Volterra, to deprive itself of such instruction? "I will fight, but I do not expect to win," he told Volterra, fearing he might not only have to relinquish hope of revitalizing the faculty but also in the process lose "a colleague from whom I would have everything to learn.... When will the happy time come when the faculty will put scientific interests above all other considerations?"[8]

Ròiti meanwhile served as Volterra's campaign manager, ferreting out information, evaluating rumors, dispensing advice, and keeping his candidate from making tactical mistakes. Nineteen professors—naturalists, chemists, physicists, and mathematicians—constituted the University of Rome's scientific faculty. Although many of them had come out publicly for Dini, Ròiti suspected that a secret ballot might tip the scales in Volterra's favor, and he advised Volterra not to request a full vote of the faculty immediately. Its members needed canvassing. Ròiti soon determined that two university stalwarts—Pietro Blaserna, professor of experimental physics, and Stanislao Cannizzaro, professor of chemistry, were ready to support Volterra, but Cerruti, who was the dean of the science faculty, was close-mouthed as to which candidate he preferred. Volterra drew this assignment.

"I think the keystone... is Cerruti," Ròiti counseled Volterra, "because if he agrees to teach mathematical physics again, many reasons brought forward by your supporters will fall, or if they don't fall, they become less important for those who love to calm their consciences with whatever excuse."[9] Appointed rector of the university that year, Cerruti had served both as secretary general to the minister of public instruction, who was responsible for all university appointments, and as a member of the Consiglio Superiore (Superior Council), an advisory group to the minister composed of professors elected by their peers. Following Ròiti's suggestions, Volterra described his family situation in his letter to Cerruti ("My mother is elderly, and all her relatives are settled in Rome, including several nephews for whom she acts as mother and whose distance is very painful to her"), asked for his support ("These and other family reasons no less important, but that I am not disclosing now, have forced me to dare to take this step toward you; sure that you will understand and sympathize with me"), and avoided mentioning Dini.[10]

Cerruti's lengthy reply offered small comfort. While the teaching, research, and other activities of the faculty at Rome would benefit immeasurably from Volterra's presence, Cerruti assured him, the matter of replacing Beltrami had turned into "a rather messy situation." A memo supporting Dini had already been circulated among his faculty, and some of them had suggested that he, Cerruti, give up the teaching of analysis to accommodate Dini and go back to teaching mathematical physics. Cerruti, however, was not inclined to accommodate this request. For years now, he had focused his research in a different direction, and he saw no need to reverse course now. He let Volterra know that he had told his colleagues as much, begging them not to create problems for him and his present situation.

Turning next to the politics of new appointments, Cerruti had little encouragement to offer. He pointed out that before Beltrami's death, Rome's science faculty had had twenty full professors, four more than the legal number. Even with one full professor less, the faculty needed the permission of the minister of public instruction to add two more professors, "were Beltrami's [chair] saved and a new [chair] added." Of course, the minister might decide simply to retain Beltrami's chair in mathematical physics, in which case Cerruti could not say how matters might end. "Many colleagues are pledged to Dini: I preferred to wash my hands of it."

Cerruti's subtext could not have been clearer: He had no intention of sacrificing his long-standing friendship with Dini for Volterra. It was bad enough, Cerruti wrote, that his own good name had been bandied about in connection with all the skullduggery surrounding the vacant chair. He added, somewhat elliptically, "I will speak to [the mathematician Alberto] Tonelli and [Luigi] Cremona about you: but I cannot go further than that."[11] He concluded by suggesting that he would not object if someone else were to personally shame the minister of public instruction into anteing up two extra professorial slots for Rome's science faculty.

Perplexed by Cerruti's reply, Volterra had the good sense to pass it on to Ròiti: "What is your impression of Cerruti's letter? Do you read something between the lines?"[12] Ròiti did not. "He is sincere, and it is clear that if he finds himself between the wall and the door—that is to say, choosing between you and Dini—he will choose Dini. . . for reasons that have nothing to do with science and teaching. Unfortunately the Roman—or rather, governmental—environment dominates this time too."[13]

But Volterra, who saw his chances of marrying Virginia slipping away, thought he had divined in Cerruti's letter an alternative road to Rome. The scenario he imagined involved Dini's going to Rome, but not as a professor of mathematical physics, and the minister's refusing to increase the number of extralegal chairs, thus forcing Rome's science faculty to hire a junior professor ("*uno straordinario*") to teach that subject—with Volterra volunteering for the position. Ròiti cut off this line of thought sharply. "For Heaven's sake, do not even for an instant be contented with a position as *professore straordinario*! Never say it again, to anybody. It would be a disastrous madness and I can't believe that someone as balanced as you could suggest such foolishness."[14] Volterra told Ròiti not to worry, but to Angelica he roundly defended the idea, pointing out that many of his colleagues had done precisely that to get ahead.

By late March, he couldn't sleep; tormented by insomnia, beset by uncertainty and anxiety over his future, he had arrived at "one of the most painful moments" in his life.[15] Even Angelica, who had gone to Rome to spend Passover with her brother's family, seemed unaccountably to have lost interest in the entire affair. "I fear that you are letting things cool down, rather than warm up," he complained in a letter to her, eager to know what Edoardo and Eleonora were thinking. Edoardo would know whom to consult in the capital about his future son-in-law's prospects—people such as the well-bred, London-born politician Ernesto Nathan, who seven years later would become Rome's first Jewish mayor. "It is right that Edoardo is acting carefully," Volterra told his mother, "but it would be good also if he acted *soon* and *with the maximum impact* if we want to avoid dragging things on. *I ask you to recommend this to him in the warmest way.*"[16]

Now all he could do was wait. He had had an offer to become the director of the observatory at Bologna, an offer he rejected out of hand, describing the city to Angelica as "extremely boring," the job as rightfully belonging "to a real and proper astronomer," and the university as "a step down from Turin." But he had nothing good to say about Turin either. Suddenly, a city he had embraced enthusiastically seven years earlier had become "insufferable" and he felt "foreign"—no friends, no pleasant memories: "Everything that surrounds me here is unpleasant and my only wish is get away from here as quickly as possible," he wrote to Angelica on April 2. Through his informal spy network, Volterra had learned that the faculty had indeed petitioned the minister of public instruction, Guido Baccelli, for two extralegal positions and that the minister had replied evasively.

"We must not give in," Volterra exhorted his mother, "and we have to fight until the very end in order not to blame ourselves that we didn't try everything or neglected anything. Therefore if Edoardo can arrange for Nathan to speak to [Guido Baccelli], it could be that [Nathan] can have some good influence."

Edoardo was apparently quite willing to busy himself on Volterra's behalf. Even the Almagià family physician, Ettore Marchiafava, a professor of pathological anatomy and clinical medicine at the University of Rome and a member of the Lincei, was corralled into lending a hand by querying Luigi Cremona, who was one of his patients. Nevertheless, no further word leaked from Rome during the whole month of April. On May 10, Angelica, who had remained in Rome, wrote to her son that Edoardo and Eleonora saw no reason not to go forward with the marriage plans, even in the absence of firm news, "because for them it [Volterra's transfer to Rome] is a sure thing." They were quite certain that the date would be no later than next October, the start of the 1900-1901 academic year.

Naturally Virginia herself still knew nothing about these plans, and no one was about to tell her—yet. The next move, Angelica explained to her son, was up to him: He must come to Rome, spend time with the family at their home, take strolls with Edoardo and Virginia, and let Virginia become better acquainted with her future husband. Virginia, wrote Angelica, had "a lot of common sense, so they [her parents] hope she will learn to appreciate your merits."

As for Volterra, he desperately wanted Virginia's consent to be "spontaneous." "It will be, for sure," Angelica assured him in the same letter. "My God, it gives me joy to see you contented! I can imagine the joy that this letter will bring you."[17]

A letter from her son crossed hers in the mail. In it, Volterra complained bitterly about the lack of definitive news from the University of Rome: "I always fear that some accidents [will turn] things for the worse, and who knows what can happen," he wrote. The uncertainty "is really terrible, and it is true we are in hell." Even the housekeeper had conspired to make his life unbearable: She had gone away the day before "leaving the lunch half-cooked, and I have no idea about [my expenses]."[18] When Volterra read Angelica's letter, his spirits soared. "I can only confirm that my feelings are still unchanged, and that it is the greatest desire of my life that this union, which would make me most happy, will occur," he assured her.[19]

Meanwhile, the rumors that had reached him about his chances of joining Rome's faculty were oddly contradictory; in the end he didn't know what to believe. Ròiti predicted that the matter would drag on for some time; Castelnuovo and another mathematics colleague, the descriptive geometer Giulio Pittarelli, foresaw a swift decision, no later than June. Volterra had also heard that Dini had privately expressed his continued interest in transferring to Rome, but Dini himself remained publicly mum. Volterra avoided confronting him directly, but Ròiti had no such qualms. "I told Dini he has a duty not to abandon the University of Pisa to its destiny, and that [Alessandro] D'Ancona should be persuaded to retire soon from [the directorship of] the Scuola Normale," Ròiti informed Volterra by letter in early May. "Dini replied that even if he were certain of becoming director of that school, he would remain unsure of the advantages of staying in Pisa."[20] Whatever the case, Volterra suspected that Dini wanted to prolong the negotiations in Rome for as long as possible.

In an angst-ridden letter on May 13 to his mother, Volterra confessed his fear that one way or the other those who did not want him in Rome would carry the day. "How is it possible to succeed with Cerruti, who certainly does not want me, and with Cremona, who has always opposed me?" he lamented. And the stakes couldn't be higher, "now that Edoardo's proposal is tied to my transfer, and if that doesn't happen everything is finished and lost."[21] Sure enough, four days later, on May 17, Castelnuovo reported that Minister of Public Instruction Baccelli had rejected the faculty's request for one additional position, on financial grounds. The decision, Castelnuovo wrote, was no surprise but it had stirred up disagreement within the faculty. He expected the faculty to revisit the matter by the end of the month, at which time Volterra most likely would have an answer "favorable to you, I hope." Castelnuovo's letter closed on an optimistic note, suggesting that he knew more than he let on: "Even Blas [Pietro Blaserna], with whom I spoke yesterday about the matter, considers our victory secure."[22]

Volterra packed his bags after receiving Castelnuovo's letter, intending to take the next train to Rome. Instead, from Turin, on May 19, he wrote to Angelica to explain his dilemma. If he showed up in Rome while the voting was in progress, it might seem to the faculty that he had come there intending to influence the outcome personally—in which case he would hurt his chances, not help them. "I think that the real question will center on whether they feel bound or not to recover the chair of mathematical physics, and on this point the voting will happen. Let's hope it goes well." He also instructed Angelica to write to Ròiti in Florence, and to consult Edoardo as well, about the propriety of going to Rome.

Angelica did not answer his letter immediately, and Volterra wrote again, two days later, demanding to know what advice Edoardo had given her: "Really I don't understand that you have not let me know anything, while I had begged you to write or even to telegraph me about [my] departure." Holed up in Turin and unable to think about anything else, Volterra had filled in the blanks on his scorecard: "Three votes can tip the scales on either side," he wrote, urging Angelica to tap the family's medical mole, Dr. Marchiafava, for information about the appointed voting day.[23] Ròiti got in touch with Volterra on May 24, after receiving Angelica's letter; he thought it best that Volterra postpone going to Rome until the faculty had finished its deliberations, but could not resist adding that "I do not begin to understand *the immense anxiety* in which you live![24]

On June 1, 1900, Virginia Almagià learned for the first time that Vito Volterra wished to marry her. Although Edoardo had told his daughter about some of the previous marriage requests he had received, the news that her forty-year-old relative had asked for her hand apparently stunned her. That evening she took the overnight train to Ancona, ostensibly to visit relatives there. She did not give her consent on the spot; neither did she turn down the proposal. Eleonora had persuaded her to go away for a few days and Edoardo had agreed to the trip against his better judgment. Having finally arrived in Rome with the explanation that he'd been called there on "family matters," Volterra was not heartened by the news that his inamorata had precipitously left town.

"Of course, somebody was very much *disappointed* by your departure— and perhaps rightfully—so much that he aroused [Eleonora's] *pity*, but if the result will be what he hopes, he will be largely rewarded," Edoardo wrote to his daughter the next day, Saturday, before leaving town himself on Sunday, June 3. He planned to spend the next three days in Genoa before returning to Rome, and he expected Virginia to be back then as well; Virginia, in other words, had until June 7 to make up her mind. In a ten-page letter posted on the 2nd, the envelope marked "*private*, for my Virginia" and the words written in a large sprawling handwriting on his personal stationery, Edoardo opened his heart to his beloved daughter.

Let's then acknowledge our faults, and learn that we should pay heed not only to those feelings that light up and spread like wildfire but also to *cooler* ones; we notice those less because they are steady....It is true, my Virginia, that we have to heed those *cooler* feelings more than those I would call *hot.* Or at least we should heed both in the same way, so it is important to weigh them up with the fine sieve of reflection....Our feelings come mainly, if not exclusively, from the environment in which we live, and this changes with the passing of time and the changes in people;...in the same way new affections replace old ones....

Focus on yourself, to think seriously about the decision you are about to take. I, as usual, don't want to influence it by any means—and whatever it is, I will respect it, as long as it is the result of serious meditation on your part. I have had many other requests [for you] (and others keep coming)....With none of these requests, though, would I be as confident in handing you over as with this man.

The only criticism I could have concerns the age difference, but this, again, more for what appears than for what is reality. The possibility of keeping you here in Rome would satisfy me, and your mother even more, who declares she doesn't want to be far from you. Think carefully about it, my Virginia: now that you are far away from the family, you are perhaps in the best position to decide.[25]

For all that he loved her and insisted that the decision was hers alone to make, Edoardo was not above using the same skills he brought to his business affairs to persuade Virginia that what was best for the family was also best for her.

His letter had the desired effect. Not long after her return to Rome, Virginia and Vito became engaged. Although he had vowed more than once to abandon plans to marry Virginia if she did not say yes "spontaneously," Volterra wooed his new fiancée with some of the single-mindedness he usually reserved for his mathematical studies. He visited her daily, always bringing her a fresh flower.

By the middle of the month, the much-sought-after transfer to the Eternal City also seemed assured. In a terse, two-line message from Ròiti, Volterra learned that the science faculty had voted unanimously to nominate two additional professors—meaning that both Dini and Volterra could come to Rome if they wanted—and that Minister Baccelli had caved in.[26] And Guido Castelnuovo, upon learning of Volterra's engagement, wrote him a short note beginning "*Caro collega* [Dear colleague]." "Permit me to use this salutation," Castelnuovo continued, "which represents the aspirations of many months, soon at last to translate itself into reality."[27] According to Castelnuovo, Baccelli had decided to nominate two additional professors if the Rome science faculty "acted in a dignified manner."[28] After the receipt of Baccelli's letter, the faculty voted unanimously to offer positions to both mathematicians, with Volterra teaching a course in celestial mechanics and Dini a course in differential geometry—though the minister noted that the transfer of Professor Volterra depended on "the favorable opinion of the Superior Council."[29]

This news cheered Volterra but then came the news that a new government had been elected, and it appeared that Niccolò Gallo, the incoming minister of public instruction, had no intention of honoring his predecessor's commitment. Volterra now wanted Edoardo Almagià's circle of well-connected friends to pass on the following message to the new minister: "The transfer of Professor Volterra must take place if the opinion of the Superior Council is favorable to the vote of the faculty of Rome."[30] The Council would approve—of that, at least, Volterra felt he could be certain.

Returning by overnight train to Turin late that June to conduct his exams, Volterra courted his Virginia by mail, reliving the few moments they had apparently spent alone together in the Palazzo Fiano's music room. "They were moments so sweet and they have left me with such an exquisite memory....The memory of this alone would be enough, my Virginia, to make me feel the terrible burden of being far from you," he wrote, at the end of a long day spent at the university interrogating students. "But I remember not only these moments. I remember so many other things....Those days, as I told you last night, passed like a dream, but they will leave an unforgettable impression on me: The strongest and the dearest that one can experience in the world!" Comparing the idyllic days he'd just spent in Rome with the day just ending in Turin had convinced Volterra "that the world had somehow changed." More likely it was Volterra, as anyone who has fallen in love could have told him, who had changed.[31]

Meanwhile, the engagement had touched off a wave of excitement among his colleagues and friends in Turin. "Hurrah, Hurrah!" shrieked Olga Segre, Corrado's wife, as soon as she opened a June 19 letter from Volterra and read the first few lines. Olga's family hailed from Ancona, and Olga knew the Almagià family well enough to gossip with her old girlfriends there about Virginia. According to Corrado, Olga had repeatedly told him that one such girlfriend "was hoping for a union between you two. And when you left last time for Rome, called by 'family matters,' how I had fervently wished that *this* was the 'family matter!'"[32] As he wrapped up his affairs in Turin at the end of June, Volterra reported to his fiancée that people from all parts of the city were heaping congratulations on him, as well as expressing "the desire to meet you on some future trip when we will have occasion to stop in Turin."[33] With his marriage only days away, Volterra's harsh opinion of the city seemed to have softened considerably.

The marriage of Virginia Almagià and Vito Volterra took place on July 11, 1900, in the palazzo that the bride had apparently been instrumental in procuring two years earlier for her family. Ettore Marchiafava acted as witness, standing in for Antonio Ròiti, whose health had not permitted him to attend the wedding. The record gives no indication of whether this was a religious ceremony. Eleonora had learned Hebrew as a child and could follow the Hebrew in her prayer book, and unlike the other members of her family, she fasted on the most important Jewish holidays. However, according to Ginetta Montecorboli, Eleonora's granddaughter, Virginia "was the least religious of all [the family], perhaps, [though] no one was really religious."[34] Volterra himself never gave any indication that he followed the Jewish laws.

The newlyweds boarded the train for Orvieto, about one hundred miles north of Rome, and the first stop on a honeymoon that would take them to France and Switzerland. The next morning, still the dutiful son, Volterra sent his mother a brief letter, also signed by Virginia, which said in part, "I have no need to tell you that I am completely happy."[35] For the first few days of their honeymoon, Volterra kept track of their daily expenditures— the price of a newspaper, a haircut, and other small purchases—reflecting the habits of someone accustomed to accounting for every *lira*. Within the week, he had abandoned the practice. Perhaps he realized that he would likely never need to worry about such mundane matters again, having acquired a rich wife and an influential father-in-law into the bargain.

But not until September would he learn that all hurdles on the road to Rome were behind him. Pietro Blaserna reported on a conversation with Minister Gallo, who had declared himself "very favorable" to Volterra's appointment and had instructed his assistant, in Blaserna's presence, to send Volterra's paperwork to the Superior Council. Ulisse Dini, it seemed, would not be accompanying his former colleague to Rome after all, Blaserna wrote, since Gallo had also acceded to Dini's appointment as director of the Scuola Normale at Pisa, in apparent accord with Dini's wishes all along. As far as Blaserna was concerned, "the matter [is] closed."[36]

CHAPTER 11

"Most Important for Our Fatherland"

In May 1899, Henri Poincaré, one of France's most celebrated mathematicians, had invited Volterra to address the Second International Congress of Mathematicians, which would be held in the following summer in Paris. For the subject of his talk, Volterra turned to the life and times of three mathematicians with whom he felt a great affinity—Enrico Betti, Francesco Brioschi, and Felice Casorati, the threesome who in 1858 had ventured across the Alps in search of colleagues with similar scientific interests in Germany and France and who had driven the agenda and direction of mathematical studies in Italy after unification.

Preparing the talk began as a labor of love, but finishing it seemed to elude him. He was still working on the talk that July, when he married Virginia. And he worked on it all through their honeymoon. From Orvieto they traveled on to the Uetliberg, above Zurich, which had mightily impressed Volterra three years earlier, during the first International Mathematics Congress. There they spent some time sightseeing, hiking, visiting with the mathematician Adolf Hurwitz, and of course working on the talk. Volterra, who remembered the 1897 congress's banquet on the summit, could now share the full sweep of the lake, the snow-capped mountains in the distance, and the twinkling lights of the city below, with the woman he loved at his side. From Switzerland, he sent a partial manuscript to the organizers of the congress and then turned his attention to completing, as he wrote to Angelica, "the well-known boring speech of which I never see the end."[1] Angelica was more interested in knowing whether or not Volterra had showed Antonio Ròiti what he'd written. "I would be pleased if he would read it and give you his opinion," she ventured.[2] Volterra hadn't really considered that option, for he now had by his side a willing confidante, muse, and helpmate, a spouse who not only doted on him but insisted, as she told her father, "that a wife must never be a hindrance to a husband; instead she must assist him, help him as much as she can in his work and spur him on in his work."[3]

By all accounts, Virginia never fell short of the standard she set for herself. "Dear Vito has almost finished his work for Paris," she assured Angelica in a letter of her own, sensing her mother-in-law's anxiety over this particular speech. "He is finishing it now while I am writing to you and I hope that by today he may be free from this labor. Besides, don't worry because *it will be a triumph*." Volterra had just read the speech aloud to her, Virginia reported, while she held a stop watch, timing it. "Later I will hear him read it on *the great day*, although Vito says that he does not want me at his lecture!"[4] Leaving their hotel on July 31, Vito and Virginia detoured to the northern Swiss border town of Schaffhausen, to have a look at the Rhine Falls, before heading for Basel, and taking a train from there to Paris to attend the congress. They arrived on August 2 in time to browse the exhibitions and pavilions of the 1900 World's Fair, the biggest tourist attraction in Europe that summer.

Paris was sunny but pleasantly cool for August. Virginia received a rousing welcome from her husband's colleagues, including "an infinite number of congratulations and hand shakings," she wrote in one note to Angelica, who had accompanied Vito to the first congress and was undoubtedly reliving it in her memory.[5] Various other celebrations surrounding the gathering had been postponed out of respect for the assassination of King Umberto I in Monza a few days earlier.

The Second International Congress of Mathematicians opened its doors promptly at 9:00 A.M. on August 6 on the fairgrounds of the Palais des Congrès. (The other sessions would take place at the Sorbonne.) Within minutes, the estimated 250 participants had taken care of procedural matters—electing Poincaré as president of the congress and approving the slate of the executive board and officers of the sections as well as the site of the next congress. Now they settled back to listen to the German scholar Moritz Cantor expand, in French, on mathematics histories since the time of Étienne Montucla, the great eighteenth-century French historian of mathematics. Volterra spoke next, on his heroes Betti, Brioschi, and Casorati. "It is to their teaching, to their labors, to the indefatigable devotion with which they pushed young Italian mathematicians towards scientific research, to their influence in the organization of advanced studies, to the friendly relations that they established between our country and foreign countries, that the existence of a modern school of analysts in Italy arose," he told his audience.[6] Recalling their historic 1858 journey across the Alps, Volterra declared that Italy's existence as a scientific nation dated from this trip. Volterra's sparing but informative references in this address to his own background and intellectual development have become an invaluable primary source about his life for many historians. The speech also served to establish Volterra's credentials as an authoritative spokesman for Italian science in the twentieth century.

Later that day, Virginia sent a telegram to Angelica and then a letter. "I would have liked to wire you this morning but my husband did not allow me," she reported, adding, "Vito remained satisfied but it took a great effort to obtain this confession." Not only had his audience erupted in cheers at the conclusion of Volterra's lecture, they had positively showered him with praise and congratulations "for his French pronunciation." Volterra added a few lines of his own, gallantly pointing out that if people had liked his talk it was due to the closing lines, which Virginia had suggested, "therefore Virginia does nothing but bring praise upon herself."[7] Writing to Volterra from Ancona, where he had taken his family, instead of attending the congress, Corrado Segre asked Volterra to send him a copy of his talk, and threatened to complain to Virginia if Volterra procrastinated, as he sometimes did. At least one of their colleagues, Segre hinted, knew how to promote his ideas. "Here I have already found on my table the address presented there by [David] Hilbert," he pointedly told Volterra.[8]

Like the Eiffel Tower, which became the iconic symbol of the 1900 Paris World's Fair, Göttingen mathematician David Hilbert's plenary talk in the same city on unsolved mathematical problems, imaginatively entitled "Mathematische Probleme," has stood the test of time, albeit in a somewhat more specialized context. In anticipation of possible linguistic difficulties posed by his talk, which he delivered in German on August 8—two days after Volterra's—Hilbert had cannily prepared and circulated a summary of his remarks in French. Reading from his prepared text, Hilbert said, "As long as a branch of science offers an abundance of problems, so long is it alive; a lack of problems foreshadows extinction or the cessation of independent development," before outlining the historical stages in the development of mathematics and the critical role that intractable problems had played in its evolution.[9] He referred to the work of contemporaries in France and Germany from Émile Picard and Poincaré to Georg Cantor and Hermann Minkowski, but not a single contemporary Italian mathematician, let alone Volterra, rated a mention. If Volterra took any notice at the time (and it's hard to imagine he didn't), he refrained from saying so publicly. Years later, however, he would roundly take Hilbert and his student the German-American mathematician Richard Courant to task for not citing him and others in their 1924 textbook *Methoden der mathematischen Physik*: "I do not understand why the only reference to functionals in this treatise is to be found in a single footnote, which merely says that this term is used in French mathematical works, so ignoring the vast literature published in England, America, Germany, Italy, and other countries," Volterra complained. He added even more pointedly:

> Still less do I understand why there is complete silence about the names of the numerous writers who have worked on the subject. And I am all the more amazed because the methods introduced and used by me from 1887 onward... were put to good use later on by one of the authors [Courant] of the treatise in question, and are moreover continually used in the treatise itself, while the second author [Hilbert] has used the notion of derivation of functionals.[10]

Hilbert wrapped up his talk by presenting ten unsolved mathematical problems drawn from arithmetic, algebra, and other branches of mathematics, for the "gifted masters and many zealous and enthusiastic disciples" in the twentieth century to work on. They had been abstracted, he said, from the complete list of twenty-three problems in the published version of his manuscript, to appear shortly in the *Nachrichten* of the Göttingen Scientific Society.[11] According to the reporter for the American Mathematical Society, Bryn Mawr professor Charlotte Angas Scott, Hilbert's talk elicited "a rather desultory discussion," although Giuseppe Peano, who attended the talk, characteristically had a strong objection to make. This one concerned Hilbert's second problem—"the compatibility of the arithmetical axioms." Peano insisted that his co-workers at Turin had already established such a system. In her report, published in the Society's *Bulletin*, Scott also remarked on the "shockingly bad" presentation of the papers she'd heard at the meeting. A litany of generic complaints, familiar to any listener who has attended such meetings, followed: "[I]nstead of speaking to the audience, he reads his paper to himself in a monotone that is sometimes hurried, sometimes hesitating, and frequently bored. He does not even take pains to pronounce his own language clearly, but slurs or exaggerates its characteristics, so that he is often both tedious and incomprehensible. These failings are not confined to any one nationality; on the whole the Italians, with their clear and spirited enunciation, come nearest to being free from them."[12] She singled out Volterra for special praise; he had, she wrote, made a point of using his platform to reach out to the specialist with a broad overview of a whole field of mathematics. The Second Congress concluded on August 11, following Poincaré's presidential address on the role of intuition and logic in mathematics. But it was Hilbert's August 8 presentation that cast the longest shadow over the direction of mathematics in the twentieth century, which may explain why few mathematicians, even today, are familiar with Volterra's lecture, let alone any other lecture, presented at that meeting.

Their professional obligations out of the way, Vito and Virginia lingered in Paris for another week, time enough to buy books and go to the theatre and the opera, visit the Louvre and Versailles, ride the elevator to the top of the Eiffel Tower, and shop for presents in the Bon Marché. After an unhurried look at Geneva, Aix-Les-Bains, and other scenic spots en route, the couple returned to Rome that September, leaving time for Volterra to nudge Pietro Blaserna to push his new university appointment through the bureaucracy of the Ministry of Public Instruction, before hurrying off by train to Pisa to give a talk about Eugenio Beltrami at the annual meeting of the Società Italiana di Fisica, a promise he had made to Ròiti months before.

Virginia remained at home, anxious for news. He drew an audience even larger than usual, he told her by letter, fleshing out the telegram he had sent immediately afterward, in which he reported that the hour-long talk had been a resounding success, much to his surprise and great satisfaction. "Since everything interests you, [I will tell you that] I didn't read it, but spoke it as if it were extemporaneous, but not hiding that it was prepared," he wrote. "[W]hen one can eliminate the reading, the remarks manage to become even more effective still"—a technique Volterra had not used with his audience in Paris. It seems he was delighted to give Virginia credit for this innovation as well. "I give my thanks to you for all the care you provided for me to work with so much tranquility and assiduity. I hope you will be content with all that resulted from this," which may have been a subtle reminder that Virginia's role as his wife included not letting their marriage vows interfere with his work. He also recounted for her benefit the business meeting of the Society, including its involvement with *Il Nuovo Cimento*, the physics journal that had become the Society's official publication several years earlier. These discussions had apparently annoyed him; he bristled at the "many small vanities, the idle chatter, and the many intrigues" of his colleagues on the journal's board, adding "Such things I am not equipped to deal with. I enjoy the part of the Society in which one speaks about scientific questions." With the excuse that he had talked too much about the meeting and about himself, Volterra, in closing, cautioned Virginia: "I write this letter for you alone"—perhaps he simply wanted to tell her, and no one else, that the desire "to embrace you again with much tenderness" was foremost on his mind.[13]

Volterra began teaching at the University of Rome that November. Fifteen students signed up for his mathematical physics course, which met three hours a week in the physics institute, situated on a secluded hill surrounded by palm trees and other tropical foliage, not far from the Santa Maria Maggiore cathedral, in Via Panisperna 89A. He had another eight students in celestial mechanics, which met twice a week in the same building, whose first floor housed the classrooms, the shop, and undergraduate laboratories; on the second floor were a library, faculty offices, and research laboratories. In his 1993 autobiography *A Mind Always in Motion*, the physicist and Nobel laureate Emilio Segrè, provided a description of attending Volterra's lectures on mathematical physics in the 1920s:

> I should add that the subsequent year [1928] I again attended his course, because he changed subject every year and from him one learned interesting notions of classical mathematical physics. Volterra's lectures were well organized and the subject matter was skillfully chosen (as I realized later), but his delivery, in a thin and slightly high-pitched or nasal voice, tended to put me to sleep. There were no textbooks, and one had to take notes; I therefore asked [Edoardo] Amaldi to write for both of us, since he wrote faster than I, and also to wake me up if I fell asleep. Volterra used to close his eyes while lecturing and somebody said that this was because, being kindhearted, he did not want to see the students' sufferings. Except for these superficial shortcomings, the lectures were profitable. One learned the mysteries of the Laplacian, Green's functions, Poisson brackets, and similar topics. It seemed sometimes that Volterra did not want to reveal the

physics underlying the equations and the analogies between different theories. After taking a course on elasticity and one on analytical mechanics, I passed the exam with the highest grades. I remember that to show my proficiency I mentioned the analogy between some elasticity coefficients and the capacity coefficients of electrostatics. I should not have done it. Volterra cut me short, remarking that he had not mentioned this in his lectures, almost as if he disliked my revealing a secret.[14]

In his first year of teaching at Rome, Volterra lectured on optics in mathematical physics and on the dynamics of heavenly bodies in celestial mechanics. He would teach both these subjects in tandem—they were his signature courses, really—until 1931, when his refusal to sign the Fascist loyalty oath would cost him his academic post.

Early in the new century, Volterra turned his attention to problems of elasticity, publishing in rapid succession between 1905 and 1907 a series of papers in Italy and France on the behavior of multiconnected elastic bodies. A region of an elastic body is said to be simply connected if any continuous curve in it can be shrunk to a point. It is multiply connected if it cannot be so manipulated—for example, if the body has a hole through it. Volterra, and the German geometer Julius Weingarten before him, found that in a multiply-connected body permanent stresses and corresponding strains could exist, even in the absence of external forces on the body, provided the body contained a dislocation, or a thin slice of missing atoms. Analogous phenomena occur in magnetostatics and in fluid flow. Volterra, as usual, was onto something big.

Shortly after being elected a foreign member of the Reale Accademia dei Lincei in 1899, Weingarten had sent Volterra a short note on the surfaces of discontinuity in elastic bodies, asking if he would present the work at a forthcoming meeting of the Lincei. As an aside, Weingarten remarked that a surface of discontinuity could not exist in an elastic body if the surface occupies a simply-connected space. Writing to Volterra again in the late fall of 1900, Weingarten announced that he had considered the case of a hollow cylinder with a circular base: "But the problem led to a transcendent equation, which has an infinite number of roots. The formulation of this equation is very difficult... and a solution founded on this equation does not have a real value." However, in the case of a cylinder whose height was very small, he had found an almost exact solution, which he called "very interesting."[15] Volterra promised to present Weingarten's note, although in his own mind he could not "understand, without demonstration, why a taut elastic body without external forces must be necessarily multiconnected."[16] Weingarten then offered an explanation, apologizing in advance if it did not meet Volterra's standards of rigor: "It is without doubt based on notions a little vague about the neutral state of a body. But this idea is generally adopted to develop the theory of elasticity."[17] He even volunteered to with-hold his note from publication if Volterra still had serious reservations. In fact, Weingarten's note, "Sulle superficie di discontinuità nella teoria della elasticità dei corpi solidi," ("On surfaces of discontinuity in the theory of elasticity of solid bodies") appeared without delay in the *Proceedings* of the Lincei, probably translated into Italian (Weingarten had offered to write the paper in French or German) by Volterra himself. In 1905, Volterra would send Weingarten a copy of his first paper on elasticity,[18] pointing out that the question they'd left hanging four years earlier, "Are there any cases where the elastic bodies are simply connected?" could finally be answered "in the negative sense, when the deformation has the property that I call 'regularity.'"[19]

Volterra's work on elasticity brought him face to face with the limita-tions of Hooke's law (which says that strain, which is defined as the amount of "give" in a material, is proportional to stress, which is the force exerted on the material) as an exact description of the history of the stresses and strains in an elastic body. In 1909, he introduced the concept of a "mixed type" of equation combining elements of integral equations and partial differen-tial equations, for which he coined the term integro-differential equations[20]; these would underlie his attempts to explain the collective memory of twists and stresses, which he called "hereditary,"* in a given body. First and foremost an analyst, Volterra was fond of saying that for him mathematics served as a useful tool for solving problems that nature presents. "There are two kinds of mathematical physics," he would tell an audience in 1912, at the inauguration of the Rice Institute in Houston, Texas:

> Through ancient habit we regard them as belonging to a sin-gle branch and generally teach them in the same courses, but their natures are quite different. In most cases the peo-ple who are greatly interested in one despise somewhat the

---

*As an example, consider the twist of a body, $m(t)$, resulting from torsion of the body, $\omega(t)$. In the simplest case, $m(t)$ depends only on $\omega(t)$,

$$m(t) = h\omega(t) \qquad \text{(Hooke's Law)}.$$

But in the real world, $m(t)$ may depend on the entire history of twists of the body

$$m(t) = h\omega(t) + \int_{-\infty}^{t} \phi(t, \tau)\, \omega(\tau)\, d\tau.$$

We can assume the effects of the twists before $t = 0$ are negligible but take into account oscillations by substituting for $m(t)$ (where μ is a constant) to obtain

$$m(t) - \mu\frac{d^2\omega}{dt^2} = h\omega(t) + \int_{0}^{t} \phi(t, \tau)\, \omega(\tau)\, d\tau,$$

an integro-differential equation. We use the modifier "integro-differential" because the equation involves both a derivative of $\omega(t)$ and an integral of $\omega(t)$. Similar integro-differential equations turn out to be important in electrodynamics, hydrodynamics, optical and elastic problems. (From: V. Volterra, "Sur quelques progrès récents de la physique mathématique," in *Lectures Delivered at the Celebration of the Twentieth Anniversary of the Foundation of Clark University* (Worcester, MA.: Clark University, 1912), pp. 66-8.)

other. The first kind consists in a difficult and subtle analysis connected with physical questions. Its scope is to solve in a complete and exact manner the problems that it presents to us. It endeavors also to demonstrate by rigorous methods statements that are fundamental from mathematical and logical points of view.

I believe I do not err when I say that many physicists look upon this mathematical flora as a collection of parasitic plants grown to the great tree of natural philosophy. But perhaps this disdain is not justified. In the evolution of mathematical physics these researches probably are to play ever an increasing part....The other kind of mathematical physics has a less analytical character, but forms a subject inseparable from any consideration of phenomena. We could expect no progress in their study without the aid which this brings them. Could any one imagine the electromagnetic theory of light, the experiments of Hertz and wireless telegraphy, without the mathematical analysis of Maxwell, which was responsible for their birth?[21]

On their return to Rome in fall 1900, Vito and Virginia had moved into a large apartment on the third floor of her parents' home, the Palazzo Fiano. Angelica moved in too, and would live with her son and his family until her death in 1916. Edoardo and Eleonora Almagià lived one floor below them, and Edoardo would stop by daily to chat with Virginia and Angelica. Meanwhile, Volterra worked hard at getting his study in order, a task that took much longer than he anticipated. "I still have all the books and papers packed in cases," he told his Turin friend and colleague Giovanni Vailati at the beginning of December, "and until they are arranged, I cannot begin to work."[22]

The Volterras' apartment opened onto a long hallway, with the living room, dining room, and bedrooms arrayed on one side; across the hall a newly built wall transformed what had been a terrace into Vito's study. In the formal living room, Virginia's harp and piano stood at attention; in their bedroom, a large painting of Moses by Guido Reni hung above their bed. Portraits of Vito and of Edoardo Almagià, painted by Virginia's sister Amelia, hung on either side of the rectangular dining-room. The dining-room had wood paneling, coming up the wall about a yard from the floor. Beyond was the red room, with walls covered in red brocade and beyond that one, the green room, with green brocade on the walls. Virginia had a small living-room of her own, where she entertained friends and family members. "The house that I knew, the one at Palazzo Fiano, was a large apartment, with nineteenth-century furniture, rather austere, but beautiful, orderly," Luisa Almagià, the granddaughter of Edoardo's brother Roberto, later told an interviewer. "Virginia was capable, and really knew how to run the house."[23]

In 1901, Virginia and Vito had their first child, Luigi, who, sadly, died shortly after birth. Another son, Gustavo, born in 1906, also died in infancy; they are buried in the Jewish cemetery in Rome, along with their grandmother Angelica Volterra. Four of their children would survive to adulthood: Luisa, born in 1902; Edoardo, in 1904; Enrico, in 1905; and Gustavo, in 1909 (the last was named after Vito's close Swedish colleague Gösta Mittag-Leffler). Each of the Volterra children was also given a Jewish name. Edoardo, who later became the first Jewish judge on Italy's Constitutional Court (similar to the Supreme Court), had a total of four names: Edoardo Abramo Alfonso Giacomo.[24] Luisa would graduate from the University of Rome with a degree in biology and in 1926 marry Umberto D'Ancona, a zoologist, who had supervised her undergraduate thesis. D'Ancona's interest in using mathematical methods to describe the fluctuations in the Adriatic fish market during World War I and Volterra's in applying mathematics to biological problems culminated in a fruitful collaboration between the two men, memorialized in D'Ancona's book, *The Struggle for Existence*, first published in German, in 1939, a year before Volterra's death. Enrico and Gustavo would also graduate from the University of Rome; Enrico immigrated to the United States after World War II, taught at the Illinois Institute of Technology and Rensselaer Polytechnic Institute, and in 1957 became a professor of aerospace engineering at the University of Texas at Austin. Gustavo, the youngest (and said to have been Virginia's favorite), practiced law in Rome.

Convocation ceremonies traditionally marked the start of Italy's academic year, and in the fall of 1901 Volterra joined a distinguished parade of professors at universities and institutes of higher learning across the land charged with presenting a learned discourse on fresh and timely topics to a captive audience of students and fellow academics. In Bologna, the embryologist Giulio Valenti began with a provocative question, "Has science become bankrupt?"; in Pisa, the geodetic engineer Paolo Pizzetti discoursed on "The influence of geodesy on the progress of the mathematical and physical sciences"; in Padua, the mathematician Gregorio Ricci-Curbastro surveyed the "Origins and development of modern fundamental concepts of geometry." Even the historian Gaetano Salvemini caught the science wave, titling his opening lecture at the University of Messina, "History considered as science." While most speakers addressed questions in their own fields of research, Volterra explicitly challenged the compartmentalization of the sciences in his inaugural talk, "On the attempts to apply mathematics to the biological and social sciences," a topic that had clearly been on his mind for some time. In a letter written about six months earlier to his former intellectual boon companion Giovanni Vailati (with whom he had kept up a lively correspondence since Vailati's move to Sicily), he asked, "What can I do to prevent removing all interest from among the cultured in other disciplines outside of mathematics?" Did Vailati think that a talk about recent attempts to apply mathematical methods to economics and biology might do the trick? Should he substitute something else? In any case, what books would Vailati recommend that he read? "You *who know everything* and are well grounded in the social sciences would be able to give me some excellent advice, and I will be very grateful if you would write me something concerning these things."[25]

Now teaching mathematics in a technical high school in Bari and as erudite and lively a Renaissance Man as ever, Vailati sent Volterra scurrying off to the library to read Francis Galton and Karl Pearson, English biometricians who had analyzed problems of heredity and evolution using statistical methods and the calculus of probability. Closer to home, the Italian economist and sociologist Vilfredo Pareto had subjected economic problems to the infinitesimal calculus, and Vailati, who had a degree in mathematics and had corresponded with Pareto, could not resist giving Volterra a quick primer on "marginal utility" and "maximum ophelimity"—Pareto's unhappy term for economic satisfaction.[26]

Volterra delivered his talk on November 4, 1901, in the assembly hall of the centuries-old Università della Sapienza, just steps from the Palazzo Madama, the meeting place of the Italian Senate. In it, he spoke "of a strong desire to find out whether the classic methods that produced such wonderful results" in mechanics worked equally well in economics and the biological sciences:

> The transition of science from what I will call its pre-mathematical era to the mathematical era is characterized thus: Elements under study are examined in a quantitative rather than a qualitative way. In this transition, definitions that suggest an idea of the elements, in a more or less vague picture, give way to those definitions or principles that determine them, offering instead a way to measure them.

Economics had made the transition from a qualitative to a quantitative discipline, he went on to argue, pointing to the writings of Pareto, Leon Walras, and William Whewell, among others, and he enthusiastically called for recognizing mathematical economics as "a scientific field in its own right." Volterra also spoke of the "application of mathematics to the biological sciences," saying, "The causes of natural inheritance in particular cases are so complex that they do not admit an exact treatment"—which explained, in part, why efforts to add a quantitative dimension to biology had not kept pace with similar efforts in economics. But if differential equations, in which cause and effect are directly linked, did not hold the key to unlocking the secrets of genetic inheritance and Darwin's theory of the origin of species, perhaps other tools did, including statistical analysis and the calculus of probabilities. And, Volterra told his listeners, if the results obtained by

Pearson, Charles Davenport, and other members of the new school of bio-metrics had not yet attained the reliability of Pareto's findings, he remained cautiously optimistic, "if only for this reason, they arouse our curiosity all the more." Volterra also shone the spotlight on the Milanese astronomer Giovanni Schiaparelli's efforts to construct a model of the organic world based on geometrical shapes (reminiscent of Johannes Kepler's attempts to inscribe the planets inside the "five perfect solids" of antiquity)—Volterra genuinely admired Schiaparelli and would later nominate him for the No-bel Prize, to no avail. Nor could Volterra resist ending his discourse on an ethnocentric note:

> If we take a quick look at the birth and development of the most original and fertile ideas that have transformed and revitalized human knowledge, we immediately see that a conspicuous number of them are due to the genius of Ital-ians....[F]rom those far-off epochs, across the centuries to today, continues unbroken the succession of Italians who have led us into the modern world, in which Italy now plays such a great part.[†]

Aside from a critical review of Pareto's *Manual of Political Economy* in 1906, which led Pareto to revise his views on utility and indifference curves, Volterra left the further mathematization of economics to others, including the American mathematician Griffith C. Evans, who studied at Rome from 1910-1912 under Volterra and in 1930 published a mathematics book for economics students (*Mathematical Introduction to Economics*).

As Volterra groomed himself to become an increasingly public face on be-half of Italian science, his curiosity about the Anglo-Saxon world increased. Although his command of English was very limited, in 1901 he visited Eng-land for the first time, accompanied by Roberto Almagià, Virginia's teenage brother, who acted as interpreter when needed. During their stay in Eng-land, they boarded at the Pension Edwards, on the outskirts of London. Roberto soon began chatting with one of the other guests, "an elderly Miss," according to Volterra, who did not take part in the discussion. "I do not know if they are exchanging eternal oaths of love between themselves or speaking about the local train schedule, because I do not understand Eng-lish and I cannot follow the conversation," he lamented to Virginia, who was pregnant with the couple's first child, and had remained in Rome. "England is the classic land of mathematical physics, therefore a visit in this coun-try with people who are working in this field is certainly advantageous," he assured her, lest she think he had gone there to amuse himself.[27] Having served on fellowship committees at home, Volterra wanted to understand how Cambridge University worked, "but to tell the truth, none of us knows what courses are offered there, or what preparation is needed in order to be

---

[†]The complete text of Volterra's 1901 discourse can be found in the appendix, trans-lated into English.

able to follow them." In short, "what they can learn there is unknown."[28] The Cambridge astronomer George Darwin (the second son of Charles Darwin) welcomed Volterra and Roberto to the university, escorted them to the Cavendish Laboratory—where Volterra met the laboratory's director, J. J. Thomson, discoverer of the electron—and wined and dined Volterra at Trinity College in the evening. By the end of his two-week visit, Volterra had a much clearer picture of the English university system (it bore no resemblance to any Italian, French, or German university he could think of) and the benefits and drawbacks of sending Italian students there. "It is peculiar how the constitution and the customs are still decidedly medieval," he told Virginia. "I don't know how wise it would be for a young Italian to do all his university studies here [at Cambridge]. He would become perhaps too much a foreigner in his own country and that would not be an advantage for him. Instead, a specialized course of study here would be very useful."[29]

Volterra and Roberto also explored London. Volterra became an expert on baby prams, thinking to bring one back to Italy; Virginia advised buying a raincoat instead. Hunting for gifts, he discovered the city's shopping emporiums, "where horses, carriages, roast meat, furniture, linen, mackintoshes, and toothpicks are sold—everything, in other words." There was nothing like them back home: "If a newborn baby were able to sign a check, he would be able to provide for himself everything he could possibly need in his lifetime."[30] With Roberto in tow, Volterra visited art galleries and museums, St. Paul's and the docks, Piccadilly Square and Oxford Street, Regent's Park Zoo, and the medieval Inns of Court. The mathematician George Greenhill, who sponsored Volterra's temporary membership in the Savile Club, one of London's most prestigious private clubs, lived in one of these inns, in a small room surrounded by a bevy of lawyers' offices. Volterra was reminded, as he wrote Virginia, of Charles Dickens's "description of these very odd Inns, but it is necessary to see them in order to get a proper idea of them," just as "it is impossible to know the English without seeing them in their homes."[31] If only he could speak to them without using sign language.

In early 1902, Angelica's brother, Alfonso, died, a loss that struck Volterra and his wife while they were still grieving over the recent death of their newborn son, Luigi. In the midst of these sorrows, Volterra completed a paper, "On the stratification of a fluid mass in equilibrium," which he sent to Mittag-Leffler for publication in a book commemorating the hundredth anniversary of the birth of the Norwegian mathematician Niels Henrik Abel. In the spring, Volterra took Virginia to Naples, for a change of air. That August, Volterra attended the Abel centennial celebration in Christiania (now Oslo), where the university awarded him an honorary degree in mathematics—which both pleased and amused him, as he had earned a degree in physics, not mathematics, from Pisa. Virginia, then expecting their second child, remained at home, but she encouraged him to take the opportunity to travel around Sweden and Norway a bit. "I have to thank you

who has pushed me to make this trip. . . .[F]ew women would have pushed
their husband to take a long trip without going along themselves and only
because most women have closed minds, which you, indeed, do not have,"
Volterra wrote, blissfully unaware that leaving his pregnant wife for a month,
with her mother-in-law and her sisters for company, might be a recipe for
domestic discord.[32]

While her husband was out of town, Virginia, who was staying in a villa
in the Alban hills, complained to her father about her mother-in-law's temper tantrums. Angelica apparently resented the fact that two of Virginia's
sisters wanted to come and keep her company in the villa while Angelica was
there. Edoardo tried to broker the peace among the women. "Do you really
want them [your sisters]?" he asked Virginia. "Between the two of us we
can talk frankly. I wouldn't want your *mother in law* to find this *excessive*,
so that she could say there is no *discretion* in your sisters' proposal." If
her sisters did come, Edoardo promised to send her a piano, although he
knew that Angelica would not appreciate his gesture. "I realize that in this
way I'll provoke all of Signora Angelica's choleric outburst; I can already
picture her flying off the handle against you and against everybody. My
poor Virginia, how many things you have to bear." But Edoardo also knew
how to wrap Angelica around his finger: "Let's see if it works to treat her
kindly and you could begin with sending regards in my name and tell her
many affectionate things, assuring her that I say them with all my heart. . . ."
Just before sealing the envelope, Edoardo thought of something else, and
he scribbled a postscript: "What *if the attitude of saving things that's so
exaggerated in you* was inherited by your children? How would this letter
of mine be interpreted in a few centuries, on the off chance that somebody
found it! How many struggles, how many divisions it could create among
future scholars, where Angelica will be handed down as a Saint?!!!"[33] How
indeed? History does not record how Virginia dealt with Angelica, but she
did save Edoardo's letter.

Having returned to his empty apartment in Rome for just a few days,
Volterra left the city by train in mid-August, headed for Sweden, with a brief
stop scheduled in Berlin. En route, he fell into conversation with a German
woman sharing his compartment, who complimented him on his fluency in
the language. Writing to Virginia from Berlin about this encounter, Volterra
took stock of his linguistic skills: "Here at least I am able to say the most
elementary things, while in England I had no other organ of speaking other
than my hands. What will happen in Scandinavia, where I know only the
article of the language?" In Berlin, he found only Hermann Schwarz and
Ferdinand Frobenius at home; Schwarz, with whom Volterra had studied
in Göttingen eleven years before, had done beautiful work there but now
lived in the past ("He is a man who has become extremely stupid, although
he is not old"), while Frobenius, with a reputation for being ill-tempered
and quarrelsome, could speak only German ("I remained half an hour looking at him, but he doesn't know how to say one word in French"). With

David Hilbert having turned down an offer to come to Berlin, and Hermann Minkowski on his way to Göttingen, Volterra sensed that a seismic shift had occurred in the German world of mathematics. "Poor Berlin school of mathematics," he remarked.[34]

From Berlin, Volterra made his way to the Baltic coast, and from there he crossed to Stockholm where his old friend Mittag-Leffler and the latter's colleague and former student Erik Fredholm awaited his visit. After spending several days there, Volterra set out for Norway, going first to Trondheim then to Bergen and finally to Christiania in time for the Abel centennial. In all, he spent two weeks on the road in Scandinavia, sleeping in small mountain hotels or in the cabins of steamships, traveling by train on the coast, horse-drawn carriages in the interior, and by boat to see the fiords up close. As he traveled, Volterra took careful note of the striking difference between the political situation in Norway and Sweden—Norway would not gain its independence from Sweden until 1905—and that of Italy. "The union of the Scandinavian countries is impossible," he told Virginia, "because neither Stockholm, nor Copenhagen, nor Christiania is ready to give up power to the others and none has so great an authority and historic importance as to impose itself upon the others." Rome, a city unmatched by any other in the world, had saved Italy from a similar fate, he said, and while it had "many defects," the city "has bestowed the inestimable gift, i.e., of having given birth to [Italy] and of keeping us [Italians] alive." Critics of Rome's bureaucracy he dismissively described as having "tiny brains," lumping his old mentor Antonio Ròiti in this category with those "who fire inflamed projectiles against Rome." Bringing his spirited defense of Rome's unique place in history to a rousing close, Volterra proclaimed, "The old-fashioned statesmanship of Sella was the only right one."[35]

Quintino Sella had been a professor of mathematics and mineralogy at Turin before exchanging academic life for the role of Italian statesman and financier as Italy fought its way to unification. Victor Emmanuel II appointed Sella to his cabinet three times as finance minister: in 1862, in 1864, and again in 1869. Under Sella's leadership, Rome's scientific halls, which had chafed under the heavy hand of papal sovereignty, came to life again. With the help of Francesco Brioschi, the new secretary-general to the minister for public instruction, Sella brought the cream of Italy's scientific faculty to Rome, transforming the capital's historic Accademia dei Lincei, then the personal institution of the pope and located on the second floor of the Capitoline palace, into a genuinely national, authentically scientific academy and moved it into the Corsini palace in Via della Lungara, which the Italian government bought for this purpose in 1883. The scientific world that Volterra inherited in 1900 when he took up his duties as professor of mathematical physics in the nation's capital was largely the creation of this visionary academician-statesman and his colleagues.

How best to safeguard Sella's legacy had lately been on Volterra's mind. Shortly after reaching Bergen, Volterra received from Sella's son Alfonso a handwritten letter penned by Sella in 1879, nine years after Rome had become the capital of Italy. Sella's letter was addressed to his mother:

> The problem for me is the following: If I keep on with the political life, I have a sacrosanct duty to accomplish—to understand thoroughly what has to be done in the various events that can happen in our country....In human affairs, it is necessary to have clear ideas....[I]f I keep on with [political life], whatever the sacrifice to Clotilde [his wife] and me, I have to follow it in such a way as to accomplish the terrible duties that are incumbent sometimes on a politician.[36]

Volterra seems to have taken Sella's words to heart; perhaps he aspired to be like Sella, to do more for his country, to be ready to sacrifice for it, if necessary. He may also have been struck by the fact that the great man, like himself, seemed to have carried on a fond correspondence with his mother. In a thank-you note to Alfonso Sella, Volterra remarked that even as a young man he had admired Alfonso's father for his outstanding qualities as a scientist, a statesman, a man who lived by his principles: "In this far country, where I now find myself, you cannot know what effect it has on me to read those lines that recall one of our greatest men, and one of the historical moments most important for our fatherland, which we love so much more when we are far from it."[37] By the end of his stay in Scandinavia, Volterra had come to believe that the purpose of his trip had less to do with honoring Niels Henrik Abel's memory and more to do with evening the odds in his homeland's favor. France and Germany, Italy's neighbors to the north, had sent scientists to the event, and "it was necessary that there should be at least someone representing Italy," he told Virginia. "I alone came, but it was a duty that Italy should be represented."[38] Volterra returned from his month-long stay in Scandinavia determined to play a more active role in his nation's future.

CHAPTER 12

"Will They Create a New World?"

Raised in the midst of his Almagià relatives and almost never out of the care of a doting mother, Vito Volterra was undeniably a family man, but, until his marriage in early middle age, he was a family man with almost none of the usual family responsibilities. Acquiring an Almagià wife both changed and did not change that situation. Accustomed to an unfettered lifestyle and thriving on professional contacts and camaraderie, Volterra seems to have spent nearly as much time on the road after he married as before. Although this would eventually create some tension in his relationship with Virginia, its immediate impact was that his wife supplanted his mother as chief confidante and correspondent. In the summer of 1902, still recovering from the death of her first baby and expecting a second, Virginia left Rome, accompanied by other family members, for what everyone hoped would be a healthful and restorative stay in the country. Volterra remained behind in the city, and in early October he wrote to Virginia that he had just begun riffling through the papers submitted by the candidates for the chair of astronomy at the University of Bologna—a position, ironically, that he had once been offered, and rejected out of hand. He had little interest in who would win first place in the competition, Volterra told her, saying, "I joined [this] Commission more for curiosity and for learning than for any other reason." Sounding very much like someone who was getting ready to play a decisive role in the organization and direction of Italian science, he added, "I am more interested in studying the commissioners, their tendencies, and the mechanism of their soul and of their mind than in dealing with the rest of it."

A veteran of many previous commissions—his committee assignments in 1902 alone included recommending Tullio Levi-Civita's promotion to full professor at Padua, ranking the applicants for the vacant chair of rational mechanics at Genoa, and selecting a librarian for the Reale Accademia dei Lincei—Volterra nevertheless seemed taken aback by the stately pace of this particular deliberating body. "Among the commissions I have seen at work, this one works with the greatest seriousness," he allowed, but the judges, it seemed to him, proceeded as if they had "leaden feet." At this rate, Volterra despaired of leaving town and joining Virginia in the country one minute ahead of schedule, but he did feel confident that he had accurately divined the thinking of the committee. "I foresee that [Michele] Rayna will be elected; but nothing has been said about it, yet," Volterra predicted, correctly as it turned out. "Naturally," he added, "I am writing this *in confidence* and *in all secrecy.*"[1]

It had been a different story earlier that year when Levi-Civita came up for promotion to full professor of rational mechanics. The outcome was never seriously in doubt; still, the matter commanded Volterra's full attention. Keen to advance the career of a man he considered a protégé and the most able of the younger Italian mathematicians, Volterra urged Senator Francesco Siacci, who chaired the commission charged with making that particular recommendation, to invoke the "clear fame" provision in the state's education statutes, instead of letting the paperwork go through the usual channels. In February, the request landed on the desk of Minister of Public Instruction Nunzio Nasi, who asked Volterra to tell him in his own words why Levi-Civita deserved special treatment. Volterra had also nominated Levi-Civita as a corresponding member of the Accademia dei Lincei; and instead of providing the minister with an accessible, non-technical appraisal of the younger scientist's accomplishments, he simply sent Nasi the same remarks he'd prepared for the Lincei, replete with language fit only for mathematicians: "He has revisited in rational mechanics the question treated by [P.] Appell, [P.] Painlevé, and by others on the transformations of dynamical equations, having taken an important and definitive step with respect to the problem. Then he took up with much success the theory of potential functions, applying to it Lie's theory of continuous [transformation] groups, and has recently finished research of great interest on stability."[2]

To this inscrutable endorsement, Minister Nasi, predictably, had no response. He turned his attention to a bill guaranteeing equal pay to male and female schoolteachers. Not long afterward, Levi-Civita secured his promotion without any special dispensation. Three years later when he did win election to the Lincei, Volterra sent his congratulations, couched in language that left little doubt about who had worked tirelessly behind the scenes to make it happen: "I have been wanting this nomination for you for a long time, because you merited it already for some time, given the very high position that you have acquired in Science."[3]

Volterra made sure he was at his wife's side in December 1902, when she gave birth to the couple's only daughter, Luisa. But by summer he was off again; his Turin colleague, professor of physiology Angelo Mosso had invited him to take part in what appears to have been a turn-of-the-century exercise in extreme science. With Virginia and baby Luisa safely in the care of relatives at the Ancona seaside, Volterra set off for Capanna Regina Margherita, a biomedical research station situated on the summit of Monte Rosa, a towering glaciated mountain straddling Italy and Switzerland. There, at some 4,600 meters (approximately 15,000 feet) above sea level, Mosso and his hardy band of researchers were studying the effects of high altitudes on humans and animals. Two other distinguished scientists— Alfonso Sella, a University of Rome physicist, and Pio Foà, the chair of pathologic anatomy at Turin—were also involved in this phase of the study.

They were investigating a theory then gaining currency in some medical circles—that at high elevations, the reduced levels of oxygen in the air stimulated the body to produce more red blood cells. As a result, it was thought, mountain living could successfully treat anemia. This was an idea, Volterra acidly noted, that had been enthusiastically embraced by Alpine innkeepers and hoteliers.

On August 7, Volterra caught the mail-coach for the Swiss town of Gressoney La Trinité. After a night at a local hotel, he set out on foot with a guide, heading for a small inn at the Col d'Olen, the mountain pass at the foot of Monte Rosa—a day's walk, he informed Virginia—where he planned to remain for several days, acclimating to the altitude ("we are above 3,000 meters") and the temperature ("it is cold here"). The hotel's accommodations were sparse—everyone slept in one large communal room on straw mattresses—but the location afforded pristine and breathtaking views of the highest peaks of the Italian Alps and bone-tingling fresh air. Only the absence of letters from home marred his stay, a deficiency he attributed largely to Italy's technological backwardness. "We are very backward here and it was easier to remain in contact [with you] from Norway than from these places although they are in Italy," he lamented. "We Italians do not understand that before building a house in any spot one needs to start with the telegraph office and the telephone." Longing for news about his infant daughter and Virginia, who was pregnant again, Volterra closed his letter with the wish that "you will be very well and that the sea will be agreeable and that you will grow fat as you promised."[4]

On August 11 Volterra resumed his ascent up Monte Rosa; after another overnight stop, he reached the peak. On the last leg of his climb, he met Sella and his group just starting down; the next day brought Mosso's latest research team, along with porters laden with food, other provisions, and the monkeys, guinea pigs, dogs, and other animals slated for experimental study.

Capanna Margherita—at the time, the highest mountain dwelling in Europe—was essentially a hut, more or less snugly constructed of double wooden walls faced with copper plates as protection against lightning. Inside, Volterra and his cohorts found several wood and coal-burning stoves and canned and fresh meat (the latter stashed outside), but no privacy. The scientists, fully clothed and wrapped in layers of blankets against the cold, slept all together on plain wooden bunk beds stacked on different levels, or on straw mattresses on the floor. Mosso's laboratory menagerie slept in the same room, making for an unappealing variation on Noah's Ark. "These animals certainly do not make the stay at the hut very pleasant, since they give off bad smells," Volterra later griped to Virginia. Shortly after he arrived, a blizzard struck the mountaintop. Supplies ran low. The stack of wood dwindled and the stoves quit working. Trying to stay warm, Volterra spent most of his time asleep on his straw mattress. He felt keenly the lack of water. The cabin's supply came from melting the ice, but in an effort to conserve the fuel used for that purpose, the researchers—Volterra included—stopped washing their hands and faces.

Although the physiological effects of oxygen deprivation are well documented today, they came as a surprise to Volterra. He found that during the day, he had trouble breathing every time he took a step or did any work; the slightest exertion shot his pulse rate up dramatically. "Afterwards, naturally, it returns to almost normal." Then, there were the headaches—faint in the morning, fierce in the evening—a phenomenon Volterra attributed to "the decrease of atmospheric pressure, which also often produces nose bleeds," but fortunately not for him. The heavy snowfall had prevented him from venturing outside, but the company inside was good, and he passed the time conversing, a bit lethargically, with his cabin mates. From Foà, one of the group's pathologists and an expert on the structure and reproduction of red blood cells, he learned that the latest experiments with rabbits seemed to contradict the notion that mountain life could help cure anemia. The scientists, he wrote to Virginia, "had found that the red blood cells increased in the ears, but remained normal in the carotids. So it is not true what the doctors paid by the hotel-keepers were maintaining."[5]

Mosso, in fact, had wanted Volterra to do some calculations on his behalf, but that would have meant remaining at the summit until the research team had finished with their experiments, and Volterra wanted no part of that. Cut off from all communication with the outside world, he painted a bleak picture for Virginia of his surroundings: "Life revolves around such a tiny space that it offers nothing to tell nor is there anything to describe other than snow and [more] snow, fog and [more] fog."[6] But five days into his stay, the wind and the snowstorm unexpectedly tapered off, allowing a group of porters and guides to bring provisions to the hut, and enabling Volterra to make his immediate departure. Guide in tow, he left Capanna Margherita that same afternoon, descending gingerly through the fresh mounds of soft snow to Capanna Gnifetti, where he spent the night. In the morning, he descended directly over the Lysglacier and the moraine to Gressoney, avoiding the Col d'Olen.

Safely below the snowline, in a hotel with a real bed, Volterra took stock of his brief but intense Alpine adventures. He had nothing but praise for the views from the summit: a panorama of the entire Matterhorn on one side, all the dazzling topography of Switzerland on the other. Now that he was no longer obliged to live in it, however, he described the Margherita hut as a firetrap, and told his wife that he shuddered to think what would happen if a blaze broke out. "In fact, the walls of the mountain descend abruptly for 2,000 meters all around, and there is only a narrow passage that leads from the Lysjoch Mountain pass to the hut, a passage that is unusable if steps have not previously been cut in the ice. Given a fire in the hut now, it is impossible to save oneself."[7] Dividing the hut into separate rooms, using metallic walls to prevent a fire from spreading seemed the best solution to Volterra, who then turned to the far more interesting topics of

how to harness wind power on the summit to generate electricity for heat
and light, and the best way to establish telegraphic communication from
there with the outside world.

Overall, his stay in the hut had not been "pleasant," and yet, as he
wrote Virginia, "I am very happy to have spent some time there, both
because I witnessed many striking things and also because it was very good
for my health to be up there." In fact, now that he had returned to a
fully oxygenated environment, he was certain that his few days high in the
mountains had been extremely beneficial. "I am, for example, completely
recovered from the light disturbance I had in Rome." The physicians there
had apparently been able to do nothing for that particular problem, and a
few days cooped up at high altitude with a band of derisive physiologists
had only strengthened Volterra's contempt for certain medical practitioners.
He wrote to Virginia:

> [You should] have listened to the physiologists—the Marghe-
> rita hut was full of them—who mocked in no uncertain terms,
> the ridiculousness of the medical prescriptions that the doc-
> tors write in Montecatin solely to justify what they charge
> for an examination. Whoever believes in things like this has
> become the victim of those who profit by taking advantage of
> human stupidity.... We criticize so much the merchants of
> human flesh; but which merchants of human flesh are greed-
> ier and more dishonest than some doctors. By good luck, I
> stopped taking that water of Tettuccio that was ruining my
> stomach, and following your advice I came here to breathe
> some good mountain air. I have to thank you also for this.[8]

(Virginia, her father, and the Almagià family physician, Ettore Marchiafava, had all given him the same advice: to work less and rest more, and
spend more time in the mountains. More often than not, he ignored their
suggestions, but while he had been sharing cramped quarters in his hut,
his civil engineer father-in-law had begun overseeing the construction of a
summer home in Ariccia for Vito and his family. Once it was completed in
1905, the Villino Volterra quickly became Volterra's home away from home,
a refuge from the hustle and bustle of the city.)

In late August, fully revived by a few days of rest and hiking, Volterra
left Gressoney to join his family, intrigued by what he had just been through,
yet skeptical that mountaineering in the Alps—"Alpinism," as it was known
to enthusiasts—"represents the most intellectual thing in the world." His
political hero, Quintino Sella, had not only founded the Italian Alpine Club,
but also preached the physical and moral virtues of hiking the slopes and
ascending the peaks. Volterra, although he grudgingly accepted the benefits
of "the physical side, perhaps," gave it as his opinion that "from the moral
side there is absolutely nothing....It is a question of paying the guide and
that's enough."[9]

Back at sea level, Volterra resumed his increasingly meteoric professional ascent. That same year, he accepted the position of administrator of the Accademia dei Lincei, a prestigious appointment that offered him a firsthand view of the Academy's inner workings, including its operating budget, and hands-on experience in dealing with publishers. As the year drew to a close, the Italian Prime Minister Giovanni Giolitti appointed him the third member of a commission that had been set up to study and propose regulations for a new technical institute in Turin. His fellow commissioners were not only renowned scientists, but also members of the Italian Senate: Stanislao Cannizzaro, at seventy-seven, the grand old man of Italian chemistry, and Valentino Cerruti, a mathematician and director of the Rome school of engineering, and a familiar face at the Ministry of Public Instruction (Volterra would join them in the Senate two years later). Cannizzaro, the committee chair, assigned Volterra the task of investigating and reporting on the organization and curriculum of other technical schools. In February 1904, he left Virginia, who had just again given birth, to deal with the household, little Luisa, one-month-old Edoardo, the wet nurse, the cook and other servants, and headed for Milan, the first leg of a trip that would also take him to Switzerland and Germany.

In Milan, Volterra combined business with pleasure. He signed a new contract with the Swiss-Italian founder of the book store, and later printing-house empire, Ulrico Hoepli, for the publication of the Academy's proceedings, visited the Polytechnic School of Milan, and best of all, had dinner one night with his old friend Giovanni Vailati in Como. The next stop on his itinerary, Switzerland's Federal Polytechnic School, brought back fond memories of his honeymoon in 1900. "Here I am in Zurich three-and-a-half years later," he reminisced to Virginia before swerving abruptly from these tender recollections to ask whether she still felt the same way about him as he did about her. "My feelings toward you have remained unaltered and all of the esteem and affection that I was placing in you," he assured her, had only grown stronger with time. "But what about your heart?" he wondered. "I feel for you all the love and all the passion [the next word is illegible] as when we set foot here last time." His next thoughts—"You haven't experienced any disappointment in this time on my account? I hope not," concluding with the wish "that the great affection that you have for me hides all of my defects from you," suggest that Virginia had detected, and perhaps even alluded to a few "defects" in her peripatetic husband's character.[10] On the other hand, perhaps Volterra, rejoicing in the birth of a new child—and a son and heir—simply felt guilty about the time he was spending away from home. There is no record of Virginia's reply.

Volterra's next stop on his fact-finding mission was Göttingen, where he planned to talk with Felix Klein, Walther Nernst, and others at the university about the development of engineering schools in Germany. If David Hilbert and Hermann Minkowski, as the mathematician Hermann Weyl once wrote, "were the real heroes" behind Göttingen's transformation into Germany's premier school of mathematics in the first decade of the twentieth century, it was Klein who "ruled over it like a distant god, 'divus Felix,' from above the clouds; the peak of his mathematical productivity lay behind him."[11] Klein's accomplishments at Göttingen were legion: In addition to turning the university into a great research center in mathematics, he had introduced new academic departments, persuaded the heads of Germany's various industrial organizations to underwrite research laboratories, and recruited outstanding scientists to lead them. In 1904, when Volterra visited Göttingen, Ludwig Prandtl had just been appointed director of an institute of applied mechanics; Carle Runge had accepted the position as chair of applied mathematics; and Hermann T. Simon had been designated the new head of the mechanical and electrical engineering laboratory.

On February 18, Volterra checked into his hotel, and knowing that every Thursday, Klein, Hilbert, and Minkowski punctually went for a brisk walk, rain or shine, he caught up with them at 3 o'clock to tag along. "In fact, yesterday it was snowing heavily, but all the same we ascended the hill, which they are accustomed to reach at 4:30 P.M., and had coffee," he wrote to Virginia afterward, adding that he and Klein then spent three hours in "important and interesting" conversation. "Klein has a special ability in explaining things with great clearness and precision and his conversation is one of the most interesting that one is able to have in Europe." Volterra subsequently pushed back his departure from the city by one day, declaring that Göttingen was "the most intellectual town in Germany." That evening, he had dinner at Hilbert's house, "where I found many professors and interesting conversation." Friday morning, he hurried off to visit Simon's engineering laboratory, which did not impress him, returned to Hilbert's house for lunch, and held more discussions with Klein in the afternoon, joining him afterward for dinner. By the time he left Göttingen, Volterra felt he had learned enough about the integration of mathematics and engineering sciences at Germany's polytechnic schools to know what it would take for Italy to catch up to the Germans. "A little good sense on the part of our mathematicians is necessary, something that is not so easy to find due to the fact that the majority of Italian mathematicians live enclosed in their little habits without much contact with the outside world and with the fixed idea that things can be done only in the way they have been carrying on for so many years."[12]

Like Klein, Volterra advocated using mathematics to address problems in engineering, applied physics, and a range of other disciplines, but, as he told Virginia, not all his colleagues shared his interest in bringing mathematical principles to bear on the training of applied scientists, much less engineers. He spent his last day in Göttingen touring the geodesy laboratory ("extremely modest"), Göttingen's Institute for Geophysics ("in which I am very interested"), and the laboratory of physical chemistry headed by Nernst (who also took Volterra for his first ride in an automobile). After a meeting with the directors of the Prussian Ministry of Public Instruction and another marathon dinnertime dialogue with Klein ("I was for seven hours in conversation with Klein, from 5 P.M. to midnight"), Volterra headed for Berlin Sunday afternoon feeling he had earned the right to spend the next day resting up.[13]

Before leaving Göttingen, Volterra had learned that Virginia wanted to dismiss the wet nurse, a certain Signora Guidotti. He did not approve, and in his next letter he told her so. "I am very unhappy and opposed to what you write me concerning Signora Guidotti. I really hope that Amelia [Virginia's sister] will not put into play the project to dismiss Mrs. Guidotti for a matter that is absolutely none of her business. This would be truly an unjust thing... about which I would be really sorry." And while they were on the topic, Virginia had not paid the midwife enough either. "I also believe that Mrs. Matilda has not been fairly compensated for her efforts with only 100 *lire*," he wrote, adding, "You have to think that she has not only worked zealously for many days, but she also found the wet nurse, for which she must be compensated." Anxious to show proper regard for the midwife, who had delivered his son Edoardo the month before, Volterra instructed Virginia to "give Mrs. Matilda a gift of clothing worth at least 50 *lire*, and I beg you strongly to not fail to do so, it being right and proper." A stern lecture followed:

> It doesn't seem right to me that people who have conscientiously labored and stayed awake during the nights for us, not be given what is due them, and it is always better to sacrifice some money rather than lose the friendship and affection of people. At the least, one doesn't want to surround oneself with an emptiness, or live in an atmosphere where people dislike you....I am really unhappy to have to write you in this manner, particularly being far away, and I fear that you will take it hard and not right what I write to you with such frankness, but....It is necessary to remember that the greatest injustices were done by those who were in complete assurance of behaving according to the norms of justice, as it is fatal in human nature to make the greatest confusion between that which is absolutely true and just and our exclusive momentary material interest. And now let us leave this unpleasant subject and turn to another....[14]

Did Volterra feel that Virginia and, more to the point, his extremely wealthy in-laws, needed a sharp reminder about their obligations to the less fortunate? The passage is interesting, both for what it reveals of Volterra's sense of social justice and for the fact that this fine moral discernment apparently did not extend to appreciating that his young wife, just past her third pregnancy in four years and left alone to cope with two infants, unhappy servants, and officious family members, might not have been in the most receptive frame of mind. The "unpleasant subject" does not resurface in their correspondence.

Berlin both dazzled and depressed Volterra—"a modern city from every point of view," he wrote Virginia, yet reminiscent of Rome with its splendid ruins of the forum and the palatine hill. One night, he went to hear the Berlin Philharmonic play Beethoven, among other composers; the next evening was spent at the theatre watching a German–language performance of Sophocles's Greek tragedy *Electra*. His days were largely spent at the impressive polytechnic school. Located in Charlottenburg, a short ride by subway, or electric or elevated train from the center of the city, the polytechnic boasted well-equipped laboratories in fields ranging from chemical engineering to engineering design to mechanical engineering, and had just opened a laboratory for internal combustion engines.

The new lab's director was Alois Riedler, a professor of mechanical engineering and a passionate believer in practical training for engineers. His arrival in Berlin in 1888 had touched off what Volterra described as "the theatre of the titanic struggle between the highest mathematics and the most advanced practice, and the latter has triumphed." The champion "of the party opposed to mathematics," he continued, was none other than "the famous engineer Riedler," whom he described as "one of the most distinctive and typical personalities, a coarse and energetic man." Volterra painted a bleak picture of the vanquished party:

> In fact, by studying Berlin as I am studying it at this moment, I see the ruins of the superb mathematical schools of the last century and in general one sees the ruins of all the intellectuality or idealism of the last century... and they produce the same effect of breaking one's heart by comparing the present to the past. This poverty of the current Berlin mathematicians! Then last night at the meeting of the Berlin [Mathematical] Society. There was not one distinguished personality and the papers presented were old stuff pulled out, which showed the absence of a modern guiding mind.

The ruins of Rome had inspired painters and poets. But Volterra could find nothing stirring to say about the remains of Germany's once-thriving mathematical community, whose current representatives he had observed "swallowing big glasses of beer after the meeting of the mathematical society." Former students of the great Kronecker or Weierstrass, most were now professors themselves at Charlottenburg, the University of Breslau, and Berlin. But what, asked Volterra, had the possessors of such fine pedigrees and academic positions, done for German mathematics? Nothing, in his opinion. "They are, however, suggestive figures because they make us think, and they are certainly great ruins. They make us think because they show that Weierstrass and Kronecker with all their genius did not succeed at anything that they hoped to accomplish. They have created neither a new mathematics nor a school of their own. Their genius has been advantageous to France and not to their [own] pupils."

Who, he wondered, was responsible for this sorry state of affairs? Perhaps Riedler, who reminded Volterra not a little of Attila the Hun, and his confederates should be held accountable for the reduced state of Berlin's, and increasingly Germany's, mathematical community. As for the victors, their future had yet to be decided, although the alternatives that lay before them seemed clear enough to Volterra. He posed the provocative question, "Will they create a new world or a [new] Middle Ages?" but stopped short of hazarding an answer. He remained intrigued by what might lie ahead for the German people, insisting that "the modern, practical, and industrial Germany that is arising on the ruins of the intellectual and idealistic Germany is a spectacle worthy of a deep study. What is better: that Germany... [filled with] the spirit of Goethe and Schiller, which imposed itself with the philosophy of Kant, which excelled with Weierstrass and [Hermann von] Helmholtz, or the present Germany that fills Europe with the production of pieces at 1 pfennig... in very bad taste and of questionable quality?"[15]

Socializing in Berlin brought a new set of challenges. Invited one night to dinner, Volterra dutifully put on a tuxedo. Much to his amazement, the other guests wore suits. Faced with the dilemma of what to wear to another party, a gala late-afternoon event to which thirty people had been invited especially to meet him, Volterra wore a suit. "In Italy it would have been ridiculous to wear tails," he wrote Virginia afterwards, chagrined to see that everyone else sported a tuxedo. The full impact of the close ties developing between Germany's burgeoning industrial caste and the engineers who were now the dominant force in its technical schools came home to Volterra as he looked around the room filled with professors, bankers, businessmen, titans of industry, and other guests. "A new world that I did not succeed in deciphering in the five hours that I spent [there]," he called it, although he thoroughly enjoyed himself, not least because he could converse in French all evening, the first time this had ever happened to him in Germany. "I haven't the foggiest idea why Mrs. [Stanislaus] Jolles had to give a grand dinner in my honor," he added, but he seemed very pleased that she had.[16]

Volterra's great fear was that Rome's school of mathematics might meet the same fate as Berlin's. "When one is there, one does not think about it, but when one is abroad and is asked who the professors in Rome are, it is a real shame to have to mention names like [Alberto] Tonelli, [Giulio] Pittarelli," he sighed, adding modestly, "the best known is Castelnuovo." If Rome would only bring in fresh, proven faculty to handle geometry and analysis, "it would work wonders."[17] (As it turned out, the task of building up a modern school of mathematics at Rome did ultimately fall to Castelnuovo, who in the years before World War I waged a long, vigorous and in the end successful campaign to bring Tullio Levi-Civita and Federigo Enriques to Rome.)

After two final stops in Leipzig and Chemnitz, Volterra returned home in early March, much relieved that his grand tour of polytechnic and technical high schools (the case in Chemnitz) was over. By June, he had sent his report to Cannizzaro, who, as a fellow professor, could not resist sending it back covered with critiques and comments. "Don't be stingy with the notes and appendices," he counseled Volterra, advising him to include as much detailed information as he could about the various polytechnics' educational objectives, courses and schedules, academic faculty, and budgetary matters. Suppose, said Cannizzaro, a senator was to ask Volterra, what there was to "emulate from foreign polytechnics in preparing the by-laws of the Turin Polytechnic, given our conditions and the means at our disposal?" Volterra had to be ready to cite facts and figures and, equally important, prepared to present his own compelling arguments "to justify" bringing such educational innovations to Italy. "Give up for a couple of weeks your favorite studies and busy yourself with an important practical subject not too difficult to solve," he told Volterra point-blank.[18]

Volterra seems to have followed this sound advice, and by June, a second Senate committee, using the commission's report as its starting point, had circulated a new bill recommending that the city's Royal Industrial Museum merge with the Royal Technical School of Engineers to become the Royal Turin Polytechnic. One year later, when the proposed legislation made its way to the floor of the Senate, Volterra was invited to speak in its support.

In what appears to have been his maiden speech to the Italian Senate, Volterra called on Italy to revamp its advanced technical school system, invoking Klein's approach in Germany as the model to emulate. "[Klein] told me (and in this he is only summarizing the current ideas in his country) that Germany's industrial movement and its miraculous economic progress are in large measure the fruits of its higher technical schools." Italy's engineering programs, in contrast, occupied a position somewhere between the pre-Risorgimento faculty of science, in which degrees in both pure and applied mathematics were required for graduation, and the modern German-style polytechnic school. The Italian curriculum, in Volterra's view, emphasized theoretical studies at the expense of applied subjects. He included mathematics in this critique, saying too much of the subject was as bad as too little. One of Germany's foremost engineers, Alois Riedler, he told the senators, had gone so far as to declare that "an excess of mathematical and theoretical studies not only is useless and takes up precious time, but is absolutely pernicious to the engineer, impairing the education of the spirit." Meanwhile, the various German polytechnic schools had replaced many of their advanced theoretical courses with extensive laboratory facilities, equipped with working machines that had been of particular benefit for mechanical engineers. In Germany, he said pointedly, a decade-long debate between pro-mathematics and pro-engineering factions over the future direction of the academic curriculum had ended in a complete rout for the mathematicians.

Volterra emphasized that it should be possible to avoid a similar debacle on Italian soil, provided that the nation's mathematicians—"and I feel the need to insist on this point despite putting myself at odds with my various friends, colleagues, and teachers" —understood the need "to reduce the teaching of mathematics as much as possible, and more than anything condense it, in few, but good hands." If his colleagues insisted upon following an opposite course, refusing to cede any ground to the technicians, the inevitable result would be that young inexperienced assistants or instructors would end up providing the instruction.[19]

Volterra's lengthy speech touched on other aspects of the bill, including the need to increase the number and kind of experimental laboratories and the choice of site for the new institution, which, he noted approvingly, complemented perfectly the industrial growth in and around Turin. The Senate voted almost unanimously to approve the bill; the Lower House followed suit, and in November 1906, the Royal Turin Polytechnic (today the Polytechnic University of Turin) would open its doors as an autonomous engineering institution. That summer, Volterra would accept the government's invitation to be the Royal Commissioner (director) and president of the new school's council of administration, although he would step down for health reasons even before the school year had begun.

Already a pillar of the scientific community in Italy, Volterra saw his reputation abroad grow by leaps and bounds during the first decade of the new century. Fresh from his tour of engineering schools in Switzerland and Germany, in the winter of 1904, he was elected a corresponding member of the Académie des Sciences de France. This was followed by an invitation to join Lord Kelvin, Norman Lockyer, George Darwin, and other eminent scientists at the British Association for the Advancement of Science meeting in August at the University of Cambridge, which also awarded him an honorary doctoral degree in science. Virginia had agreed to accompany him to England, leaving the children with their nannies and family at the seashore. They went first to the Third International Congress of Mathematicians in Heidelberg, before heading for Cambridge. There, as guests of King's College, they indulged in a week of "running from one place to the next"; an "English custom," wrote Volterra to his mother, that required "changing of clothes I don't know how many times a day, especially [for] Virginia, who is always very smartly dressed."[20] From Cambridge, the couple went on to stay at the Pension Edwards, where Volterra and his brother-in-law Roberto had boarded three years before.

By then, Virginia was eager to return home to their two children ("It's been 24 days that I am far away from my adored little ones"), while Volterra maintained that Virginia had greatly exaggerated her maternal feelings ("but on this point I am truly not willing to discuss it because I feel too much their absence"). Virginia's desire to include the children on their trips, and her husband's insistence that they be left behind, would become a running point of contention between them. At the Edwards boarding house, time seemed to have stood still. As Volterra noted, "we found the same Miss that we had left three years ago and in the same position in which we left her. Among the others, we have met again the wife abandoned by Mr. Baedeker [this was most likely a joking reference to the Baedeker travel guides, an indispensable reference for many tourists of that era], who has taken us again under her protection pointing out all the colors and the directions of London's omnibuses."[21]

In March 1905, in his last official act as prime minister, Giovanni Giolitti appointed forty-three new Senators of the Kingdom of Italy, mostly former members in the Lower House. Of those appointed, four were recognized for their high scientific standing: Enrico D'Ovidio, professor of mathematics at Turin; Emanuele Fergola, director of the Capodimonte Astronomical Observatory in Naples; Augusto Righi, professor of physics at Bologna; and Vito Volterra. In an article about the new crop of senators, one of Italy's popular magazine's described him—"Illustrious physicist, professor at the University of Rome, and mathematician, is the Israelite Vito Volterra, also of the Lincei."[22]

Volterra's appointment reaffirmed the Risorgimento tradition of the scientist-statesman in the service of king and country. Years later, Castelnuovo, who had an armchair view of his colleague's efforts to advance the cause of science in Italy in the first quarter of the century, recalled that Volterra had probably been preparing himself over a long period of time to assume this new role:

> More than twenty years needed to elapse after the *laurea*, between austere meditation and rigorous lectures dictated by the professorial chairs at Pisa, Turin and Rome, before he felt the need, even the duty to assume a directive role in the Italian scientific movement. And the delay was advantageous, both because twenty years of study had enlarged his views in all the fields of the physical sciences, and because the universal authority which his discoveries had acquired served to smooth out the obstacles that intervene with every new initiative.[23]

The commitment and drive that Vito Volterra ("illustrious physicist... mathematician... Israelite") would bring to these varied enterprises would continue unabated until 1925, when Italy became a dictatorship under Mussolini.

CHAPTER 13

"A Political Man"

"Between you and me, I can really say that you have not only the most beautiful mind, but still more the most beautiful soul among our mathematicians."[1] This was Volterra's old mentor and high-school teacher, Cesare Arzelà, now a professor at Bologna, happily acknowledging the congratulations that his former prize pupil had sent him in June 1905, on the occasion of Arzelà's receiving Italy's Royal Mathematics Prize. Arzelà, who shared that year's award with Volterra's longtime friend and colleague Guido Castelnuovo, seemed dumbstruck by the honor, and, after commenting that he doubted that his work merited such recognition, he turned solicitously to another topic: Volterra's health. The stomach problems that had plagued Volterra off and on for the past few years were evidently well known to his inner circle, and now Arzelà earnestly urged him to test the waters of the Montecatini mineral springs in Tuscany. Their curative benefits were reputedly excellent; even more to the point, said Arzelà, "everyone goes there." He told Volterra that he really needed to mingle more, now that he had become a senator.

The exact nature of Volterra's medical problems are not clear, but it hardly comes as a surprise to learn that the intense and often competing demands of his private and public life had produced classic symptoms associated with overwork, pressure, and stress. Conscientious to a fault, a born perfectionist and workaholic, a renowned academic, a devoted husband, father, and son; an Italian senator; a rising, if still relatively youthful, elder statesman of Italian science; and, as his nation's popular press had not hesitated to remind him earlier that year, a Jew—he was an obvious candidate for almost any of the conditions demurely grouped in those days under the rubric of "digestive upset."

Not long after receiving Arzelà's letter and shortly after the birth of a new baby, Enrico, Volterra reluctantly consented to try a medical regimen. He did not go to Tuscany, but instead went to the well-known Bohemia spa town of Karlsbad, located in what is now the Czech Republic. At the time a playground for wealthy and titled Europeans, the town sits in a narrow valley that runs between two lines of hills covered with pine trees. Its reputation as a health resort rested on the local mineral springs, which annually attracted legions of invalids seeking cures for various physical ailments. Volterra rented a small room with a beautiful view of the valley in a guesthouse in the Hirschensprungzeile Haus "Belvedere" area, located on the left bank of the Tepi River. His "cure," like that of his fellow patients, consisted of drinking and bathing in the mineral springs several times a day. Ettore Marchiafava, the family's doctor in Rome, had urged him to go there; Virginia and his colleagues seconded the idea. "I don't doubt that a prolonged stay at Karlsbad, followed by a dose of Alpine air, will make the annoying disturbances that tormented you in the winter disappear," Castelnuovo had assured him, pointing out that his young son had recently been "ordered to bathe in the sea."[2]

Although pleasantly surprised by Karlsbad itself, Volterra wrote to Virginia that he had little faith in the doctors overseeing his case. "It is strange that, although I don't trust one bit that race of people who are the physicians, everybody ends by falling into their trap and letting them fleece and suck us dry....The trouble is once in the hands of these vampires, they have no desire to cure the patient." He was convinced that the main objective of the medical staff was to prolong the cure as long as possible.[3] Nevertheless, he followed his prescribed course of treatment to the letter, breakfasting in the designated café and obediently downing glasses of mineral water on schedule. Reflecting on his own willingness to adhere to what he considered a somewhat tyrannical routine, he thought he understood why his fellow sufferers at the spa might find it appealing to blindly follow the doctors' instructions. "The advantage of a medical treatment is that there is no need to think anymore, but we automatically do what has been prescribed." It was his experience that many people really preferred to let others do the thinking for them, seeing "the advantage, or feeling the need, not to think at all in life." For such individuals, the goal is to "choose that career in which, once settled, they can go ahead by the irresistible force of things."[4] In future years, he would return to this theme, at a time when it was no longer his health but Italy's that was in question.

As his first week of the spa treatment drew to a close, Volterra felt so much better that he actually started praising the quality of the water and making arrangements to spend several weeks at St. Moritz in the Swiss Alps. As his physical state improved, his view of tidings from home soured a bit. He wrote to Virginia that while he was eager to have her join him in Switzerland, he was adamantly opposed to her proposal that she bring at least one of their three children. She had written to him that the children's physician, a Dr. Ronchi, had approved the idea, prompting him to reply, "Ronchi and doctors in general say only what they believe will make you happy."[5] No doctor in his right mind, he added, would sanction uprooting such young children (Luisa, then not quite four, was the eldest) from the warm waters of the Adriatic to the cold slopes of the Alps for such a frivolous reason. Virginia refused to abandon the plan. Vito stuck to his guns. They had discussed this many times before, he reminded her, and he would not be bullied into saying yes now, much as he longed for his wife's company. Let all the doctors in Rome oppose him, Volterra pointedly told Virginia; he would never permit it, "in the name of common sense."[6] Once again Virginia replied that she could not see her way to remaining "calm" if obliged to leave all three children behind, and, once again, Volterra refused to back down. "I don't have to tell you again how contented I would be if you were to come with me...but the idea of bringing some of the children, if you will remember, *was excluded* even before I left Rome." Summing up, he added, "I have not changed in the slightest my opinion, as is my habit, after all."[7] Ever the professor, Volterra never quite rid himself of the tendency—which is not uncommon in the profession—to address cherished intimates, starting with his smart, strong-willed, and significantly younger wife, as if they were intelligent but untutored graduate students. In the end, Virginia and the children remained at the Almagià villa at Palombina, a sandy beach area north of Ancona, while Vito went to St. Moritz by himself.

En route, he made a side trip to Munich, where his arrival coincided with a full-scale production of *Das Rheingold*, the first opera in Richard Wagner's *Nibelungen* cycle. Volterra caught a performance of it, which he greatly enjoyed, at the Prinz Regent Theater.[8] Arriving at St. Moritz, the highest village in Switzerland's Engadine valley, he spent the next five weeks breathing the brisk pine-scented air, and strolling mornings and evenings along the hamlet's many footpaths, often joined by his Turin colleagues Corrado Segre and Pio Foà, who were also staying at the hotel. Meanwhile, a steady stream of letters from Virginia kept him briefed on developments at home—her father's newfound passion for motoring (too reckless), baby Enrico's weight (too little), a gift of pet donkeys for the children (too extravagant), and the ongoing construction of their villa in Ariccia (too many delays). If Virginia was hoping that these reports would spark a conjugal dialogue, she was not disappointed. Edoardo Almagià had enthusiastically embraced that new-fangled contraption, the automobile, writing to his daughter that he dreamed of buying a car "of 24, 40, 50 horsepower [that could] fly across the cities of Italy, and even abroad."[9] (His fervor was sidelined, but only temporarily, a few weeks later when he injured his shoulder in a car accident.) His son-in-law, however, was unimpressed. "As usual I don't begin to understand why people take long trips by automobile when one can go by train much more easily....it will be a custom that disappears just as the custom of traveling by bicycle has ended. In any case it will be necessary at least to have a chauffeur who knows how to drive and not drive alone; otherwise one really becomes a slave of the automobile."[10]

As for baby Enrico's weight problem, Volterra—who perhaps by now considered himself something of a lay medical expert—blamed it on the coastal air, insisting that the climate interfered with the wet nurse's ability to produce milk. Enrico's apparent failure to thrive, he maintained, dated "from the moment that you [Virginia] went to that place." The pet donkeys immediately became a pet peeve; such extravagances, he told his wife, had no place in the life he envisaged for his children. "I wish for them to become simple and serious people and not people who love elegance and sport," he protested to Virginia, who had serenely maintained their children were not being spoiled in his absence. "It seems to me they are not being educated according to my desires," he continued, adding, "If you wish me to be calm you will make this kind of amusement end."[11]

It is clear from his correspondence that Volterra had never much liked the Almagià family compound in Palombina. Rejoining his wife and children there in early fall only increased his eagerness to see how *Il Villino Volterra*, then under construction with his money—not Edoardo's—was coming along. After only four days, he headed for Ariccia, one of the dozen hill towns surrounding Rome, to personally supervise the progress on the house. Its location had been carefully chosen—situated just off the Via Appia on the highest spot in the village, it stood on a promontory commanding views of both the deep gorge that divided Ariccia from neighboring Albano and of the plain below that stretched as far as the sea. Yet nothing about the work under way there pleased him—not the location of the cesspool, which had been put under the dining room and bedroom windows; not the stairs, painted in a color that made the ceiling look too low; and definitely not the decorative motifs of garish flora and "horrible little children" that had been applied to the walls by "apprentice house-painters." "I am not exaggerating anything," he wailed, wondering how anyone could live in rooms painted with indecent-looking "bandy-legged little angels," or flashy-colored fruits.[12] Workmen also greeted him at their apartment in Rome, where he dined surrounded by dust and disorder. But this, he told Virginia, not a little pointedly, "was nothing compared to those disturbances that caused me so much suffering in the four days that I spent at Palombina."[13]

* * * * * *

Volterra marked the start of the new year by returning to Sweden, which he had last visited in 1902. On the recommendation of Volterra's longtime Swedish colleague Gösta Mittag-Leffler, King Oscar II had invited him to present a course that winter at the University of Stockholm on the integration of partial differential equations. En route he learned that he had been awarded the Civil Order of Savoy, a decoration based solely on scientific merit, which pleased him much more than the knowledge that the honor carried with it an annual pension. Arriving in Stockholm, he took up residence in the Grand Hotel, an upscale establishment whose clientele enjoyed access to the era's most modern amenities, including a post office, telegraph office, and coffee bar. In these congenial surroundings, Volterra began working on his lecture series. On February 3, he gave his first talk: an overview of the rapidly shifting terrain in mathematical physics, whose Olympian certainties were being shaken to their foundations by Einstein and Lorentz's revolutionary new formulations of space and time, Planck's startling characterization of the light quanta, and Michelson-Morley's demonstration of the nonexistence of the supposedly all-pervasive ether. These and other developments signaled the end of the purely Newtonian universe, and while

Volterra did not downplay the challenges all this posed for his field, he was confident that new approaches and solutions would be found. Acknowledging "that mathematical physics is going through a period of crisis," Volterra conceded the necessity of

> abandoning certain ideas in order to follow new ones....But, even if some concepts that we now have on the nature of natural phenomena and some fundamental principles have to be shaken by some new facts and discoveries, a part of mathematical physics has a good chance to save itself from the shipwreck.
>
> In fact, it [mathematical physics] represents, perhaps in a rough way, but certainly in a very simple way, a great part of the known natural facts, it [mathematical physics] connects them together and it has an indisputable practical utility.[14]

He had no sooner finished this first lecture than he set about preparing the next, a pattern he would follow throughout his five-week visit. As he wrote to Virginia, "As soon as I have finished [one lecture], it becomes necessary that I start to write the next, because the following Saturday after my lecture I must deliver the manuscript, which is then printed the following week." Lectures two through eleven would focus on special topics of his own choosing, including the application of partial differential equations to such classical physical phenomena as the propagation of heat, hydrodynamics, and elasticity. He would devote three lectures to this latter subject, with the emphasis on his theory of the deformation of elastic solids, amply illustrated with three-dimensional models of cylinders in various states of distortion, which he had brought from home.

In his spare time, Volterra toured the countryside, sometimes by sleigh (for which he seems to have had a higher regard than the automobile), and attended concerts, operas, and student parties. His official status as a guest of the monarchy guaranteed that he would be swept into the social whirl of the Swedish capital. He dined with writers, journalists, politicians, and just about everybody with a connection to either mathematics or to Mittag-Leffler, who entertained Volterra lavishly and often at his house at Djursholm, just outside the city. On at least one occasion, he stayed there overnight in order to spend the following morning in his host's library, reputed to be one of the finest mathematical libraries in the world. The two probably compared notes on their respective collections; Volterra's own personal library contained several thousand volumes on science and the history of science, including nearly 500 priceless tomes printed before 1500.

The Grand Hotel's first-class restaurant attracted a cross-section of Stockholm's professional elite, and Volterra frequently found himself dining there with, among others, the Italian chargé d'affaires in Stockholm and the secretary of the Russian embassy. They in turn made it their business to introduce the distinguished visiting scientist to colleagues in other embassies.

It was Volterra's first prolonged exposure to political and diplomatic circles outside Italy (where, in any case, he had mostly associated with scientist-parliamentarians like himself), and the experience undoubtedly contributed to his success as Italy's leading spokesman for science in the first decades of the twentieth century. What he learned from these sophisticated public servants in Stockholm seems to have increased both his social confidence and his ability to move easily among the academic, scientific, and political worlds. Already accustomed to thinking of himself as an ambassador for the cause of Italian science, he found his view of how best to fulfill that role expanding. In a letter to Virginia about yet another diplomatic gala that he had attended at the Swedish Ministry for Foreign Affairs, he wrote that he had caught himself thinking, "I must remember once again that I am a political man, a thing that very seldom comes to my brain!"[15]

Back in Italy, Virginia, then expecting another child, had taken over Vito's role as sidewalk superintendent in Ariccia. She arranged for an electrician to wire the country house, for the builder to install custom-made bookcases in Volterra's study on the second floor, and for the installation of a heating system. Possibly in a bid to avoid a replay of the summer's somewhat testy correspondence, she had asked her husband to write her more descriptive letters. In late February, he obliged with an account of a day filled with heady socializing, commencing with an audience with the Swedish king. The affable monarch (who would die the following year) "spoke to me in Italian... thanked me for having come here, and afterward he spoke of the *Acta* [the influential mathematical journal founded years earlier by Mittag-Leffler] and told me that he loves mathematics." Volterra was then turned over to the Crown Prince, who "spoke about the antiquities of Rome, about earthquakes, and volcanoes." That afternoon, accompanied by the Italian chargé d'affaires, he made a formal call on the Swedish foreign minister ("he spoke a little about politics") and then met with the secretary general of the Swedish Foreign Affairs Office ("a young man who told me that his greatest desire would be to spend his honeymoon in Italy.")[16] On another occasion, intrigued perhaps by the name, Volterra ventured alone to a party sponsored by the Concordia Society, a club dedicated to promoting "women's studies" and "universal peace or the brotherhood of nations." The evening turned out to be a marathon of speeches, food, toasts, and songs, and the enthusiastic participants did their best to make Volterra feel at home, speaking to him in "an Italian that is really awful." Although he failed to see any point to the group's existence ("It seems to me impossible to take seriously stuff that in Italy would seem ridiculous," he harrumphed to Virginia), he did not beg off when he received another invitation a few weeks later, explaining "there will be good music and a supper."[17] Still, he much preferred the party held in honor of the Danish critic, writer, and parliamentarian Georg Brandes, where the conversation flowed along literary and political lines.

Italy was just then going through one of its not-infrequent periods of political turmoil, and Volterra, newly attuned to the significance of such developments, wrote his father-in-law to ask what he thought about the country's new prime minister, Sidney Sonnino. In his reply, Edoardo Almagià said that he foresaw a short tenure in office for the new leader. "I don't believe that the present ministry, although composed of good elements, can do better than the one it succeeded," he told Volterra. Assessing the increasingly chaotic state of Italian politics, the astute and politically savvy guardian of the Almagià family and fortune, added, "The defect is a Parliament in which there are many groups with sinister tendencies . . . and a country that still remains in a state of infancy. . . . So we must still rely on our [word ineligible]. . . and pull on with the men that the country can give us."[18] (Sonnino's government lasted all of four months; Edoardo Almagià died in 1921, barely missing Benito Mussolini's rise to power.)

If Volterra needed further confirmation of just how unsettled things could become on the home front, it arrived in the form of another letter from Virginia. She had enclosed photos of the two oldest children in costume— Edoardo decked out as a *bersagliere* (marksman), and Luisa striking poses as a nun. The sight of his daughter in a habit seems to have appalled Volterra, although the irony of the situation was not lost on him either. "To be entirely sincere, I did not like Luisa's costume as much as [Edoardo's]," he chided Virginia gently. "I was even more displeased to learn from dearest mother that Luisa carries herself [as a nun] with perfect. . . naturalness." Volterra almost never spoke publicly about his personal life and certainly always thought of himself as a patriotic citizen of Italy; nevertheless, the thought of his own daughter in the garb of an institution that had spent a millennium hounding and defaming her ancestors clearly struck a nerve. He managed to find some humor in Luisa's "monkish tendencies," telling his wife that "perhaps we should be grateful that it doesn't jump into Mrs. Massari's [possibly a reference to the governess] head to dress Edoardo as a priest!"[19] In the meantime he immediately purchased and shipped off a colorful and traditional Swedish costume for his little daughter to wear.

By early March, Volterra, having absorbed at least as much new knowledge in Sweden as he had imparted, was ready to return home. On March 10, he presented his last lecture, and marked the occasion with a quick note to Virginia: "I am satisfied with the lectures I gave, which were published little by little, with no small labor on my part."[20] The hard work had paid off handsomely; the first edition of *Leçons sur l'intégration des equations différentielles aux derives partielles* (Lectures on the integration of partial differential equations), published in French in Uppsala, had gone to press by the time he had returned to Rome. Volterra remained in Stockholm just long enough to make the obligatory round of farewell courtesy calls, starting with the Crown Prince and the minister of foreign affairs ("Here they look at the formalities much, much more than in Italy," he wrote to Virginia) and to attend Mittag-Leffler's sixtieth birthday gala. That evening he caught the train for Paris, arriving just in time to present a scheduled lecture before the Paris Academy of Sciences, which had elected him a corresponding member two years earlier. After a week that included a memorable dinner at the home of the mathematician Emile Borel, where he met Marie and Pierre Curie, who had shared the 1903 Nobel Prize in physics for their work on radioactivity, Volterra came home, bursting with energy and full of plans to launch an organization dedicated to the promotion of science in Italy.

Back in Rome that spring, he nominated his venerable Senate colleague Stanislao Cannizzaro (with whom he had collaborated on the Turin Polytechnic report a few years earlier) for the Nobel Prize in chemistry, and spent time over the next several weeks working quietly behind the scenes to enlist support for the selection of what would have been the first Italian Nobelist in chemistry. He was not successful (Guglielmo Marconi would share with Ferdinand Braun the Nobel Prize in physics in 1909), but this effort to enhance the profile of Italy's scientific establishment helped refine ideas that he publicly unveiled for the first time that summer in Milan at the fiftieth anniversary of the Società Italiana di Scienze Naturali (The Italian Society of Natural Sciences).

In an impassioned speech before the gathering on September 15, Volterra called upon his colleagues to endorse the establishment of a new association dedicated to the advancement of science in Italy. He pointed out that the idea had a partial precedent—the Congresso degli Scienziati Italiani (the Congress of Italian Scientists), which had briefly flourished in the country between 1839 and 1875. Its members had met a dozen times in cities throughout Italy, attracting respectable turnouts that ranged from a low of about 420 scientists in 1839 to a high of nearly 1,200 in 1844. Founded during a period of political upheaval and foreign domination in Italy, the gatherings had derived much of their intellectual vitality from the unrequited undercurrent of Italian patriotism that permeated the meetings. Unification, ironically, had ended up robbing the organization of its political importance; after many struggles to persist in the new political environment, the organization ceased to be after its twelfth meeting in 1875, in Palermo.[21]

While not hesitating to invoke the memory of this nineteenth century body, Volterra drew a sharp distinction between what he described as its old-style "scientifically aristocratic" character and his own vision for a new organization that, he insisted, "must be scientifically democratic." Inspired by the venerable British Association for the Advancement of Science and similar organizations on the Continent, and in the United States and Australia, Volterra envisioned a federation of scientific organizations resting on "a broad base, which can extend its roots freely throughout the country and embrace all those who, full of good-will, love science... those who have directly made a contribution to it [and] those who simply desire to inform themselves of what others have discovered."[22] This Italian Society for the Progress of the Sciences (Società Italiana per il Progresso delle Scienze, or simply SIPS) would function as a corrective to excessive specialization in the sciences, fostering what today would be called an interdisciplinary environment in which scholars from different fields could meet and exchange ideas; it would also offer opportunities for students and young scientists to explore new avenues of research. Volterra assumed that "the rapid intellectual development of Italy" would sustain SIPS as it shouldered "the advancement and the popularization of science" in the years ahead.[23] He conceded that his vision of opening the organization to the widest possible membership carried some risks, but called them "worth running... in order to create something youthful, vital, and modern, provided that courage and good-will are not lacking."[24]

Not everyone shared Volterra's optimism. When he learned about his colleague's grand plans several months before the Milan meeting, the normally ebullient Sicilian mathematician Giovanni Battista Guccia sounded an uncharacteristically grim note. In a letter to Volterra, he predicted that SIPS might easily die a thousand deaths before realizing its potential as a catalyst for improving public awareness of science and the scientific mission. "I will say immediately that the idea of founding [it]... is excellent," he wrote Volterra. But, he asked, had Volterra identified the right "man" with the "energy" (he capped both nouns for emphasis) required to steer such a complicated enterprise safely to shore. "If you have '*in pectore*' [in your heart] this man, this energy, if you really *had found it in Italy*, and will be willing to let me know *who is it*, in this case I will be able perhaps to say it that the 'idea' is capable of being carried out with success."

But he was skeptical that such a person existed and felt compelled, "with my usual frankness," to warn Volterra that only disillusionment lay in store. Volterra might think him "pessimistic," but Guccia, with characteristic bluntness, offered to explain why history was about to repeat itself:

Well then, according to me, the praiseworthy attempt of an Italian Association for the progress of the Sciences is destined to abort, or at the very least follow the same fate of that which died in 1875 with the Palermo congress, in which the most eminent men of Italy convened and in which wonderful speeches were uttered, full of promises for the future of the association, but that, unfortunately, contained... politics: a microbe that in Italy enters throughout and... kills everything! in a particular way science! Politics! Here is the great enemy of Science in Italy! Here is why institutions that prosper and flourish in other countries cannot catch on in Italy! Let someone try to inaugurate a Congress or... a Library without the intervention of His Excellency the Minister and patriotic speeches concerning it: containing which is this: that without Victor Emmanuel, Mazzini, and Garibaldi there would not have been (for example)... the wireless while in these said lectures the progress, the future, the needs of Science are relegated to the last seat, well then, 'and to

finish'!....I repeat that a man of great authority and of phe-
nomenal energy would [need to] rise, who, *free from any ties
with politics*, wishes [to] and can resolutely confront the difficult problem, that is: to separate, in Italy, science from
politics.[25]

Except for the political connections conferred by his Senate seat, Volterra's own qualifications bore a curious resemblance to Guccia's job description, although this does not appear to have been Guccia's intention.
His private reservations had no effect on the Milan meeting, which voted
unanimously to adopt Volterra's plan. Before the gathering broke up, an
organizing committee had been formed, and by January 1907, a circular
describing the new organization was in widespread distribution. Later that
year, the Society held its first meeting in Parma, having duly elected Vito
Volterra as its first president. Guccia, who died in 1914, did not live long
enough to see SIPS absorbed into the Fascist state (and Volterra's name
erased from its history), but he would surely not have been surprised.

Volterra's public triumph in Milan coincided with an intense private
sorrow. Shortly before the conference, Virginia had given birth to a new
baby who, like the couple's first child, died soon afterward. This little boy
had been named for Volterra's Swedish colleague and good friend, Gösta
Mittag-Leffler, who, upon hearing of the child's death, wrote to Volterra, "I
admire you to have been able to do a piece of work as remarkable as that
which you speak about [studies relating to the distortions of elastic solids] in
the sad period that you are going through."[26] In Milan too, Volterra seems
to have sought consolation in his work and professional obligations, but at
the end of the day, he freely expressed his terrible sense of loss and isolation
in letters to Virginia. "You cannot imagine the immense sadness of being
this far away," he wrote, seemingly terrified by the thought that he had no
one with whom to share his loneliness. "I don't know if you understand and
feel the same, but for me it's a huge, absolutely unbearable suffering."[27] The
anguish of losing a second child filled him with anxieties about the rest of
the family: Would Virginia please write to him in greater detail about their
three children? When did she expect her father to come to Rome? Had she
arranged for the family doctor to come to the house to make sure the other
children were in good health?

From Turin, where he felt compelled to travel next to deal with an
avalanche of administrative and academic duties as Royal Commissioner of
the newly established Turin Polytechnic, Volterra's litany of unhappiness
continued. "It's a largely depressing thing to be a public man and to be
more or less mixed up in politics and to have to think about public administration that doesn't interest me while the family and studies...fill [me] with
satisfaction."[28] Meanwhile, unanticipated problems arose, forcing him to
postpone his return home by several days. On September 23, still grieving
for his son and missing his family, Vito sent his wife a lengthy letter that

said in part: "I had dreamt of marrying you and to never leave you, while instead circumstances always keep us apart. My ideal would be to make the two of us become one in thoughts, in feelings, in everything, but how much time have we been truly together and united? Very little according to my wishes." For the first time in his exceptionally busy life, he had experienced a "horrible sense of loneliness," he said, adding, "I feel the absence of you and of your closeness." However, Virginia could make it up to him and drive away his pervasive sadness by "staying very close to me and never leaving us, not even for a moment." He went on to suggest that she put his needs ahead of their children's. "They always have your care and your presence and your closeness while I am deprived of it and I am suffering for a long time," he said, not troubling to downplay his misery or his sullen mood. Knowing that his letters often ended up being read by his mother, who lived under the same roof, Vito again reminded Virginia of his wish that she not share his letters with the family. "They are written for you alone and no one certainly will be able to identify themselves with how I feel." He added, "I am writing to you with the same trust and with the same abandon as [when] people say— or better still—whisper, the most intimate and sweetest words."[29]

Volterra's tenure as Royal Commissioner did not long survive this trip. When he returned to Rome, suffering intensely from a recurrence of stomach problems, Edoardo Almagià, the family patriarch, took matters into his own hands. In a letter to the family's physician, Dr. Marchiafava, he wrote at the end of October 1906:

> My daughter Virginia wrote me that the state of health of her husband is not improving at all. On the contrary, she notices the harmful effect caused by the [many] occupations and jobs in which he is involved; and those are for him a serious concern when they have the character of a task entrusted, I would say, to his judgment, such as the one assigned [to him] concerning Turin.
>
> You know very well the state of his health and you can advise him about the direction he should follow. But you should also know the strength of his character and the care with which he devotes himself to every mission assigned to him, in order to understand that it is necessary to dissuade him from every occupation for a certain period of time. In this way his health could take advantage of the cure that you are going to prescribe, and of course you'll recommend rest above all. Oh, if you could have him exonerated from the commitment he took on in Turin! His Excellency [Giovanni] Giolitti entrusted it to him and could also agree to withdraw it! It would certainly be a holy deed to do for Volterra![30]

Confident that Marchiafava would follow his instructions to the letter, Edoardo then wrote a stern note on the same topic to his son-in-law, prudently omitting to mention that he had primed the doctor beforehand. Evidently concerned that Vito might berate Virginia and Angelica for undue interference in his professional affairs, Edoardo curtly informed him that they had no choice but to go along with the physician's advice. In the face of all this, Volterra capitulated, and wrote to Marchiafava (who in turn passed the letter on to Prime Minister Giolitti), formally asking that he be temporarily relieved of the duties of Royal Commissioner. Under doctor's orders, he spent several months in Ariccia, now furnished with a well-stocked library and a wonderful trove of marble relics purchased from local farmers. By the start of 1907, he had resumed his university duties and myriad extracurricular activities. But he never resumed his leadership role at the Turin Polytechnic, and officially resigned from his post there early that year.

He had recovered sufficiently by the summer to immerse himself in a wealth of activities relating to the upcoming inaugural SIPS meeting that September in Parma, at which he would be elected the Society's president. He delivered the event's opening speech "Science at the Present Moment and the New Italian Society for the Progress of the Sciences,"* reporting afterward to Virginia, "I remain very satisfied by the outcome of the speech, even if only the first rows could hear it," a problem he blamed on acoustics in the theatre. He followed this with a reminder of a promise she had evidently made him before he left Rome: "I would be very happy if you were able to come or at least come before the end [of the meeting] to go together to Venice," he wrote, adding," I hope that on this point you will not have difficulty as you promised me."[31]

Virginia did make the trip, and it was perhaps during what seems to have been a tranquil interlude in their lives that the couple purchased eight small oil paintings of Venetian street scenes by the artist Ermo Zago, which hung in their dining room in Rome for many years.

With his health and spirits more or less restored, Volterra plunged with gusto into a new round of institution-building. In 1906, he had been elected president of the nation's first scientific organization for physicists, the Italian Physical Society (IPS); in 1907 he was named dean of the science faculty at the University of Rome, a position he would hold until 1919; and in 1908, he helped found an organization devoted to promoting oceanographic research in the Mediterranean. He ended his two-year term as IPS president that same year, but would continue to play a vigorous role on the executive board for many years. Indeed, his involvement with this organization, or at least its forerunners, dated back to his student days at Pisa, when his physics professor Riccardo Felici recruited him to join the editorial board of Felici's journal *Il Nuovo Cimento*. When the Italian Physical Society was formally established in 1897, the journal became its official publication. As

---

*His inaugural lecture appears in an English translation in the appendix.

the IPS president, Volterra worked hard to boost both the quality and quantity of the journal's articles, and to promote an active schedule of meetings for the Rome chapter. As he stepped down in 1908, the Italian experimental physicist Antonio Garbasso, who had taken his degree in physics at Turin during Volterra's tenure there, thanked him profusely on behalf of his assembled colleagues "for what he had done in order to give new life to the Italian Physical Society."[32]

Volterra's prominence in the IPS typifies the multifaceted roles that a select circle of Italian mathematicians began to fill during this period. Their statistical edge (mathematicians outnumbered physicists by four to one in Italy's universities before World War II), and their keen interest in the application of mathematics to real-world phenomena, meant that scientists like Volterra and his younger colleagues Levi-Civita and Castelnuovo came to occupy a unique niche in their country's physics community, which at the time was almost exclusively composed of experimental physicists. When theoretical questions arose, these gifted experimentalists frequently turned for assistance to those among their mathematical colleagues who had both the training and the inclination to address problems at the interface of math and physics.

As a result, in the first decades of the twentieth century, Volterra and like-minded colleagues became the primary spokesmen in Italy for the radical new ideas of Albert Einstein, Hendrik Lorentz, Max Planck, and Arnold Sommerfeld, among others. Like Volterra, Levi-Civita came to spend as much time at physics as at mathematical meetings and to devote considerable time to the newly formed SIPS. When the Society met in Padua in 1909, Levi-Civita presented "Sulla costituzione delle radiazioni elettriche" ("On the constitution of electrical radiation"), an illuminating critique of recent theoretical contributions by Einstein and Lorentz, among others. Besides Volterra, the audience included the émigré theorists Max Abraham and Ludwig Silberstein, whose work Levi-Civita also discussed, as well as the renowned Italian experimentalist Augusto Righi and his younger colleague Orso Mario Corbino. A recent addition to the physics faculty at Rome, Corbino (who was well acquainted with the topic) had already heard Levi-Civita give a similar talk earlier that month at the IPS. Clearly impressed by the mathematician's ability to talk physics to physicists, he had promptly written him a long letter, praising "the brilliant way, the lucidness with which you posed the problem of such capital importance for modern physics."[33]

Corbino then went on to describe, in some detail, his own remarks on the same subject at an IPS meeting in Rome. He reviewed the discovery by Hans Geiger, Ernest Rutherford, and others of the bafflingly different forms of radioactivity dubbed alpha and beta rays, and the theorists' efforts to explain these and related experimental findings. The experimentalists, he said, had uncovered these startling new phenomena; now the mathematicians' turn had come "to transform their marvelous mechanism in order to adapt it to the changed needs of the theory. We physicists can only wait and hope that the bonds with the theoreticians will become more intense and effective."[34]

Levi-Civita could not have agreed more. Within a decade, his courses at Padua would focus increasingly on topical issues in theoretical physics: analytical mechanics with application to thermodynamics and relativity (1912–1913); kinetic and statistical theories with application to quanta (1913–1914); the mechanics of continuous media from both the classical and relativistic point of view (1915); analytic mechanics, again including the Newtonian potential and relativity (1916); and the electromagnetic field (1918).

Largely at Volterra's behest, Rome's IPS chapter began to sponsor group discussions on special relativity and, a decade later, general relativity, which Corbino, Castelnuovo, Silberstein, Abraham, and others regularly joined. Indeed this group played a major role in making Einstein's powerful and discomfiting theories known not only in Italy, but also in Spain and Portugal. In 1909 Volterra prepared to venture considerably farther afield. That summer, he accepted an invitation to participate in the twentieth anniversary celebration of the founding of Clark University, in Worcester, Massachusetts. (As always, he wanted Virginia to come too, but he could not persuade her to accompany him.) In issuing the invitation, Clark University physicist Arthur Gordon Webster had asked Volterra to deliver a series of lectures on mathematical physics, adding handsomely that "the opportunity to hear a contribution from such a distinguished source does much to ensure the success of our celebration, at least from the point of view of the department of physics"[35] (which in large part consisted essentially of Webster himself). As he began contemplating his talks, Volterra asked Castelnuovo, who had recently spoken in Rome on the four-dimensional universe of Hermann Minkowski, if he could look at the younger colleague's lecture notes. "This [trip] will not be restful for you," Castelnuovo predicted, "but you do well not to let slip an opportunity to visit some cities in the United States, and to bring over there the voice of Italy." Although Castelnuovo seemed doubtful that his lecture would interest Volterra's American audience, he dutifully sent him the outline he had saved, "in part recopied for you today."[36] Unlike pure mathematics, which was developing by leaps and bounds in the United States during the early part of the twentieth century, theoretical physics got off to a late start in America. On August 4, Volterra boarded the steamship *Regina D'Italia* in Naples, eager to bring a taste of the new physics to the New World.

CHAPTER 14

"A Professor in America"

A modest institution with aspirations to become a major center for scientific research in the United States, Clark University in Worcester, Massachusetts, orchestrated a series of international conferences in 1909 to mark the twentieth anniversary of its founding. In a major coup, its president, Granville Stanley Hall, a leading psychologist, successfully recruited Sigmund Freud and Freud's then-disciple Carl Jung to speak at the event's psychology conference. (Freud's five lectures on psychoanalysis during his first and, as it turned out, only visit to the United States, would attract a diverse audience, including Emma Goldman, the high-profile anarchist of her day, who conspicuously planted herself in the front row.) The university's sole bona fide physicist, Arthur Gordon Webster, had lined up an equally impressive roster for the physics and mathematics conference; the speakers included some of the finest minds in science, among them the 1907 Nobel laureate in physics Albert A. Michelson, the 1908 Nobelist in chemistry, Ernest Rutherford, and Vito Volterra, who had agreed to give three lectures.

Volterra's reputation in American academic circles preceded his arrival in the United States, thanks in part to a course of lectures given by the Harvard mathematician Maxime Bôcher, subsequently published as *An Introduction to the Study of Integral Equations* in 1909 in the Cambridge series of small volumes in mathematics and mathematical physics. In producing what one reviewer called "the first connected account of the subject,"[1] Bôcher, who remembered Volterra very well from his student days in Göttingen (he had earned his doctorate there in 1891 under Felix Klein), had credited Volterra with being the first to discover "a very remarkable and important relation between the theory of integral and of algebraic equations,"[2] and presented English-language audiences with the first comprehensive overview of his contributions.

Although the intrepid Volterra had traveled widely throughout Europe, the North American continent was terra incognita to him. Rarely happier than when an opportunity to take a trip presented itself (travel, as he once explained to Griffith Evans, his first American student, helped him "to relax my brain and to think about different things"),[3] Volterra was in a mood of high anticipation as he sailed out of Naples on August 4, intending to write his lectures for Clark University on board the steamship. The next morning, when the ship docked for the day at Palermo, physicist Orso Mario Corbino, who had recently been appointed to the faculty at Rome, joined Volterra on board; later they toured the city and visited the Physics Institute at the University of Palermo, where an assistant proudly showed them a microscopic diamond, apparently fabricated in one of the university laboratories. At exactly eight that evening the Italian liner raised anchor, heading for the Straits of Gibraltar and the open ocean beyond.

As the ship steamed across the Atlantic, Volterra wrote out his talks, setting a goal of ten pages every day. Evenings were spent getting to know some of his fellow first-class passengers, among them the boat's chaplain—a professor from Como who had been promised a bishopric that never materialized—the brother of an Italian general, Italy's royal commissioner for emigration, and the ship's doctor and chief officers. But by the fifth day, the novelty of the transatlantic voyage had begun to wear thin. In the privacy of his cabin, Volterra kept a running diary, which he mailed *in toto* to Virginia, after disembarking two weeks later in New York:

> August 9, 1909: All land has disappeared; we are on the ocean, very calm water, it looks like a lake. August 10: Nothing new, one sees only sky and water. I work every day, writing the usual ten pages....August 12: One sees flying fish. Life on the boat is extremely boring....August 16: I finish writing the lectures....August 17: Good weather....The boat contains 600 emigrants....They are all from the south of Italy and extremely dirty, but in good financial condition. The majority of them have either already been in the United States and are returning there, or they have been urged to come by friends or relatives. Many emigrants travel first class. At least so-called first class, because the accommodations are third class. The clerical part of the ministry...fill the boats with priests, nuns, chaplains for the emigrants, but they have little luck. It is enough that the emigrants don't boo the chaplains! The anarchists on board publish a paper...but I don't see it....Now that I have finished writing the lectures, I sleep all day on the ship. I intend to protest about the awful service too....In the afternoon, awful weather....One remains on the bridge until 10 P.M. in the rain because it is boring to stay in the cabin. The smell of the passengers is so unbearable in the halls that it is almost impossible to go there. August 18: We are in communication with the coast by wireless telegraph, and our arrival is announced for 9 P.M....I always feel hungry....The Commissary tells me that on board ships, in first class, one eats too much and many people suffer from indigestion. Here there is no danger from this....My health is excellent and by now this unbearable journey has ended.[4]

Volterra landed in New York on August 19, reservation in hand for a room at the Fifth Avenue Hotel, in midtown, only to discover that the establishment had been torn down in the meantime. He settled for a room with a shared bath on the seventh floor of Holland House, another hotel on the same block. "I like New York," he wrote to Virginia, after spending his first afternoon taking in the sights from a tour bus. "Here the buildings are very tall, at least twelve stories, and they touch the clouds."[5] After a visit to Wall Street the following morning, he headed for the city's less glamorous districts, including Chinatown and the Lower East Side, the likes of which he had never encountered before. "These consist of the Italian, the Chinese, and the Jewish quarters," all of them so dirty they are impossible to describe and one right next to the other," he informed Virginia.[6] The stark contrast between Little Italy, in lower Manhattan, and upper Manhattan, "so clean and beautiful," amazed him. Many of the Italian Americans he encountered casually were transplanted Neapolitans and Sicilians, who worked long hours for low wages, shining shoes and sweeping streets—the same jobs that earlier waves of immigrants to America's shores had held. While praising his countrymen's industry and higher standard of living in America, Volterra recoiled from other aspects of life in Little Italy, which he attributed "to the proximity of so many bad elements. Certainly the crime rate of the Italians here equals that of Sicily, with crimes and blackmail," he told Virginia. How Volterra could reach such a sweeping conclusion, having just barely landed in New York, is not hard to explain: he devoured the local Italian-language newspapers. And just as instantly, he had become an expert on immigration ("there are 600,000 Italians: no Italian city has as many... as New York"), wages ("the stone workers [make] about 30 *lire* per day: the minimum salary is about 7.50 *lire* per day"), and labor laws. "Nobody has to work more than 8 hours per day," he reported, with more enthusiasm than accuracy.[7]

One of the Italian papers, noting the arrival of the distinguished visitor from the old country, reported that he had toured the city with one of the priests he had met on board the ship. "It is not true," Volterra complained to Virginia, enclosing the offending news clipping with his letter. "I always toured the city alone."[8] In fact, within five days of his arrival in New York, Volterra was boasting that he had mastered the subway system, the elevated trains and streetcars, and even the ferryboat schedules—all without ever speaking a word of English. His easy command of French did not help. As he reported to Virginia, "The French language is useless....the most necessary language to know is English." Still, he was delighted to find that every corner he turned seemed to be inhabited with "waiters, barbers, workers, and peddlers" speaking Italian.[9]

On August 24, having been advised by Webster, his host at Clark, that the university would not reopen until September 1, Volterra collected his belongings and set out on a grand tour of the American northeast, commencing with a visit to Niagara Falls. On the New York side of the falls, he donned a woolen bathing suit and rubber raincoat (unfortunately no photo commemorates his brief time in this singular attire) and followed the guide along the narrow passageway that separated the cascading waters from the wall of rock behind them, before taking a short boat ride directly under the waterfall. After spending the night at a German hotel, which afforded him the opportunity both to converse in that language and enjoy generous portions of food, he crossed the border into Canada on a Lake Ontario steamship bound for Toronto. From there he traveled along the St. Lawrence River to Montreal, in the French-speaking province of Quebec, having spent a total of two days on the Canadian waterways. Delighted to once more be able to speak French, Volterra visited the University of Montreal, finding the campus empty except for one kindly engineering professor who showed him the laboratories. He debated briefly whether to go on to the province's capital, also named Quebec, or return directly to Boston, and the prospect of being able to continue speaking with North Americans in a common tongue won. But even here, in a city "totally French," Volterra realized that "in this moment the most important [language] to know is English." But try as he might, he could not manage to speak it, not even a few words. "Occasionally one finds somebody, especially from the Chicago area, who speaks German," he wrote Virginia, "but it is not so often as I expected. How happy I would be if I were able to speak an English word, but I don't know any. I don't even know how to say 'eat' or 'drink.' While the German words remain in my mind, of the English words not one remains."[10] Still, he never let his shortcomings in this area impede his enthusiastic explorations of the North American continent.

From Quebec, Volterra made his way back to Cambridge, Massachusetts, where he spent the morning of September 2 visiting the Massachusetts Institute of Technology ("interesting and imposing"), and the afternoon socializing with the Webster family at their home in Worcester, an hour's train ride away ("an excellent welcome"). Dining that evening in the Brunswick, a hotel located directly in front of MIT, he feasted to his heart's content on local farm-fresh foods ("excellent meat, butter, milk, tea, eggs"). "Here they eat an enormous amount of fruits," he informed Virginia, marveling, or perhaps shuddering, at the thought that an American breakfast "starts with fruit" —something no proper Italian would have considered acceptable back then or perhaps even now. "My health is very good," he assured Virginia, who seems to have fretted continually in her letters about his well-being.[11]

With five days remaining before the start of the celebrations at Clark, Volterra made a quick visit to Harvard, then took the overnight train to Washington, D. C., finding it "a beautiful and quiet city." He estimated the local population at about 200,000 inhabitants, less than half that of Rome, which surprised him, given the city's stature as the nation's seat of government. Touring the city's "imposing spots," he climbed the steps of the Capitol, viewed the Washington Monument, and admired the Library

of Congress, which he made a point of returning to in the evening specifi-
cally "to see its system of illumination."[12] From D.C., he continued on to
Baltimore, Maryland, and Johns Hopkins University, before heading back to
New York, where he visited the Metropolitan Museum of Art and Central
Park, and watched the annual Labor Day parade on Fifth Avenue,

Traveling alone and unable to talk to anyone in English, Volterra began
musing about the languages in which he communicated so fluently back
home in Europe and how much longer they would be useful in America. "It
is certain that at the present time the most indispensable language seems
to be English," he wrote Virginia.

> French, as I already wrote to you a few days ago, is un-
> derstood only by cultivated people. It is becoming a little
> like Latin. Nor does it seem to me that the French people
> have the power of expansion. Germans speak their language
> themselves, but it does not seem to me that others wish to
> speak it. English remains, which is the language of North
> America and of a great part of Asia where it extends into
> Japan, China, and Oceania [i.e., present-day Australia and
> New Zealand].[13]

On September 7, when Volterra presented his first American lecture
at Clark, by pre-arrangement he spoke in French. Many in his audience,
which was made up mostly of U.S. scientists and mathematicians, would
have received some training on the Continent and would have been sufficiently familiar with the language to at least follow the main points; in any
case, "Sur quelques progrès récents de la physique mathématique" ("About
some recent progress in mathematical physics") was couched largely in the
attendees' universal language of mathematical notation. After thanking his
hosts for the opportunity to "touch on some points that have been the
aim of my personal research," and pledging to resist the urge to do more
than "consider the evolution of some ideas," Volterra launched into his talk,
which addressed in part the identical nature of mathematical approaches
to electrodynamics and elasticity: "If you ask any mathematician if in his
mind he makes a distinction between the theories of elasticity and those of
electrodynamics, he will tell you that he cannot do it, because the types of
differential equations he encounters, and the methods he must use to solve
the problems that are introduced, are entirely the same in the two cases."[14]
He then proceeded to support this assertion with a variety of examples that
illustrated the interplay between mathematics and natural phenomena, including the use of the calculus of variations to derive Maxwell's equations of
electrodynamics, and Minkowski's four-dimensional treatment of electrodynamics, where he both acknowledged and closely followed the lecture notes
provided to him by Castelnuovo. "[The lecture] went very well," he wrote
Virginia later that day. "At least I was very pleased." Both Michelson and
Rutherford had come to hear him speak.[15]

The fast pace of life in America clearly appealed to Volterra; his letters are filled with enthusiastic accounts of a conference agenda that had him racing from lecture to lecture during the day, while looking forward to more talks each night. Not everyone shared this opinion: the great Rutherford, then the Langworthy Professor of Physics at the University of Manchester, and the scientist who first divined the existence of the atomic nucleus, told Volterra that the English (Rutherford actually hailed from New Zealand) viewed the American way of life as "too intense," and that he preferred, when he could, to lie on a sofa and smoke his pipe." Still, Volterra's positive impressions were only reinforced by what he perceived as the informal dress code followed by academics in America. "Here, life is much simpler than in England," he told Virginia. "At parties, people go indifferently in white ties or in morning dress!"[16]

It was in this relaxed and genial frame of mind that Volterra presented his remaining two lectures, taking pains in both to acquaint his American audiences with the intellectual achievements of his own countrymen. In a lengthy discussion of the theory of elasticity, he cited the work of the mathematicians Betti, Beltrami, and Somigliana, among others, described his own research involving "the equilibrium of multiconnected elastic bodies,"[17] and reviewed the recent experimental work of physicists Corbino and Giulio Trabacchi to experimentally verify the distribution of the internal tensions predicted by Volterra's own theory of dislocations. When the presentation of honorary Clark degrees to the conference speakers formally brought the event to a close on September 10, Volterra had every reason to feel elated with his first experience of America. Not only had his lectures been well received, but Michelson and E. H. Moore, Michelson's colleague in the mathematics department at the University of Chicago, had both invited him to lecture there the following summer. The well-traveled mathematician had accepted on the spot. "I hope that we will come back together, you and I,"[18] he wrote to Virginia. As it turned out, Volterra did not go to Chicago in 1910, although he would visit the city when he returned to the United States in 1912; and he would come back once more to America in 1919. But he came alone both times.

On September 14, Volterra returned to Europe aboard the steamship *SS Kronprinzessin Cecilie*, having visited both Yale University, in New Haven, Connecticut, and Columbia University, in New York City, before boarding the huge liner, bound for Cherbourg. "[America] is a great country," he wrote to Virginia on the eve of his departure,

> and what constitutes the principal difference from our Europe is that it is a country in the process of being formed, while ours has already been made for centuries and remains crystallized. Here converge all the races of the world, millions and millions of people. Many millions more will be able to come, and these millions of men, of different natures,

characters, and races, are on the way to forming a new race with particularities and characteristics of its own. America nourishes itself with this multitude of men, which it digests, forming blood of its own blood as well as its own flesh, and it increases and lives with an extraordinary and indescribable prosperity.[19]

Volterra returned home with vivid stories about life in America and a determination to see it again. "I have a strong desire to become a Professor in America,"[20] he would later disclose to his wife, and he particularly liked the idea of working at Rice, in Houston. There is no record of what Signora Volterra thought about this idea; but it was not to be. A Volterra would eventually become a professor in America, but it was Vito and Virginia's son Enrico, who would become a professor of aerospace engineering at the University of Texas at Austin, many years after his father's death.

In contrast to Volterra's sea voyage from Europe, which had followed a leisurely southern route, the North Atlantic crossing on the return took only six days. Aside from having to prepare a few words that he would deliver as the outgoing president of the SIPS at the organization's upcoming meeting in Padua, Volterra found he had "nothing to do," except relax. He kept getting lost on the ship, which seemed to him more like a large first-class hotel than an ocean-going liner. On one occasion he turned a corner and stumbled upon the dressmaker's quarters; another time he walked by mistake into the tailor's workshop. Surrounded by a seemingly infinite array of private and public rooms—the latter included a dining room especially designed for children—he actually gave up counting how many rooms were set aside for writing, reading, coffee, conversation, and smoking cigars. "I have never before seen and could not have imagined anything like this," he confided to his diary, shortly after settling into his quarters. He found that he enjoyed relaxing each afternoon in a chaise lounge on the deck. It was in this posture, bundled up and napping after lunch, that he was roused one day by one of the ship's stewards, who wanted to call his attention to the "icebergs....blocks of ice that are moving southward in the ocean....One sees also many whalers, who fish for whales in the surrounding area, and we passed very close to a rowboat from a whaler," he wrote in his diary. Before he knew it, the voyage had ended. In one of his last diary entries, dated September 19, Volterra wrote, "It fact, it seems to me that I embarked yesterday; it even seems that I have not for a moment left the Holland House Hotel...."[21]

Disembarking at Cherbourg, Volterra spent the night in Paris, continuing on the next morning to Padua, where the third annual meeting of the SIPS was already under way. He arrived in time to greet the attendees, to pass the president's baton to Giacomo Ciamician, professor of chemistry at the University of Bologna, and, on September 25, to declare the meeting officially closed. From there, he went directly to the country house in Ariccia, where his family was waiting to welcome him, and got down to completing a series of notes on the use and application of integro-differential equations in cases involving the theory of elasticity—work he had set aside several months before.

Arriving back in Rome toward the end of the year, he was pleased to find a letter from Federigo Enriques, who hoped to entice Volterra to lend his name to the list of organizers planning the International Congress of Philosophy, to be held in 1911, in Bologna. Volterra's name, Enriques felt sure, would command the attention of scientists both at home and abroad. There was one delicate matter that Enriques wanted to bring up, involving Giuseppe Peano, Volterra's old sparring partner from Turin. Enriques had invited Peano to head up the section of the congress dealing with logic and the theory of science. "I add that—before making this invitation [to you]—I made a point of consulting with Peano (whom the organizers of this section have already asked to participate) and that like me, he would be happy to have you accept."[22] A good citizen of science, Volterra lent not only his name to the enterprise, but quickly wrote to Max Planck in Berlin, inviting him to be a participant.

In 1910, Volterra did indeed return to the Americas, but this time the journey took him south of the equator. To commemorate the hundredth anniversary of Argentine independence, the government in Buenos Aires staged a lavish series of centennial celebrations, including an International Exposition and an Inter-American Scientific Congress, which Volterra attended that summer as an official representative of Italy's Ministry of Public Instruction. In mid-June, he boarded the *Re Vittorio* in Genoa, learned with pleasure that the ship would dock at Barcelona as well as at Rio de Janeiro and Santos in Brazil, and set to work on the talk he planned to deliver, which he drafted at his standard rate of 10 pages a day. "I am beginning to write the lecture... on the new concepts of space, time, and mass, taking advantage of my previous lectures," he wrote Virginia, as the liner steamed toward the Straits of Gibraltar. For his audience at the University of Buenos Aires, Volterra fashioned a talk that blended the latest scientific research—Minkowski's four-dimensional universe and Max Abraham's rigid sphere model of the electron—with the newest fashion in science fiction—time-travel as imagined by English author H. G. Wells, whose best-selling novella *The Time Machine* was then all the rage in Europe.[23] After finishing it, he began preparing a short communication on the "Functions of lines, integral and integro-differential equations" for presentation at the centennial's Science Congress.

Disembarking in Barcelona, Volterra visited the University of Barcelona, where he discovered that the mathematics faculty offered a course on integral equations. Had he been able to find the professor who taught the course, he later told Virginia, he would surely have kissed his cheek. When the ship dropped anchor one evening off a tiny Portuguese island west of the Canary Islands, Volterra and several other passengers, along with the ship's engineer, opted to take a quick boat trip to shore. They were on their way back when the crew of the tender ferrying them to the liner veered off course, apparently intending to rob them. A man of obvious experience in these matters, the engineer pulled a revolver out of his clothes and persuaded the crew to return them to the ship. Volterra seemed unfazed by the midnight adventure; one wonders what Virginia thought when he related the incident to her. A few days later, the rising heat and humidity signaled to the passengers that they were approaching the equator; as they crossed the iconic imaginary line, all the sirens on the ship sounded. Star-gazing at night from the deck, Volterra beheld for the first time the constellation of the Southern Cross. When the tropical weather and choppy water prevented him from working, he found relief in the novel *I Moncalvo* ("The Moncalvos") written in 1908 by Enrico Castelnuovo, the father of his colleague Guido. "Try to find a copy and do read it," he advised Virginia. "I am sure it will interest you."[24] It is not hard to understand the book's appeal for Volterra: Set in Italy thirty years after unification, it relates the story of a family caught between the traditional tenets of Judaism and the desire to become assimilated Italians.

When the boat docked at Rio de Janeiro, Volterra mailed a batch of letters to Virginia, including this rather lyrical description of his first impressions of Rio:

> The city, as everybody says, is one of the most beautiful in the world, as is the bay that is surrounded by strange and picturesque mountains, while the town itself extends along the shores and climbs to the heights....However, recently the town was completely rebuilt and a very long, beautiful, and completely new road was built. The architecture is very strange, unlike any I have ever seen: all the houses are colored and with ornaments completely original. In any other place it would be completely incongruous, but here the effect is very beautiful. The vegetation is tropical, with magnificent plants and trees everywhere. Especially interesting are some very long alleys with palm trees taller than I could have ever imagined.[25]

While at sea, Volterra had received no messages from Virginia. Arriving in Buenos Aires, he was happy to find her telegram until he saw that it contained only one word: "Benissimo." The meaning of "very well" was clear enough; still, almost a month had elapsed since his leaving Italy, "and all the news received reduces to a single word! It is a bit too little," he protested, in a letter that he dispatched almost immediately.[26] With almost a week at his disposal before he was scheduled to give his lectures, Volterra explored the city and attended some of the centennial events, but the days passed slowly. "My impression of the country I will tell you verbally," he wrote Virginia on two separate occasions, perhaps fearing that his letters might be opened en route.[27] Before leaving Argentina, Volterra traveled around the interior and the region south of Buenos Aires, but waited to pour out his true feelings about the country until he had safely boarded the *Regina Elena* for the return passage.

> If I had to write a book about Argentina...I would entitle it "The Country of Mud." Everything there is mud, materially, morally, and intellectually....Take away from a Latin race all the beautiful, ancient, and noble traditions; take away any taste and any feeling for art; take away the aptitude toward any scientific and philosophical speculation; give them a soft and disagreeable language that deforms any thought and makes it bombastic, exaggerated, clumsy, and inaccurate; let this race freely expand over fertile soil that gives forth its products without effort to the most lazy, and the monster that will be born...will be called the "Argentine Republic...." The people who go to Buenos Aires have only one aim: to make money and to make it as soon as possible....The scenes of disorder that I witnessed were worse than any I ever saw in Italy or in France. Nobody is able to organize anything; people lack discipline; everything is organized at the last moment. The Argentines, moreover, have an unlimited pride, especially now after the recent celebrations....At a reception at the Senate where I had been invited, the President of the Senate, a man who struck me immediately as frivolous and rather stupid, said, when he learned that I was Italian, "Ah, Italian—I am very happy." With the idea of paying me a compliment, he added: "We consider Italy almost an extension of ourselves!" I felt obliged to answer that this was not the case, but rather that it was "old" Italy that considered Argentina as one of our offshoots![28]

What was it about Argentina that provoked such an outburst from Volterra? In the decades just before and after the turn of the twentieth century, Italian emigrants to Argentina accounted for almost half of the country's European-born population, and they had a large hand in shaping the culture of the country. What Volterra didn't say, probably because it was already clear to Virginia, was that the vast majority of those Italians were poorly educated, poverty-stricken people from southern Italy and Sicily. Like most Italians who hailed from north of Naples on the Italian peninsula, Volterra did not have a high opinion of southern Italians. The language he derided so viciously was not Spanish, the national language, but *Lunfardo*, a kind of lower-class Argentine *lingua franca*, composed of words drawn from many languages including what Volterra obviously considered sub-literate southern Italian dialects. But this alone may not be enough to explain his almost visceral loathing of aspects of life in Argentina; he had written with approval and even admiration of the Italian immigrants, also largely from Sicily and southern Italy, whom he had seen and talked with in New York City. While it is easy to label Volterra a snob, it is equally hard to discount all of his observations, particularly when considering the divergent paths followed by Argentina and the United States in the twentieth century.

* * * * * *

Toward the end of October 1910, Griffith Conrad Evans, a twenty-three-year-old Harvard Ph.D. in mathematics who had written his doctoral dissertation on Volterra Integral Equations, arrived in Rome to begin a two-year postdoctoral fellowship with Volterra. (Other foreign students would follow, including Réné Gateaux, a young Frenchman who died on the battlefield in the early days of World War I, Szolem Mandelbrojt, and Joseph Pérès, who became a collaborator and a close friend). Their first meeting, which took place at the Rome Institute of Physics on Via Panisperna, led to a warm and lasting friendship. Fifty years later, Evans could still recall vividly the invitations to "come to lunch [at Ariccia] when I could... which I could often," and the Roman sculptures on display there. "Volterra's house was a small stone castle, which he and Mrs. Volterra had made into a museum by buying up the relics of antiquity that constant building in the neighborhood... brought to the surface." He also retained lively memories of the goat tethered in the garden of the Rome School of Engineering, dining on the bushes while students streamed inside the building to attend Volterra's lectures on integral and integro-differential equations.[29]

In Evans, Volterra found a bright, ambitious, and eager Italianophile, a willing disciple who taught himself Italian and immediately set about contributing papers in that language to the *Proceedings of the Accademia Nazionale dei Lincei* on problems close to Volterra's heart, and who was proud to consider himself an honorary member of Volterra's family circle. In his study of economics' evolution from a qualitative to a quantitative discipline, economist E. Roy Weintraub describes this period in Evans's career as "the marker event in his intellectual life," and the six papers he wrote under Volterra's tutelage in Rome as "the defining intellectual themes in his mathematical life."[30] Their paths would continue to cross after Evans returned to the United States.

Meanwhile, Volterra's honors and foreign invitations continued to accumulate. In 1910, he had been elected a Foreign Member of Great Britain's Royal Society, and the following year he was named a Foreign Associate of the National Academy of Sciences, the most prestigious scientific organization in the United States. Late that same year, he accepted an invitation to give a series of lectures and seminars on the functions of lines at the Sorbonne in Paris. Arriving there in mid-January 1912, Volterra checked into the Hotel Lutetia, about a fifteen-minute walk from the Sorbonne. Although Volterra observed that it was more expensive than other hotels on the Left Bank, he was struck by its comfort and charm, impeccable service, and interesting guests, and he resolved to keep his accommodations for the entire six weeks. Indeed, for the next twenty-six years, until he became too ill to travel abroad, the Lutetia would remain Volterra's home away from home in Paris. No sooner had he unpacked than he began urging Virginia to join him, at least for a week. "It would be a diversion for you and it would certainly please me greatly," he wrote, clearly not relishing the idea of a long stay in the City of Lights by himself. In a new letter the next day, he tried a different approach: "I assured everybody that before my departure you would be coming here." "Everybody" in this context meant France's top tier of mathematicians—Paul Appell, Émile Borel, Jacques Hadamard, Paul Painlevé, Emile Picard, and Henri Poincaré—all of them colleagues of long standing.[31] After more weeks of silence on this subject from Virginia, Volterra wrote, with some irritation, "I am waiting for *precise* [he underlined "precise"] information about the day of your arrival, and I advise you not to delay it because it is a real embarrassment not to know how to respond to invitations that cannot be left unanswered." (That final phrase is underlined twice.)[32] On February 12, he wrote again, vowing to telegraph her the next morning if he heard nothing. When daybreak brought no word from her, he sent the telegram, informing her that he had accepted several invitations in her name. Responding a few days later, Virginia blamed her inability to join Volterra on a chronic respiratory condition. "I am sorry that you always have a cough and a sore throat," replied her husband, who could not resist adding that the many visitors she regularly received at home had undoubtedly overtaxed her strength.[33] In the face of this domestic stalemate, Volterra played his final and possibly his only trump card; on February 13, he wrote to his father-in-law, then in Alexandria, Egypt, on business. Sizing up the situation, Edoardo Almagià wrote a letter to both Vito and Virginia, but appears to have rather pointedly mailed it to Vito in Paris:

> I received with pleasure letters from both of you with the date of the 12th from Paris and of the 13th from Rome, and from what Virginia writes to me I will hope that this letter of mine will find both of you in Paris. . . .I find it right [to think] that Virginia has also gone to Paris to witness and perhaps to collect also the honors given to her husband. . . .I wish now to hear that Virginia has satisfied Vito's wish and she is already in Paris. . . .I am sure that such a little journey will be very advantageous also to you, my dear Virginia.[34]

Virginia joined Volterra in Paris on February 25, toward the end of his stay.

While Volterra was in Paris, his French colleague Jacques Hadamard invited him to lunch to meet Edgar Odell Lovett, the president of the newly established Rice Institute in Houston, Texas. When Lovett, who had headed the department of mathematics and astronomy at Princeton University before accepting the job of Rice's founding president, disclosed that he was still looking for professors for his new university in Texas, Volterra recommended Evans on the spot. "But don't tell Evans if you see him," he later cautioned Virginia. Lovett also invited Volterra to participate in the inauguration of his Institute that October, offering $1,000 in exchange for a speech about how the educational system worked in Italy. Once again, Volterra swore Virginia to secrecy, although it appears that he had already made up his mind to accept the invitation.[35] In February, he again spoke to Lovett about Evans and gave him reprints of his protégé's papers to read. In one of last official acts before leaving Europe, Lovett took the train to Rome to see Evans; by mid-March, Evans had been offered, and accepted, an appointment as assistant professor of mathematics at Rice (he later moved on to the University of California at Berkeley, where he became chairman of the mathematics department).

Between January and March 1912, Volterra gave fourteen lectures at the Sorbonne, which attracted a steady attendance of professors and students, but by the time he had delivered the eleventh, on February 21, he realized that he had enough material for a dozen more. "I don't know how to condense so much stuff into only three lectures," he groaned, the *cri de cœur* of teachers everywhere.[36] He also reached an agreement with the publishing house Gauthier-Villars to bring out his Sorbonne lectures as a volume in its series of monographs on the theory of functions. The rising young French mathematician Joseph Pérès, introduced to Volterra by a mutual colleague, collected and edited them, and corrected the proofs for publication, an effort that prompted Volterra to praise his editor's "intelligence" and "zeal" in the preface when they were published the following year.[37] Teaching at the Sorbonne spared Volterra the pain of listening to his colleagues in the Italian Senate hail Italy's determination to establish a colonial presence in Africa, even at the cost of waging war against Turkey. Eyeing the far-flung colonial holdings of the other European powers and fired by dreams of a long-vanished imperial past, the government in Rome had set its sights on Libya, a territory of the crumbling Ottoman Empire. In September 1911, after spurning several overtures for negotiation and compromise, and turning a deaf ear to offers of concessions from Turkey, Prime Minister Giolitti declared war, and a month later Italian troops had bombarded the Libyan city of Tripoli. In the months leading up to the September declaration, Italians of all stripes had embraced the notion of armed conflict as a matter of national honor and pride. When the writer and rabid Italian nationalist Gabriele D'Annunzio famously declared in 1911 that "Africa is only the whetstone on which we Italians shall sharpen our sword for a supreme conquest in the unknown future," his words resonated with the mood and aspirations of many Italians. To Volterra's dismay, these ardent new nationalists included his own cousin Roberto Almagià, his Uncle Alfonso's son, who somewhat ironically, in view of his father's misgivings decades ago about the young Vito's career aspirations, had followed his eminent cousin into academia and become a renowned geographer at the University of Padua.[38] Known in the family as "Il Ligure" after the Via Liguria address of the Rome apartment that he had inherited from his father (the same apartment where Volterra's mother had stayed when she went to Rome to see about getting Vito married), Roberto publicly defended Italy's imperialist aims. "Tell 'Ligure' to put water in his wine," Volterra wrote sourly to Virginia, not amused by his cousin's jingoism. A few days later, he let fly with a much harsher judgment: "Tell 'Ligure' that in his search for applause, he should not say things that may embarrass him later on, and that he should not support nationalist illusions that in the end will only bring about damage... and ridicule"[39]

To calm his nerves, Volterra had vowed not to write about politics in his letters home, a promise he found nearly impossible to keep. How could "our government," he agonized in one letter, be considered "civilized" when it "goes on a crusade and helps the Banco di Roma?" (The Bank of the Holy See, through which the Vatican carried out its business dealings, had obtained mining concessions in Libya, from which it was anxious to realize a profit.) Under these circumstances, he added cynically, "a senator might also aspire to 'the Pontifical Throne.'"[40]

When Volterra read in the paper that Italy's Parliament had reopened on January 22, he told Virginia he was "very glad not to be there where I would be obliged to hear the most disgusting and awful oratory," adding in the next breath, "For God's sake tell 'Ligure' not to quote verses by [D'Annunzio] who is a shame and a real dishonor."[41] But Roberto Almagià was hardly alone in feeling a sense of resurgent national pride. Writing to Volterra from Rome, his colleague Orso Mario Corbino complimented him on a rave review he had read of Volterra's Sorbonne lectures, and went on to say, "In this reawakening of national life that infuses everything, the homage paid to you in this marvelous capital is for us a new source of pride and joy."[42] Volterra, however, felt only apprehension and shame. "I have been reading the news from Italy," he finally admitted to Virginia,

> which makes me very sad. The best thing would be to not open the newspapers, so as to know nothing of what is happening there. I hope that soon after I return, we shall be able to go to Albano. This will be restful for me, and also I shall not have to remain in the middle of all this with ideas that run so counter to what almost everyone else is thinking. If only the Italians could for once open their eyes, they would see how little they are thought of elsewhere in Europe and in the United States. We are in a bad period, and it will not be short.[43]

That fall, with a foreboding sense of the gathering storm on the European horizon, Volterra returned to the United States. His 2,000-mile itinerary was clearly meant to acquaint him with as much of America as possible: From New York, he traveled by train to New Orleans, going from there to Houston, where he had accepted Lovett's invitation to lecture at

Rice. He followed this up with more talks at the University of Illinois and the University of Chicago, before returning briefly to New York City to deliver a paper at a meeting of the American Mathematical Society. He then headed for Princeton University, which had named him the Louis Clark Vanuxem Lecturer, to give three lectures on the theory of permutability; finally, he hastened to Harvard, where he lectured on functionals. In the midst of his ongoing love affair with the United States, news from Europe arose once more to depress him. The disintegration of the Ottoman Empire, which had emboldened Italy to declare war on Turkey, now threatened to engulf the Balkans, where Serbian nationalists and their regional allies had mobilized with the aim both of driving out the Turks and throwing off the hegemony of the Austro-Hungarian Empire. "I am very worried by the news of the war in the East," he wrote Virginia from his room at the University Club in Urbana, Illinois, where the local foreign-language newspapers were reporting on the conflict in the Balkans. "It is an enormous danger for Europe. A little beginning of fire is enough to cause a great fire in Europe. The war of Italy against Turkey was a first spark, now the fire is in the East. Let us hope that it will not spread."[44] Two years later, the fire in the "East," now burning out of control, had ignited the conflagration known as World War I.

CHAPTER 15

"Empires Die"

In July 1915, fifty-five-year-old Vito Volterra, newly commissioned as a lieutenant in Italy's Army Corps of Engineers, reported for duty at the Central Institute of Aeronautics in Rome, an institution that he had helped found in 1908. He was met there by the Institute's military director, Major Gaetano Arturo Crocco, who put to Volterra a problem that he described as entirely new in the annals of artillery. Crocco, an aeronautical engineer who had designed a number of airships for civilian as well as military use, had it in mind to install 65 mm-caliber mountain guns on Italy's M-type dirigibles. He asked Volterra to calculate the firing tables for blasting this weaponry from the air to a target on the ground. Decades later, Crocco (who after World War II would go on to become Italy's leading space and rocket scientist) could still recall that Volterra needed only a few weeks to solve this airborne-armaments problem and devise what Crocco called "a great innovation in the history of ballistics."

A year later, the concept had moved from the drawing board to the staging ground. At Campi Bisenzio, an air station near Florence, Volterra, Crocco, and two of Crocco's longtime associates prepared to test the validity of Volterra's calculations. The team had obtained encouraging results in the lab, but this was not the same thing as firing a powerful gun from a zeppelin inflated with combustible hydrogen. When a group of French officers who had come to Italy expressly to see Crocco's latest aeronautical innovation in action expressed doubt that the dirigible's skin would survive the shock wave set off by the cannon blast—with obvious consequences for Crocco and any others who would be test-firing the gun from the airship—the major was sufficiently unnerved to consider postponing the dry run while he consulted once more with the mathematical savant from Rome. "I went back and asked Senator Volterra once more to review his solution of the problem and its various unknowns. [He] meditated on it for a few days and said to me, 'Nothing bad will happen, but the phenomenon is interesting: I will be on board with you.'"[1]

Two months after the Serbian nationalist Gavrilo Princip assassinated Archduke Franz Ferdinand, heir-apparent to the Austro-Hungarian Empire, and his wife, Duchess Sophie, in Sarajevo, on June 28, 1914, Europe turned into a bloody battlefield. With deadly precision one national domino after another toppled into warfare: Under the pretext of retaliating for the murder of its royals, Austria-Hungary declared war on insurgent Serbia and sent troops against Belgrade; Russia, coming to the aid of its Slavic ally, mobilized against Austria. The Empire's chief ally, Germany, swiftly declared war on Russia, and also marched against France after the French, bound by a not-so-secret treaty to Russia, refused to accede to German demands that they remain officially neutral. Germany's determination to open a second front on its western flank required the invasion of Belgium, a noncombatant, and when this occurred in the dead of night on August 4, Great Britain, bound by treaty to defend Belgian neutrality, declared war on Germany. After a century marked by unprecedented progress and what many had come to regard as an irreversible spirit of enlightenment, the heart of darkness opened in Europe. The continent was at war.

Incredibly, Italy was not initially drawn into this catastrophe. Its one-time alliance with Germany and Austria-Hungary had fallen apart the year before, leaving the nation a free agent. Not sure which side to join or whether to enter into an alliance at all, and anxious to command the best possible terms if it did, the government in Rome proclaimed its neutrality in August 1914 and spent the next nine months negotiating with both sides. Italian public opinion was divided, mostly between those who were convinced it was in Italy's best interests to stay out of the conflict, and those who urged intervention on the side of the Allies, as the coalition of Britain, France, Russia, and their allies came to be called. In this ongoing national debate, Volterra sided whole-heartedly with the interventionist camp. He was as appalled by Germany's naked aggression against helpless Belgium and mobilization against France—the nation he venerated above all others as Italy's inspiration, role model, and soul mate—as he had been by Italy's trumped-up war with Turkey over Libya in 1912. While ideologically consistent, this position placed him in the company of some strange bedfellows, including the self-proclaimed poet-patriot Gabriele D'Annunzio, whose rabble-rousing support of Italian imperialism in Africa Volterra had denounced so vehemently only two years earlier. If he perceived any irony in supporting a cause that D'Annunzio and Italy's Nationalists, for very different reasons, had also ardently embraced, he gave no hint of it. "In my opinion Italy must take its place at the side of its Latin sister, France, and of its allies against Austria and Germany," Volterra wrote an old colleague in Paris, soon after Germany invaded Belgium. "It is her role and her mission. She must not fail in it."[2]

Using his prestige and standing as the dean of Italy's scientific community, Volterra waged an active campaign in support of the country's intervention on the Allied side. In this he was vigorously abetted by Italians who shared his view that only the defeat of Germany and Austria could safeguard hard-won democratic freedoms in Italy and elsewhere, as well as by those who could not have cared less about the future of constitutional democracy in Europe. Chief among them was the large, growing, and unfailingly vocal Nationalist Party, which saw the opportunity to wrest the Italian-speaking territories of Trieste and Trent from a defeated Austria-Hungary as the first transcendent step toward the creation of a Greater Italy. Speaking to a cheering crowd in early May 1915, on the fifty-fifth anniversary of the start of Garibaldi's drive for Italian unification, D'Annunzio proclaimed his vision of "an Italy that shall be greater by conquest, purchasing territory not in shame but at the price of blood and glory....After long years of national humiliation, God has been pleased to grant us proof of our privileged blood."[3] Mingling with the deliriously aroused listeners was Volterra, who told his wife afterward "that the celebration of this morning went very well, and it was very beautiful."[4]

In the midst of these popular demonstrations, the government of King Victor Emmanuel III, with the concurrence of key parliamentary leaders, concluded that an alliance with Britain and France (who had promised the Italians not only Trent and Trieste but additional territories in the event that the Central Powers were defeated) offered the best deal for Italy. But neither the Italian public, which remained deeply split on the war, nor the parliamentary rank and file knew that Italy had made a secret agreement to come in on the Allied side when Prime Minister Antonio Salandra called Parliament back into session on May 20, 1915, to rubber stamp a motion supporting the government "in the event of war." Three days later, Italy entered the conflict. When word of the government's covert machinations finally did leak out, it did irreparable damage to Parliament's credibility as the country's constitutional authority, a point that was grasped by, among others, Mussolini, who would exploit it to the hilt in 1922 when he became prime minister.

"So we are at war; long live Italy!" enthused Carlo Somigliana,[5] one of Volterra's oldest schoolboy friends, on the eve of the vote in Parliament. Italy declared war on Austria on May 23, 1915; it would wait until August 1916, however, to declare war on Germany.

Volterra lost no time in volunteering for military service. He was inducted that July into Italy's Army Corps of Engineers and almost immediately assigned to assist Crocco with his aeronautics and armaments research. For the next three years his war work would both consume and invigorate him. The peripatetic professor who had happily written to his wife a few years earlier about the hectic pace of academic gatherings in America now threw all his energy, industry, and force of intellect into the Italian war effort, as he rushed from meeting to meeting in cities throughout Italy, hurried off to military consultations in Paris, participated readily in hazardous field tests of untried equipment, and braved the action at the front lines to monitor and carry out demonstrations of new military technologies. Thriving on a schedule that would have exhausted a much younger, physically fitter man, he was convinced that he had never done more worthy or important work in his life.

His preoccupation with the European conflict left Volterra little time for home and family, but there was one important exception. In early 1916, his eighty-year-old mother, who had enjoyed generally robust health throughout her long life, fell ill with a chronic fever. "I wish to receive news daily by telegram about mother, and I beg you... not to conceal anything from me,"[6] wrote Volterra to his wife, rather in the manner of a general ordering up bulletins from the front. It developed that Angelica had contracted an aggressive bladder infection—a serious matter at her age in those pre-antibiotic days—and on March 4, 1916, she died of the complications, with her beloved son, the wife she had chosen for him, and her four grandchildren by her side. The redoubtable matriarch of the Volterra clan was laid to rest in the Jewish cemetery in Rome, near the graves of her two infant grandchildren, Luigi and Gustavo. The burden of looking after Angelica in her last illness had fallen almost entirely on Virginia, who had left off her volunteer work in a military hospital to care for her ailing mother-in-law. Her three sons, reacting perhaps to the death of the grandmother who had been a constant presence in their lives, and to their father's prolonged absences, had become so disruptive and quarrelsome that they were almost unmanageable, she reported in her letters; this, too, seems to have taken its toll on her. "Take courage," Volterra exhorted her, after a hasty visit home that April. "I was terribly sorry to see you so depressed; it is necessary to force oneself to [be brave.]"[7]

A few days later, he spelled out in greater detail a credo that he had apparently adopted as his philosophy of life: "It is important always to take care of one's health, to have a firm character, and to take an interest not merely in one thing at a time, but in many different things."[8] Indeed his convictions about the rewards of stoic sacrifice in wartime had taken on almost epic proportions. When he learned in late May that his French colleague Jacques Hadamard had lost his son at the horrific Battle of Verdun, he wrote to Virginia that the condolences he had sent in both their names was really "a telegram of admiration for the heroic end of his son." What can Virginia have thought when she read this meditation from her husband: "What a pity that our sons are not yet old enough to go to war." Edoardo, the eldest, had just turned twelve.[9]

In another letter to his wife in early June, Volterra again admonished her to stay calm and not worry about his safety: "For my part, I have never for an instant lost the most complete serenity," he told her, adding, "I am of the opinion that I would not lose it under any circumstance in my life."[10]

Four days later, Volterra made good on his pledge to join Major Crocco and his associates in conducting the first canon tests aboard a dirigible pumped full of flammable hydrogen. The airship had been tethered to an embankment at the Campi Bisenzio airfield; the visiting observers, who had tried unsuccessfully to talk Crocco out of what they perceived as an insanely risky experiment, had prudently stationed themselves well out of range of any rogue detonation. From inside the dirigible, Crocco, as he later recounted, "personally fired the first shot, using *one-quarter* of the cannon's charge. The covering of the dirigible wasn't affected at all. Not even at *half charge*. At *three-quarters* of charge there was a slight shudder; and not much happened at full charge. Nothing exploded." Volterra's firing tables had passed their first test with flying colors. By then, the wary spectators had begun to edge closer, eager to try their hands at firing the cannon. Early in July, Volterra's tables had their first in-flight test, and fifty rounds of cannon fire, aimed at spaced targets on the ground in accord with his exacting calculations, similarly found their mark.[11]

After these phenomenally successful trials, Volterra spent the remainder of the summer commuting between Rome and Campi Bisenzio, calibrating barometers and other instruments, refining his tables, and carrying out more experiments aboard armed and airborne dirigibles. In July he was participating in one such test on board a zeppelin designated Dirigible Number Seven, flying some 5,000 meters (approximately 15,000 feet) above the air station, when an accident occurred. As the crew struggled frantically to bring the damaged airship under control, Volterra calmly monitored the variations in the zeppelin's motion during the emergency descent. For conduct that Italy's Minister of War, General Paolo Morrone, would describe as "cold-blooded coolness" under fire, Volterra earned a mention in the army's dispatches ("posted and read to the soldiers the following day," he proudly reported to his wife) and a promotion to captain. Three months later, in Venice, Italy's Minister of Justice, Vittorio Scialoja, presented him with the War Cross for bravery, an event that Volterra would subsequently describe as one of the most beautiful in his life.[12] Describing Volterra's wartime conduct some years later, Morrone would recall that "in addition [to his other accomplishments], he has the merit to have been the first to propose helium gas for dirigibles." If Volterra's idea of inflating zeppelins with stable helium rather than incendiary hydrogen had been widely adopted, the history of modern aviation might have been quite different.

Volterra never wavered in his conviction that the "war to end wars" was an entirely just cause, or in his optimism about its outcome. Like many pro-war Italians, he had been convinced from the outset that the hostilities would be short-lived, and that Italy and her soldiers would emerge covered with glory. "The war is going very well," he assured Virginia over and over,[13] although the news from the Italian front, where the troops had been struggling unsuccessfully for a year to break through the Austrian lines, told a far different story. Emilio Lussu, who served with the crack Sassari brigade on the Italian front for four years, recalled in the book he later wrote, *Sardinian Brigade*, that in late spring 1916, his brigade had been fighting on this front (the Carso) ever since the beginning of the war,

> and we were heartily sick of it. Every inch of ground reminded us of some engagement or of the grave of a fallen comrade. Over and over again we had captured trenches from the enemy, but the situation remained just the same. There were always more to be taken. Trieste seemed still as

far away as ever, lying as though weary in the summer heat, between us and the sea.[14]

(After World War I, Lussu was elected to Parliament, but in 1926 he was expelled from the Lower House on "anti-Fascist" charges and subsequently imprisoned. Escaping to France, he later joined the Italian Resistance, and at the end of World War II returned to Italy, where he served as a Senator in the newly constituted Parliament as a member of the Socialist Party.)

As her husband continued to declare the conflict in Europe all but over and to condemn Italians who expressed apprehension about the future as "enemies of their country," Virginia returned to her volunteer work with wounded soldiers. "By now it is a completely lost game for the Germans," Vito assured her, shortly before Italy finally entered the lists against Germany in August 1916. "Don't worry and don't let the children become accustomed to being afraid."[15] He had been dismayed by the collapse of yet another Italian government that June, which he regarded as "a triumph for the neutralists and for the pro-Austrians."[16] If only, he lamented, the country could rely on the university students and other young people who had marched in the streets in May 1915, demanding Italy's entry into the war. Unfortunately, most of those stout-hearted patriots were now at the front or otherwise engaged in the war effort. "At the present time all the best constituents are fighting the war and at home there are only old people, invalids, or even worse," he maintained, glossing over the inconvenient truth that not all of his colleagues felt as he did. Tullio Levi-Civita, whom Volterra greatly admired as a scientist and mathematician, and whose career he had consistently championed, was a Socialist and firmly committed to the Party's position of non-alignment. (The Italian Socialists had split bitterly over the question of Italy's involvement in World War I; among those expelled from the Party for advocating an end to neutrality was a charismatic and ambitious journalist and former pacifist named Benito Mussolini.)

That fall, Volterra moved his military research operations to the Italian front near Trieste, on the Adriatic side of the peninsula, where he had been asked to assess the effectiveness of a new technology for detecting the location of enemy guns. In mid-November, he received orders to report to Paris, where he caught up with his Sorbonne colleague Émile Borel as well as Paul Painlevé, who had given up all of his academic posts a few years earlier and become a full-time politician. Like Volterra, these consummate mathematicians had thrown the full force of their energies and expertise into the war effort and were now in charge of the French Department of Technical Inventions, Studies, and Experiments, which they had founded expressly to organize, coordinate, and promote national defense work. (Painlevé, who would become France's Minister of War in 1917, was at the time Minister of Education, Fine Arts, and Inventions for National Defense; he would go on to serve twice as the nation's Prime Minister.) Joining them in the French capital, Volterra plunged into a round of meetings at the l'École Normale and various scientific laboratories, including briefing sessions with Italy's ambassador to France, and meetings with Madame Curie and General Pétain, the hero of the Battle of Verdun. He then set off with Painlevé and Borel for a visit to the French front "to study firsthand the military apparatus of the Allied forces," an expedition he breezily described in a letter to Virginia as "a tour in the provinces for a few days."[17] It was the first of many such trips he would make to England and France over the next two years.

Volterra's almost total immersion in his war work did not preclude his following with intense interest the election of a new president of Italy's Accademia Nazionale dei Lincei, a position he coveted. Upon learning that Francesco D'Ovidio, a distinguished philologist, had won the election, Volterra did not bother to hide his pique and disappointment from his wife. "What you [Virginia] write concerning the Lincei does not surprise me....I had nobody on my side since everybody was excessively sluggish." Worse still, he continued, "the Academy will carry on in the usual way without anything new. I don't know if I will ever reach [the presidency of the Academy], but if I do, it will be at such an age that I too will have become old and therefore incapable of accomplishing anything."[18] (Volterra's fortunes would improve in 1920, when he was elected vice-president of the Lincei; he went on to serve one term as president, from 1923 to 1926.)

Inspired by the example of the French Office of Technical Inventions, Volterra returned to Italy with the idea that the nation's Ministry of War ought to establish a similar entity, headquartered in Rome. Such a government agency, with the authority to coordinate activity among Italy's various military boards, its widely dispersed commissions of inventions (whose functions were somewhat akin to that of the American patent office), and the nation's several hundred scientific and technical institutes, would surely act as a catalyst for "a mobilization of the scientific resources of the country," spurring the design and development of technological innovations with military applications. Volterra also recommended that representatives from Italy's European allies, England and France, be stationed in Rome to facilitate a rapid exchange of information.[19] The government was receptive to these proposals, and in March 1917, the Ministry of War authorized the establishment of the *Ufficio Invenzioni*, with Senator Volterra at the helm; shortly afterward, Volterra enlarged the agency's scope to include cutting-edge research on war-related problems.

As the number of Volterra's high-level, and presumably confidential military assignments increased, he had less and less to say about his war work in his letters. "Nothing especially new here," he wrote from Paris that April, "but the papers are full of news and you will learn it naturally by reading them."[20] One current event that could not have escaped Virginia's notice was America's entry into the war in April 1917. President Woodrow Wilson's decision to come to the aid of Britain and France inaugurated a period of intense communication and collaboration between Volterra's *Ufficio* and America's wartime science advisory board, the National Research

Council (NRC), founded and directed by the American astronomer George Ellery Hale. Volterra's delegate in Washington, D.C., Giorgio Abetti, a solar astronomer who had worked at Hale's Mount Wilson Observatory, near Pasadena, California, sent Volterra a steady stream of reports on such topics as aerial photography, magnetic and electromagnetic communications with submarines, and oil and gasoline.

In 1918, to strengthen and expedite the exchange of information between Italy and the United States, Italy's military mission in Washington, D.C., asked American officials to send a scientific attaché from Hale's National Research Council headquarters to Volterra's agency in Rome (now renamed *Ufficio Invenzioni e Ricerche*, in acknowledgement of its expanded research functions). Hale hand picked S.L.G. Knox, a seasoned engineer with extensive ordnance experience, for the job. That July, Knox sailed to Europe, with Hale's assurance that an Italian-born Columbia University physicist, Gioacchino Failla, would be joining him shortly as his interpreter (Knox, somewhat oddly, in view of his selection, did not speak Italian). Knox was in France when Failla arrived in August and presented his credentials, only to discover that Volterra refused to acknowledge that his diplomatic passport constituted proof of his identity and security clearance. When Knox, who returned to Rome the next day, tried to bring Failla along on a series of laboratory visits, Volterra told him that "that was impossible as they had received no official notice of his assignment from their military mission, as they had in the case of the others from the NRC." Even after Knox pointed out that he would be helpless without his translator, Volterra "would not budge, although very polite, of course." Knox cabled Washington to take action on the case, pointing out that in view of the fact that he had come 5,000 miles with "the full knowledge and consent of the Italian ambassador," he found this official conduct inexplicable. Knox could never make out whether Volterra was acting under instructions or on his own initiative, but he continued to treat poor Failla in a manner that the perplexed American envoy could only describe as "most extraordinary... I took Failla in to interpret for me and Volterra refused to talk to him, persisting in using French, in which I am conversationally very limited."[21]

The problem may have come down to a matter of personal style. Knox, a plain-spoken civilian engineer, with a typical American's can-do attitude and distaste for formality, undoubtedly felt that the Italians could afford to disregard a minor irregularity on his and Hale's personal say-so. Volterra, who had wholeheartedly embraced the American way of life on his two trips to the United States, may have seen things differently on his home turf in wartime. He may very well have felt that while breezy informality and seat-of-the-pants diplomacy were charming and appropriate in New York, Chicago, and Houston, a pillar of the Italian scientific establishment—a captain, moreover, in Italy's army—was honor- and duty-bound to respect the bureaucratic proprieties. Although matters improved after Volterra apparently received official clarification of Failla's status and, perhaps more to the point, discovered that his émigré *paisan* had written a well-received book on radium and radioactivity, Knox suspected that other factors might also have been at work. He wrote to Hale, "I had been warned by our military and naval attachés here that there was a lot of prejudice against so called 'bastard Italians'—American citizens born in Italy who had not come back here to fight."[22]

By the time Hale received Knox's letter in early October, both he and Volterra were in London, attending the first of three inter-allied scientific conferences convened by England's Royal Society. Although soldiers were still dying on the battlefield, it was obvious that an Allied victory was only a matter of time, and much of the discussion centered on the shape of the international scientific community in the postwar era. Hale had assured President Woodrow Wilson that the gathering had no intention of issuing hostile proclamations aimed at enemy scientists, but neither he nor Volterra had any real objections to make when the majority of delegates approved a resolution that prohibited scientists from the Central Powers from joining "international" scientific associations. The French and Belgian representatives, whose countries had suffered enormously during the conflict, had pushed hard for such a declaration, which was also strongly endorsed by their colleagues from elsewhere on the continent. Most delegates from the United States and Britain, while perhaps not feeling quite so strongly about it, were more than willing to go along. Two weeks after the Armistice on November 11, a second inter-allied conference got under way in Paris, and Volterra found himself elected a member of the meeting's executive committee.

At the close of World War I, the future of Italian science hung in the balance, and it is instructive to compare Italy's situation at this juncture with that of the United States. In America, the National Research Council had won its spurs organizing research for the war effort, and after the war Hale, along with other leading American scientists, did his best "to circumscribe the role of the federal government in the development of science."[23] In Italy, on the other hand, the government held all the cards, for better or—as Volterra's late Sicilian colleague Giovanni Battista Guccia had once presciently suggested—for worse. In July 1919, Italy joined France, England, Belgium, and the United States in inaugurating the International Research Council. Hale was named president and Volterra the vice president of the new organization, whose statutes excluded the defeated nations' scientists from the privilege of membership. (The ban was rescinded in 1926, clearing the way for German mathematicians to participate in the International Congress of Mathematicians held in Bologna two years later.) By the following summer, ten more countries had pledged their allegiance to the Council's rules about the international exchange and promotion of scientific ideas. Like the other signatories, Italy agreed to set up an integrated hierarchy of scientific organizations at home; the *Unione Matematica Italiana* (Italian Mathematical Union) was formed in 1921, following the model of existing Italian organizations for astronomy, chemistry, and oceanography. Other disciplines would soon follow. In 1923, a royal decree brought the *Consiglio Nazionale delle Ricerche* (National Research Council), composed of representatives from each national scientific union, into existence.

* * * * * *

In the summer of 1922, a new face appeared at the gates of Rome. It belonged to a brilliant and cocky graduate of the University of Pisa and the Scuola Normale who had just taken his doctorate in physics at the precocious age of twenty. Years later, the physicist Franco Rasetti, who was Enrico Fermi's fellow student at Pisa, would remember him there as "an extraordinary student, who at the age of nineteen knew more than all of his professors."[24] Fermi's outstanding gifts had already caught the attention of the scientists at the University of Rome's Physics Institute. After graduating from Pisa he had hastened back to his hometown, bearing a letter of introduction from his awed thesis advisor, which he presented to the Institute's director, and Volterra's longtime friend and colleague, Orso Mario Corbino. Corbino, an experimentalist, was Italy's best-known physicist. Like Volterra, whose name had become synonymous with Italian mathematics in scientific circles around the world, Corbino yearned for a physicist whose experimental and theoretical aptitude would do for physics what Volterra had done for mathematics. In Fermi, whose own advisor had been stumped by his student's papers on relativity (Levi-Civita was one of the few people in Italy capable of understanding what the young man was talking about), Corbino thought he might at last have found the appropriate person.

Although he couldn't match Volterra's record as an organizer and founder of scientific organizations, Corbino, who was also a Senator of the Kingdom of Italy, had served as Italy's Minister of Public Instruction, and had done important work in the field of magneto-optics earlier in his career. In his biography of Fermi, the physicist Emilio Segrè describes Corbino as "an excellent speaker" who "liked to arrange promotions, transfers, and the like, and usually succeeded in these attempts." Corbino had one other quality, Segrè remembers: "He was as shrewd and clever as anybody I have ever met, but what distinguished him from many other equally ardent academic politicians was the loftiness of his purposes and the sureness of his judgment."[25] Certainly, Corbino was deeply impressed by Fermi, and they began to meet and talk almost daily about physics, plans for future work, postgraduate opportunities, and, increasingly, politics.

Italy had survived World War I only to surrender its democratic institutions to Benito Mussolini. The war had left the country with soaring inflation, monstrous debt, rampant unemployment, and widespread labor unrest, as well as a disenfranchised and increasingly disaffected officer corps. The Socialist Party, at the time the largest in Italy, was split between its left wing whose rhetoric had become ever more militant in the wake of the Bolshevik Revolution in Russia, and its moderates, who found themselves hamstrung by both the Party's strident leftists and their own inability to form a coalition within Italy's faction-riven Parliament. Into this breach stepped Mussolini, backed by a motley alliance of nationalists, militarists, old-line conservatives, hard-line rightwingers, and Italians desperate to embrace almost any alternative to the prospect of a Communist-style takeover. The future *Duce* had begun his political career as a Socialist, writing for one of the Party's newspapers. Expelled by the Socialists in 1914 for his pro-war stand, he took his revenge in 1919 by founding the National Fascist Party, which rapidly became both his personal and political base. He was elected to Parliament as a Fascist deputy from Milan in 1921 and lost no time in capitalizing on the social and civic turmoil that was tearing the country apart.

In October 1922, following months of street violence, Mussolini's militant supporters, the Blackshirts, announced plans to march on Rome. Fermi happened to be in Corbino's office at the Rome Physics Institute on the morning of October 28. As his future wife, Laura, later wrote in a family memoir, "This time they did not talk of physics....Corbino was preoccupied with the political situation. He did not like the profession of violence made by the Fascist leader, Mussolini." The cabinet had drafted a decree declaring Italy to be in a "state of siege"; the document, which would have authorized the army to use force, if necessary, to quell the street demonstrations of Mussolini's backers, awaited the king's signature. When Fermi asked Corbino if he thought the king would sign the decree, Corbino, who was convinced that armed intervention would only produce more violence and most likely lead to civil war, replied, "I think there is a chance that the king may not sign the decree. He is a man of courage." "Then there is still a hope," Fermi said. "A hope?" Corbino replied. "Of what? Not of salvation. If the king doesn't sign, we are certainly going to have a Fascist dictatorship under Mussolini."[26]

Corbino's prediction, of course, came to pass. Within three years the one-time journalist and ex-pacifist had transformed Italy's besieged parliamentary democracy into one-man rule. As a Senator, Volterra seems only to have wanted to distance himself from the chaos and corruption in Parliament in the months leading up to Mussolini's takeover. "The reopening of the Senate oppresses me; it is of no interest to me at all, like all politics, in which there is neither anything beautiful nor comforting," he wrote to his wife in 1921.[27] The start of the year found him back in Paris, presiding over the meetings of the International Bureau of Weights and Measures, which had elected him its president (a position he would retain until his death in 1940). Shortly before leaving Rome, he had suffered an angina attack, and there had been some question of whether his physician, Ettore Marchiafava, would allow Volterra to travel. Volterra recovered, but the family doctor urged him to quit smoking cigarettes. From Paris, he wrote to Virginia that he was making a point of not reading the Italian-language newspapers, which in any case, "have very little new to say about Italy." He added, "Happy the countries that have no history!"[28] Was Volterra thinking about the United States, which he had visited, once again with great pleasure, in 1919, when he made this comment? It was his last trip to America.

"I traveled today with a trainload of Fascists,"[29] Volterra laconically wrote to his wife on October 26, about a recent trip from Rome to Florence. Decked out in their black shirts, his swaggering fellow passengers were among the waves of Fascist militia who would soon converge on the capital in anticipation of Mussolini's impending arrival from Milan, where he had been waiting for the king's formal invitation to form a new cabinet. On the morning of October 30, Mussolini arrived in the Eternal City with a mandate to govern Italy. It was two days after Corbino's bleak conversation with Fermi.

Although Volterra was certainly appalled by the Fascist triumph in the Italian capital, he initially seems to have sought refuge from "all politics" in a new scientific investigation. That autumn, he began studying the phenomenon of liquid jets, a well-known problem in classical hydrodynamics. His work formed the subject of a short research note, which he presented at the 1923 Liverpool meeting of the British Association for the Advancement of Science. Edoardo, his oldest son, now nearly twenty, accompanied his father to the meeting. (By 1926, Volterra had developed enough material to offer a course of lectures on the theory of liquid jets at the University of Rome, which he also gave in 1929 at the University of Cluj in Romania, and again, in 1930, at the Institut Henri Poincaré in Paris.)

His public reticence about the state of affairs in Italy was short-lived. In the spring of 1923, he joined with a handful of other senators and university professors in voicing strong objections to the new rules that Mussolini's Minister of Public Instruction, Giovanni Gentile, had imposed in the field of education. Mussolini had given Gentile, a prominent Italian philosopher who taught at the University of Rome, a mandate to overhaul the country's educational system, which he had exercised, in part, by abolishing the teaching of natural history and combining the teaching of physics and mathematics at the junior-high-school level. Responding to a letter from a university colleague obviously in sympathy with his own views, Volterra wrote, "It is a grave moment for the fate of higher education, and it would be useful for those who are interested in [its fate] to express their thoughts with haste and efficiency. . . .It has been noticed that the University Association [a nationwide union of university professors] (I don't name names for obvious reasons) has been too concerned with material interests so that according to public opinion and the government, its voice today will seem objective and calm as is necessary." In short, most of Volterra's Italian colleagues were not about to risk their positions for principles. Neither, as it turned out, were their country's politicians.[30] A month later, at a sparsely attended meeting of like-minded senators in Rome's Palazzo Madama, Volterra took aim at

Gentile's secondary school reforms, charging that they would have the effect of "diminishing and lowering the scientific culture" at the pre-university level, by denying entering university students the "technical knowledge indispensable for dealing with advanced studies... [which will do] harsh damage to the functioning of the university." (In fact, Gentile, an exponent of the idealist school of philosophy, had also proposed dispensing with the university system altogether.)[31]

Later that summer, Volterra, in his capacity as the new president of the Lincei, formed a commission to examine the impact of Gentile's reforms on the country's school system. Although Gentile modified his stance on some provisions of his bill, the results, from the physicists' point of view, proved to be disastrous. Against the backdrop of these events in 1924, the academicians of the Lincei's class of physical, mathematical, and natural sciences unanimously selected Volterra to be the first president of Italy's National Research Council.

Not all of Volterra's colleagues shared his almost visceral antipathy to the Fascist regime. In a letter to astronomer George Ellery Hale, the Florentine astrophysicist Giorgio Abetti, who had served as Volterra's emissary in Washington during World War I, spoke for many Italians, including highly educated professionals like himself, when he described Mussolini's path to power as "quite a revolutionary one but so well arranged that there were no disturbances at all. Mussolini the leader of the fascisti... was elected 'premier' by the king and formed quite a new ministry." New elections would be held in the spring, he added, and the results of those should eliminate "the bad elements... which in recent years have failed so miserably at running the State, at least that is our hope."[32]

Still, Abetti's views were an exception to the general feeling among Italy's scientists. Recalling the political atmosphere in Rome during his stay there in 1924, a postdoctoral student of Levi-Civita's, the Dutch geometer Dirk Struik wrote, "The scientists I knew were almost all antifascists... although their antipathy or skepticism towards the Mussolini regime was not, for so far as I could see, a militant one. An exception was Volterra." Initially, some of Volterra's colleagues were willing to adopt a wait-and-see attitude. One of them was Rasetti, who decades later said: "In the first few years... Fascism didn't seem very bad. In fact, a large class of Italians welcomed it, because the Communists were very powerful and practically disorganized all industrial production, disorganized the railway traffic. So at that time Mussolini seemed a fairly reasonable dictator. The first act that really disgusted the more reasonable people was the [Giacomo] Matteotti murder... in 1924. But in 1922 it still looked as if it would become a reasonable dictatorship; after 1924, people lost hope in that." Matteotti, a Socialist deputy in Parliament, had denounced Fascism with much venom there, and afterward told a colleague that he had written his own obituary. True to his prediction, he was stabbed to death by Fascist vigilantes in June of that year.[33]

Fermi's revulsion over the Matteotti murder was so great that he considered permanently leaving Italy. Anxious to keep their stellar colleague on Italian soil but fully sympathetic to his concerns, Corbino and Volterra took the lead in making sure that Fermi had ample opportunity both to forge scientific contacts abroad and to advance professionally at home. In winter 1923, Fermi went to Göttingen to do quantum physics work with Max Born. When he returned to Rome, Corbino saw to it that he was given a paid position at the university teaching mathematics to biologists and chemists. The following spring, Volterra was instrumental in helping Fermi obtain a Rockefeller-financed International Education Board Fellowship to study theoretical physics at the University of Leiden in Holland. As the board's principal talent scout in Italy, Volterra also saw to it that two other talented young physicists—Rasetti, Fermi's friend from Pisa, and Enrico Persico, from Rome, received Rockefeller support.

It was one of the signal ironies of Volterra's later years that the near-total collapse of the Italian social and civic order from which he had benefited all his life coincided with the emergence of precisely the kind of dynamic Italian physics community toward which he had worked throughout much of his career. During the 1920s, his versatile circle of senior mathematicians forged close personal as well as professional ties with this rising young generation of Italian physicists, whose acknowledged star was Fermi. Laura Fermi, who married Enrico in 1928, wrote this account of one of their frequent gatherings at the home of Guido Castelnuovo, the head of Rome's mathematics faculty:

> In the Castelnuovos' small parlor ten or twelve middle-aged people were sitting in a circle on green plush chairs....Most men wore beards; most women were clad in black. The men were the greatest Italian mathematicians of that time: Volterra, Levi-Civita, Enriques...Saturday after Saturday they gathered at the Castelnuovos with their wives and children, to spend a few hours in informal chatter among congenial friends. They talked of the latest happenings in the faculty of science: of births and deaths; of marriages and flirtations; of faculty policies; of new discoveries and theories; of the rising stars in physics.[34]

In the fall of 1927, less than a year after he had discovered the quantum mechanical statistical laws now known as "Fermi Statistics," Fermi, then twenty-six, was appointed to Italy's first designated academic chair in theoretical physics, at Rome. A second such chair, newly established at Florence, went to Persico. That same year, Franco Rasetti transferred from Florence to Rome, where he became first assistant to Corbino, as well as a new regular at the Castelnuovo gatherings. In September, the members of this close-knit group turned out in force for an international physics meeting that the Italians hosted at Lake Como to commemorate the centennial of the death of Alessandro Volta, the Italian physicist who pioneered the development of the electrical battery, and whose name was posthumously bestowed on the electrical unit known as the volt. The cost of the conference was borne by the Italian electrical industry, the province of Como, and a subsidy from the state (which Mussolini, after considerable prodding, had agreed to at the last minute). In complete contrast to the animosity that had prevailed toward Central Europe's scientists at the end of World War I, the conference drew a cross section of the greatest names in physics from throughout Europe, including Max Planck, who came from Berlin, along with Ernest Rutherford, Niels Bohr, Hendrik Lorentz, Arnold Sommerfeld, and Robert Millikan, America's best-known physicist. The most prominent absentee was Albert Einstein. He did not like the idea of entering Fascist Italy, but his scruples were not widely shared. When the conference ended, most of the participants went on to Rome, where they had been promised an audience with *Il Duce* himself. Volterra, however, was not among them; he had continued on to Paris, where he was to chair a meeting of the International Bureau of Weights and Measures.

His open breach with the Fascist state had begun three years earlier. Weeks after Matteotti's murder in June 1924, Volterra had joined with twenty other senators in casting a vote of no confidence in Mussolini, as the head of the government. The annals of the *Atti* of Parliament for June 24 and 26 also reveal that more than two hundred senators gave their blessings to the party in power, and that only a tiny handful (six) abstained, rather than cast a vote either way. Fortified by this lack of political opposition, Mussolini introduced over the next two years a series of laws that abolished what was left of parliamentary democracy in Italy, including his 1926 Decree on Public Safety, which essentially transformed the nation into a police state. The regime criminalized public opposition to Fascism, imposed widespread censorship, issued new restrictions on travel and emigration, and explicitly sanctioned the use of violence to suppress dissent.

By then Volterra had essentially severed his connections with Parliament. The failure of the 1924 no-confidence vote marked one of the last times he ever set foot in the Senate, and he withdrew to the relative security of his science and family life. The Volterra family had lost a mainstay in 1921 when Virginia's shrewd and formidable father, Edoardo Almagià, died, but 1926 brought a fresh face to the family and a new colleague to Volterra. Unlike her mother, who had been privately educated, Luisa Volterra, as an emancipated young woman, had enrolled in the University of Rome (her three younger brothers would ultimately matriculate there as well), where she majored in biology. She fell in love with Umberto D'Ancona, the young marine biologist who was supervising her thesis. They married in 1926, and soon afterward, Luisa's father and her husband embarked on a scientific collaboration that took Volterra, then sixty-six, into entirely new territory.

Although he had devoted part of his 1901 inaugural lecture at Rome to pioneering efforts to apply mathematics to the biological sciences, Volterra had never attempted to carry out such research himself. But when his son-in-law, who had been studying the fish markets between 1914 and 1923 in three Italian cities on the Adriatic—Fiume, Trieste, and Venice—"asked me," Volterra later recalled, "if it were possible to give a mathematical explanation of the results that he was getting in the percentages of the various species in these different periods,"[35] Volterra jumped at the opportunity to apply his beloved differential equations to D'Ancona's problem. His first paper on the topic, "Variations and fluctuations in the number of individuals in coexisting animal species," appeared in 1926 in the *Memorie* of the Lincei, and he would subsequently expand on this initial work in more than thirty papers, books, and articles. His last paper on the subject, "Calculus of variations and the logistic curve," was published in 1939 in *Human Biology*.

Volterra wrote a number of these publications in collaboration with D'Ancona. In the course of his statistical analyses of Adriatic fish markets, the young biologist had come upon a surprising fact: during World War I, when most commercial fishing ceased, the few catches that were made showed a dramatic rise in the number of sharks, ray, and other predator fish, compared to their pre- and postwar ratios. But what was the relationship between armed conflict and the number of fish in the sea? In framing this problem, Volterra posited two coexisting fish species, one of which preys upon the other. The mathematical model that he developed explained D'Ancona's empirical observations in this way: The predators (in this case, the sharks) ate the bluefish (the prey). The lull in commercial fishing during the war increased the sharks' food supply and, in the absence of any significant human catches of sharks, initially allowed more shark offspring than usual to survive. But, as the growing shark population outstripped the number of bluefish available for food, the shark numbers fell, concomitantly allowing the population of bluefish once again to rise to pre-war ratios by the end of the war. Volterra's model could also account for D'Ancona's pre-and post-war percentages: the fisherman, he observed, "by disturbing the natural conditions of proportion of two species, one of which feeds upon the other, causes diminution in the quantity of the species that eats the other, and an increase in the species fed upon."[36] As the historian of science Sharon Kingsland has pointed out, "Volterra considered his analysis to be part of evolutionary biology, an attempt to investigate, along mathematical lines, the day-to-day interactions of organisms as a first step toward a fully mathematical, general theory of evolution."[37]

Only after his 1926 paper appeared did Volterra discover that the Ukrainian-born physical chemist Alfred J. Lotka had already carried out similar predator-prey calculations, modeled on an interacting parasite-host system. Today, the predator-prey model that the two scientists arrived at independently is universally known as the Lotka-Volterra theory, and it can be found in the opening chapters of virtually every textbook on theoretical ecology.

It is hardly fanciful to suppose that as Volterra mathematically modeled "the laws of the struggle for life by a group of species living in the same environment in such a way that some devour others,"[38] his mind was occupied with the increasingly ugly predations of Mussolini's Italy. The previous spring, at a pro-Fascist rally in Bologna, Gentile had floated the idea of creating a Fascist academy of arts and sciences, whose establishment would both celebrate the achievements of the nation's Fascist-leaning artists, scientists, and intellectuals, and, not incidentally, serve as a counterweight to the anti-Fascist tendencies of the Lincei, over which Volterra continued to preside. The University of Rome philosopher did not stop there. With the obvious backing of the Mussolini regime, he used the Bologna meeting as a forum for promulgating "The Manifesto of the Fascist Intellectuals." Its opening statement, a defense of the historical need to overthrow "the demo-socialist" political tradition, made it clear that the battle lines were being drawn. There followed a ringing endorsement of a number of popular Fascist themes: the necessary use of violence by the black-shirted youth gangs, the supremacy of the Fatherland, the glorification of war, and the urgent need to restore greatness to the nation. On April 21, the traditional anniversary of the founding of Rome, the Gentile manifesto, along with the names of several hundred signatories, was published in newspapers across the country.

Ten days later, the Neapolitan philosopher Benedetto Croce, countered with a manifesto of his own. Although he had supported Mussolini in 1922, Croce had apparently undergone a change of heart; his document advocated acceptance of a universal culture, not one confined to a particular political system (Croce would later describe Fascism dismissively as "a parenthesis in history"). He too collected the signatures of several hundred professors, artists, and other intellectuals, including those of Levi-Civita and Castelnuovo. Volterra also signed, although he had made it clear in the past that he was no great admirer of Croce. The fact that the document was not only anti-Fascist but that more than twenty percent of those signing it were Italians of Jewish descent, was not likely to endear either Volterra or his fellow Jewish Italian intellectuals to Mussolini.

It was under these conditions that Volterra, in his capacity as president of the Lincei, delivered an emotion-filled tribute to Italy's scientific community at the Academy's annual convocation in June 1925. Speaking in the presence of the king and queen, who were celebrating their twenty-fifth anniversary on the throne, he invoked World War I as a defining moment in the nation's history in which "the unity of the homeland" transcended nearly all other considerations, including those of the nation's scientists, who had willingly carried out

> patient and obscure work that required the sacrifice of many hours and the deferral of many quests for new truths, in the service of more practical and urgent [matters]. This work was

accomplished with tenacity, courage, and faith. . . .in every place where science had something to offer the war effort— among the hardships and the dangers of the front, on the sea, and in the air, as well as in offices and laboratories. The entirety of this work. . . is a testament to the high level of the contributions made by Italian science.[39]

Was Volterra making one last effort to remind Italy's nominal rulers of how much the nation owed its patriotic scientists, many of whom were now as united in their distaste for Fascism as they had once been in support of their country in wartime? If so, his remarks fell on deaf ears: Levi Civita's Dutch student Struik, who attended the gathering as Levi-Civita's guest, could still remember decades later how bored the sovereign had looked while listening to the speeches, and how afterward the audience had surged forward almost desperately to shake the hands of the royal couple. "I preferred the food and the sherry," he added, "and I was happy to see that Levi-Civita also kept his distance." It was the last public speech Volterra would ever give in Italy.[40]

The regime had needed three years to come up with the idea of an academy to showcase the pantheon of Fascist intellectuals; it needed another three to choose the proper candidates for the new Accademia d'Italia and to secure the site Mussolini had set his heart on—the Palazzo Farnesina, directly across the street from the Palazzo Corsini, home of the Lincei. (The acquisition of the Farnesina marked a highly visible first step in the chain of events leading up to the Lincei's formal annexation by the Academy of Italy in 1939.) Not long after Gentile's pronouncements in Bologna, the regime's campaign to sideline Volterra as Italy's spokesman for science began in earnest. In early 1926, a story appeared in several newspapers announcing Volterra's decision to step down as president of the Lincei, immediately prompting his friend Carlo Somigliana to urge him to reconsider on the grounds that his resignation would harm the Lincei and was certainly "not justified on your part."[41] Volterra assured Somigliana that the stories in the papers were false, but he had in fact drafted, although not yet sent, a letter of resignation. "The news [of my resignation] was not quite correct but certainly suggestive," he wrote to another concerned well-wisher. "But what will happen in the future is very difficult to predict and talk about now." Before he sent the letter, he crossed out "not quite correct" and substituted "premature."[42] In May 1926, he finished out his three-year term of office and tendered his resignation. At the end of the year, Volterra's term as president of the Italy's National Research Council also expired; it too was not renewed, which came as no surprise. Three years later, in February 1929, Italy's 1909 Nobelist, Marconi was officially sworn in, with much pomp and ceremony, as the Council's new president. Eighteen months later, he became president of the Accademia d'Italia as well.

The list of science candidates submitted to Mussolini in 1929 for final approval for the Accademia d'Italia included Fermi and the mathematician Federigo Enriques. Like most of the other mathematicians of Volterra's circle, Enriques was Jewish; in politics, he went his own way and would ultimately join the National Fascist Party in 1933. (Fermi, then twenty-eight, joined the Party nine days after he was nominated for Academy membership.) At the last minute, the government deleted Enriques's name from the list of candidates and substituted the name of another mathematician, Francesco Severi, in its place. The most obvious, and probably the likeliest, explanation for the change is that Mussolini's regime did not want Jews in its Academy, and in fact none were ever admitted. It is also possible that some of the scientific circles to which Italian Jews, especially mathematicians, belonged, were so clearly identified with the old intellectual order in Italy that, in the case of Enriques at least, Mussolini felt compelled to seek out a candidate who had fewer ties with the past.

Like Castelnuovo and Enriques, Severi was an internationally known algebraic geometer, but he had never been part of Volterra's circle. He did not share the group's breadth of scientific interests, and he cared little for history or their other intellectual pursuits. In private, he had frequently lashed out at Castelnuovo and Enriques, complaining that their work was "overrated" and his own "underrated." His political evolution from Socialist to Fascist paralleled Mussolini's (following the Second World War, he would become a devout Catholic). After Mussolini became dictator, Severi regularly sent him reprints of his articles, richly annotated. He had bucked the regime in signing Croce's manifesto rather than Gentile's, but this was a detail the government found it possible to overlook. Overall he made the most of his opportunities; and his appointment to the Accademia d'Italia made it clear that in the eyes of Italy's new rulers, Vito Volterra was no longer titular head of the Italian school of mathematics.

Volterra spent the winter months of 1929 in Paris, lecturing on the mathematical theory of biological fluctuations at the Institut Henri Poincaré; the lectures would later be published in the series "Cahiers scientifiques," under the direction of the French mathematician Gaston Julia.[43] By then, he had spoken on the subject often enough to know that the audience would be entertained by an early allusion to the coefficients of greediness exhibited by the predatory fish, and would then settle back in their seats and listen with pleasure and interest to the rest of the talk. Back in Rome, the newspaper *Il Messaggero* carried a short announcement of Volterra's inaugural lecture at the Institut. It had probably escaped the attention of the Fascist censors, "who people say are practicing [control] over anything that could cast a favorable light on the famous signers [of the Croce petition]," Levi-Civita wrote him in February.[44]

Unbeknownst to him, Volterra had already caught the authorities' attention some months earlier when he visited Rome's police headquarters to apply for a new passport. The resulting paperwork, which he never saw, surfaced in the records long after World War II, and it provides a fascinating look into the idiosyncrasies of Mussolini's bureaucracy. At least three different officials weighed in on Volterra's request: the first duly noted that the petitioner was "a liberal opponent"; the second helpfully jotted in the margin that he had been "among the signers of the Croce manifesto"; and the third found it prudent to add that Senator Volterra belonged to the group of "opposing Senators." Ultimately the document landed on the desk of *Il Duce* himself, and he gave the application his personal imprimatur, appending "Si" with a flourish and boldly scrawling "M" across the middle of the page. Quite possibly, Mussolini was hoping that this celebrated and tenacious thorn in his side would leave Italy for good—a vain hope as it turned out.[45]

Volterra was once again at liberty to travel, but as he quickly discovered, being a Senator did not protect him from being embarrassed and harassed by the regime. While the government could not easily strip him of his title as Senator (according to parliamentary rules still in effect, senators could neither be arrested without authorization of the House to which they belonged, nor removed from office during their lifetimes), it did find other ways to hound him. In the spring of 1930, officials in the Ministry of the Interior summoned Volterra to a meeting; when he arrived, accompanied by one of his sons, he immediately took the position that members of the Italian Senate could only be questioned by their peers, that is, other senators. Brushing aside these protests, the ministry official handed Volterra a copy of the Italian-language anti-Fascist magazine, *Libertà*, and told him that he was suspected of having provided information for one of its articles during his previous year in Paris. Volterra was excused after emphatically denying that he had done any such thing, but the episode left a bitter taste in his mouth. Later that year, he and Virginia were traveling by train from Rome to Paris when Italian border guards ordered the couple to leave their sleeping compartment and disembark in the middle of the night, on the pretext that Volterra's passport had expired. Several hours later, they were permitted to resume their journey, but on a much slower train, with no sleeping accommodations.

The regime did not find it so easy to marginalize Volterra and his like-minded colleagues within the fraternity of Italy's leading scientists. In 1931, the Accademia d'Italia announced its intention to award the first Mussolini Prize in science. Fermi, a member of the awards commission, urged that the award be given to a mathematician, inasmuch as Italy's scientific credentials rested in part on its achievements in the field of mathematics. The three he named for the committee's consideration were Volterra, Levi-Civita, and Guido Fubini, a specialist in analysis at Turin. Not surprisingly, Fermi's suggestions came to nothing; somewhat more surprisingly, the next name put forward was also that of a Jew, and a militant anti-Fascist to boot—Giuseppe Levi, a professor of human anatomy at Turin. After two ballots, the committee approved his name and asked Filippo Bottazzi, the Neapolitan physiologist who had nominated him, to prepare a report on Levi's work. At this point, the government stepped in; Bottazzi's report was filed, but it was never read in the Academy's general assembly. The regime was determined not to award the Mussolini Prize to a scientist who, in its words, was both "Jewish and [a] signer of the noted manifesto, so-called, of the intellectuals" when it had at hand an "old black shirt"— the Himalayan explorer Filippo de Filippi, who had conveniently come in second to Levi in the committee vote. The die was cast in April 1931, when Mussolini gave the order, "either De Filippi or nobody" (a comment he wrote in blue pencil on the margin of an April 19 memorandum). The Academy's general assembly approved de Filippi's name almost immediately, and Mussolini presented him with the prize two days later.[46]

In November, a few months after flexing its muscle over the Mussolini Prize, the government announced that all of Italy's university professors would be required to sign a loyalty oath professing allegiance and devotion to king, country, and the Fascist regime. Alone among his colleagues in mathematics throughout Italy, Volterra refused to sign, a decision that brought to an end a university career of nearly fifty years. Back in Paris the following winter, out of reach of the Italian censors, he sent a long letter to his former postdoctoral student Griffith Evans, now a world, and seemingly a lifetime, away at Rice Institute in Houston, Texas. Volterra wrote:

> I am now unemployed in Italy because the Fascist government dismissed me from the position as professor that I carried out for half a century, because for my principles and my conscience I refused to take the oath of loyalty to the Fascist regime. Such an oath would have moreover limited my political liberty as Senator that was guaranteed to me by the statute of the kingdom. In short, today I am no longer a member of the Italian university.[47]

Out of Italy's 1,250 university professors, twelve ultimately refused to sign the oath and left their teaching posts. Of the dozen dissenters, two were scientists—Giorgio Errera, a chemist at the University of Pavia, and Volterra. His colleagues Levi-Civita, Castelnuovo, and Enriques all signed it; refusing would have meant the loss of their jobs and their livelihoods. "If there had been hundreds of refusals... it would have carried weight with public opinion and abroad," one historian of the period has noted.[48] Three years later, when the loyalty oath was extended to all the members of academies and other learned institutions across the nation, ten academicians refused to sign, one of whom, again, was Volterra. A member of just about every Italian scientific society when Mussolini became dictator in 1925, he belonged, ten years later, to none. In 1936, the Vatican's Academy of Sciences, which had expanded in size and scope, made him a member.

Deprived of his university position and stripped of nearly every other professional and academic affiliation, Volterra in his last years became *persona non grata* in his own country. Among many other petty humiliations, mail addressed to him at the Lincei after his dismissal was pointedly returned to senders, not forwarded to his permanent residence in Rome. "A famous Italian Academy," it was said nearly thirty years later, "foreswore and pretended to ignore this eminent Italian, systematically returning the scientific papers that scholars from every part of the world were sending to his old address at this same Academy, always with the same stupid and ridiculous motivation dictated by a servile bureaucratic insensitivity: 'Returned to the sender: the addressee is unknown.'"[49]

One of the few relatively unchanged aspects of Volterra's life was his freedom to travel abroad (the regime may still have cherished the notion that he would leave and not come back). He spent the better part of the next several years in a kind of voluntary exile, working for the most part in Paris, with trips and short stays elsewhere in Europe. In 1936, he and his younger French colleague Joseph Pérès issued the first volume of their joint work, *Théorie générale des fonctionelles*. The following year, he and Virginia visited Egypt, in the company of the conductor Arturo Toscanini and his wife, whom the Volterras had met by chance in Alexandria. He and his wife became virtually inseparable during these years; Virginia, in the words of one family intimate, "created a sort of niche for her husband, so that he had only to eat, drink, sleep, and go out, because he was very active... you could [almost] say that those two people were one person."[50]

Although Mussolini was undeniably a bully and a brute, and his regime was not, to say the least, particularly well-disposed toward Italian Jews, they were generally left alone as long as they did not actively oppose the dictatorship, and, in fact, more than a few were fairly prominent Fascists. By the late 1930s, however, Mussolini and Hitler had signed a number of "friendship treaties," and Italy had come under increasing pressure to follow Germany's lead in introducing anti-Jewish legislation. Overt Anti-Semitism became an official component of the Fascist state in July 1938, when the regime issued its *Manifesto of Italian Racism*, which declared, in part, that "the Jews do not belong to the Italian Race."[51] The new laws had no sooner been issued than Enriques, Levi-Civita, and their younger Jewish colleagues were dismissed from their university posts, from the Lincei, and from the host of other Italian learned societies to which they belonged (Castelnuovo had retired from university life in 1935). At a stroke, the prolific and vigorous mathematical and mathematical physics community that Volterra had labored for more than half his life to build in Italy was demolished. Even though his refusal to sign the loyalty oaths had already cost him his academy memberships, the vice-president of the Lombardy Institute of Science and Letters nevertheless felt compelled to send him a curt one-sentence letter, informing him that he could no longer take part in the Institute's activities because "you belong to a non-Aryan race."[52]

The racial laws, which mandated that Jews could not attend public schools or universities, serve in Italy's armed forces, work as journalists, teachers, or notaries, use libraries, or own more than a limited amount of land, hit Italy's entire scientific community hard. By the 1930s, Italian Jews held seven percent of the appointments in the state university system, where they were represented far out of proportion to their overall numbers—there were at the time about 40,000 Jews in Italy out of a total population of forty million. The story goes that Luigi Fantappiè, who had been one of Volterra's better students in the 1920s, showed up on Volterra's doorstep that summer, gloating over the regime's anti-Semitic legislation. "How is it possible," Volterra is reported to have said later, "that I did not have the presence of mind to throw him down the stairs?"[53]

For the ailing, embattled mathematician, the racial laws came almost as a deathblow; he would never have believed, he told more than one associate, that the appalling anti-Semitism of the past would rear its ugly head again in his lifetime. His health, which had begun to decline in the wake of the loyalty-oath crisis, now went rapidly downhill. By the time he and Virginia left Paris for the last time in the summer of 1938, it was clear he would never again leave Italy. His devotion to his work and his mental acuity never faltered; in May 1940, the month he turned eighty, he put the finishing touches on a paper dealing with the equations that describe the history of elasticity in a body—research that had been a leitmotif of his career. "I don't need to tell you of the clarity of his thinking," Virginia wrote to a family member, before adding, a few lines down, "However, in these last days...he is very depressed."[54] He died less than six months later and was buried at the top of a knoll in the local cemetery in Ariccia, not far from his beloved country house.

His headstone, engraved with the name "Vito Volterra," bears no epitaph although Volterra had in effect produced his own a few years earlier, along with his final judgment on Mussolini's Italy. A well-wisher had asked him to autograph a photograph that, in the fashion of the day, had been made into a postcard. "Empires die," Volterra wrote, "but Euclid's theorems keep their youth forever."

In the end it fell to Guido Castelnuovo to rescue the ideals of Volterra's circle from the rubble of Italy's ill-fated pact with Nazi Germany. One of the first tasks he tackled at the request of the post-war government was the restoration of the Accademia Nazionale dei Lincei. (The Lincei of his generation had ceased to exist in 1939, following its formal annexation by the Accademia d'Italia.) On October 17, 1946, standing before the entire membership of the reestablished Lincei, Castelnuovo, in his capacity as the reconstituted organization's president, spoke at length about his late colleague, saying in part:

> Any sign of weakness on his part, a minimum gesture of consent toward the party in power would have returned to him

his influence and honor and would have placed him again in the mainstream of Italian science. But he did not bend...he preferred...remaining faithful to the ideals that had inspired his entire life, the ideal of liberty, of respect for others' opinions, of love between individuals and among peoples. So closed in October 1940, amidst the indifference of the ruling classes and the silence of our press, the life of one of the greatest sons of Italy, whom we don't know whether to admire more for the breadth of his creativeness or the nobility of his character.[55]

Samuel Giuseppe Vito Volterra, born in the year of Italian unification and the liberation of Italy's last Jewish ghettoes, died on October 11, 1940, at his residence at Via in Lucina 17, thirteen months after Germany's army, the *Wermacht*, crossed the border into Poland and ushered in World War II.

Epilogue

Vito Volterra lived just long enough to witness the wreckage of the brave new world into which he had been born in 1860. He did not live to see a new era rise from the ashes of the old—the restoration of human freedoms and parliamentary democracy in much of Western Europe; the new prominence accorded science and technology in the aftermath of the Manhattan Project and the onset of the Space Age and the Cold War; and the post-war revitalization of the science and mathematics community in Italy. Nearly all of his friends and family survived the war and the horrors of the Nazi occupation of Rome and northern Italy, including Naples, in 1943. Italians by and large had no stomach for Hitler's Final Solution, and managed to thwart the Germans' efforts to round up and deport Jews in occupied Italy through a combination of non-cooperation, obstruction, and outright sabotage. It is estimated that eighty to ninety percent of the country's Jews (including several thousand who sought safety there from elsewhere in occupied Europe) survived the Holocaust.

A few of Volterra's more senior associates did not live long enough to experience the rise of Fascism. Antonio Ròiti, who had been the young Vito's mentor, and professor, and a trusted confidant of both mother and son throughout his life, died in 1921, the same year as Virginia's father, Edoardo Almagià. On December 18, 1921, Volterra opened a meeting of the Lincei's class of physical, mathematical, and natural sciences with an announcement of Ròiti's death and—in accordance with Ròiti's explicit instructions that there be no commemoration or last honors—adjourned the session as a gesture of mourning.

After her husband's death, Virginia's mother, Eleonora, lived out the rest of her life in Rome and passed away in 1932. Tullio Levi-Civita, the dean of relativity theory in Italy, died in 1941, a year after Volterra. Enrico Fermi won the Nobel Prize in physics in 1938, went to Stockholm to collect it, and with his wife, Laura, whose family was Jewish, continued on to the United States, where he would spend the rest of his life. In 1942, he oversaw the first controlled nuclear chain reaction in an atomic pile he had constructed under the stadium at the University of Chicago, and soon after he became one of the leaders of the Manhattan Project to build the atomic bomb.

Italy had declared war on the United States in December 1941, three days after the Japanese attack on Pearl Harbor. After the Allies liberated

Sicily with only token opposition in August 1943, the rest of the country
erupted in political turmoil, culminating in the overthrow of Mussolini, the
dissolution of the Fascist Party, and the establishment of a new government
under Marshal Pietro Badoglio. In September, the new government surrendered to the Allies; the Germans retaliated by occupying Italy as far south
as Naples. The Allies entered Rome in the summer of 1944. The following spring, Mussolini and his mistress were captured and shot by partisans.
Several weeks later, the German army in Italy gave up the fight, and the
entire peninsula was liberated.

After the war, Francesco Severi, whom the Fascists had installed as a
kind of puppet doyen of Italian mathematics, found himself under investigation by the "Reconstitution Commission," which had been established in
1944 to restore the functions of the Lincei. The commission's specific mandate was to look into allegations of wartime collaboration against members
who had been accused of taking an active part in Fascist political life or who
had remained loyal to Mussolini after he was deposed in September 1943.
Severi was subsequently absolved of any criminal activity in a report that
concluded that he "had not received from fascism anything more" than was
his due as a distinguished scientist.* His "moral rectitude," the commission added, was never called into question. Some mathematicians remained
skeptical, nevertheless.

As for other Jewish members of the Volterra circle, Federigo Enriques
and Guido Castelnuovo lived through the Nazi occupation of Rome, changing hiding places frequently. In 1941 Castelnuovo organized a clandestine
university for Jewish students in Rome, operated in conjunction with the
Fribourg Institut Technique Supérieur, a private school, in Switzerland. He
and Enriques taught there with a number of other professors, both Jewish
and non-Jewish, until the occupation forced the school's closure in the fall
of 1943.

After the liberation of Rome, Castelnuovo, then seventy nine, was called
out of retirement to breathe new life into Italy's pre-Fascist scientific organizations. Appointed general commissioner of Italy's National Research
Council and president of its mathematics committee, he played a major role
in reviving and reconstituting both organizations and contributed also to the
rebirth of the Lincei, whose president he became in April 1945. He held the
post until his death seven years later, and was named a Senator for Life in
Italy's Parliament in 1949. The Institute of Mathematics at the University
of Rome was named for him in 1953.

Volterra's first cousin Roberto Almagià, the geographer who had been an
enthusiastic proponent of Italian nationalism in the years before World War
I, rose to become the head of the Italian school of geographers in the interwar
years. In 1938, he became a guest of the Vatican library and published

*Francesco Severi to Beniamino Segre, October 15, 1945, Beniamino Segre Papers,
private collection.

under the pseudonym of Bernardo Varenio; after the war he resumed his academic career at the University of Rome. He headed the National Research Council's committee for geography, geology, and oceanography from 1945 until his death, in Rome, in 1962.

Virginia Volterra remained in Rome after her husband's death. When the racial laws prohibited Italian Jews from having servants, she had asked the family's librarian, Gualda Caputo Massimi, to come to the house less frequently, fearing that the police, who kept surveillance on the palazzo, would call Mrs. Massimi down to police headquarters for questioning. In 1943, after the Germans occupied Rome and began rounding up the city's Jews for deportation, Massimi took Virginia into her home and concealed her there for more than a month. Later, her son Edoardo, who had joined the resistance movement in Rome, came and found a secure place for Virginia with a group of nuns in Via Vicenza, close to the city's main train station.

After the war, Volterra's four children resumed their professional careers in the law and the sciences; all are now deceased. (It might have pleased Volterra to know that a number of his direct descendents now reside in the United States.) Their mother lived out her days in Rome and died in 1968 at the age of ninety-three. Virginia Volterra never lost her lively spirit or her streak of independence. The story goes that after the war, she refused to greet or even acknowledge individuals like Severi, who had ostracized her husband after the racial laws were passed. With magnificent disdain Virginia would behave as if these former colleagues of her husband did not exist. She never offered any explanation for her behavior. Those who remembered the old days understood.

APPENDIX A

Sir Edmund Whittaker, "Vito Volterra, 1860-1940"

"Vito Volterra" by Sir Edmund Whittaker is reprinted with permission of the publisher, The Royal Society, from *Obituary Notices of Fellows of the Royal Society of London*, 3 (1941), 691-729.

VITO VOLTERRA

1860–1940

VITO VOLTERRA was born at Ancona on 3 May 1860, the only child of Abramo Volterra and Angelica Almagià. When he was three months old the town was besieged by the Italian army and the infant had a narrow escape from death, his cradle being actually destroyed by a bomb which fell near it.

When he was barely two years old his father died, leaving the mother, now almost penniless, to the care of her brother Alfonso Almagià, an employee of the Banca Nazionale, who took his sister into his house and was like a father to her child. They lived for some time in Terni, then in Turin, and after that in Florence, where Vito passed the greater part of his youth and came to regard himself as a Florentine.

At the age of eleven he began to study Bertrand's *Arithmetic* and Legendre's *Geometry*, and from this time on his inclination to mathematics and physics became very pronounced. At thirteen, after reading Jules Verne's scientific novel *Around the Moon*, he tried to solve the problem of determining the trajectory of a projectile in the combined gravitational field of the earth and moon: this is essentially the 'restricted Problem of Three Bodies', and has been the subject of extensive memoirs by eminent mathematicians both before and after the youthful Volterra's effort: his method was to partition the time into short intervals, in each of which the force could be regarded as constant, so that the trajectory was obtained as a succession of small parabolic arcs. Forty years later, in 1912, he demonstrated this solution in a course of lectures given at the Sorbonne.

When fourteen he plunged alone, without a teacher, into Joseph Bertrand's *Calcul différentiel*: he does not seem to have had access to any work on the integral calculus at this time,

691

and when he attacked various special problems relating to centres of gravity, he discovered for himself that they could be solved by means of an operation (integration) which was the inverse of differentiation.

His family, whose means were slender, wished him to take up a commercial career; while Vito insisted on his desire to become a man of science. The struggle between vocation and practical necessity became very acute: and the family applied to a distant cousin who had succeeded in the world, to persuade the boy to accept their views. This man, Edoardo Almagià, who died at the age of eighty in 1921, was one of the most celebrated civil engineers and financiers in Italy in the latter part of the nineteenth century : as a contractor of public works he constructed many important railways and harbours at home and abroad, including the harbours of Alexandria and Port Said in Egypt: the proprietor of vast estates in Tuscany and the Marches, he was renowned for his charities: and it was in the course of excavations at his palace in the Corso Umberto—once the Palazzo Fiano-Ottoboni—that the discovery was made of the sculptures of the Ara Pacis of the Emperor Augustus, which are now among the treasures of the Museo delle Terme in Rome.

His interview with his young relative turned out differently from what the family had expected. Impressed by the boy's sincerity, determination and ability, the older man threw his influence on the side of science, and turned the scale. Professor Roiti offered a nomination as assistant in the Physical Laboratory of the University of Florence, though Vito had not yet begun his studies there: it was accepted, and now the die was cast. The young aspirant entered the Faculty of Natural Sciences at Florence, following the courses in Mineralogy and Geology as well as in Mathematics and Physics. In 1878 he proceeded to the University of Pisa, where he attended the lectures of Dini, Betti and Padova: in 1880 he was admitted to the Scuola Normale Superiore, where he remained for three years: and here, while still a student, he wrote his first original papers. Under the influence of Dini he had become interested in the theory of aggregates and the functions of a real variable, and he gave some examples [3] [1] which showed the inadequacy, under certain circumstances, of Riemann's theory of integration, and adumbrated the developments made long afterwards by Lebesgue.

In 1882 he graduated Doctor of Physics at Pisa, offering a thesis on hydrodynamics in which certain results, actually found earlier by Stokes, were rediscovered independently. Betti at once nominated him as his assistant. In 1883, when only twenty-three years of age, he was promoted to a full professorship of Mechanics in the University of Pisa, which after the death of Betti was exchanged for the Chair of Mathematical Physics. He now set up house in Pisa with his mother, who up to that time had continued to live with her brother. In 1888 he was elected a non-resident member of the Accademia dei Lincei: in 1892 he became professor of Mechanics in the University of Turin, and in 1900 he was called to the Chair of Mathematical Physics in Rome, as the successor of Eugenio Beltrami. In July of that year he married Virginia Almagià, one of the daughters of the distinguished relative who had first made it possible for him to follow a scientific career. She had inherited intellectual brilliance from her father, and great beauty from her mother, and as the wife of Vito Volterra took upon herself all the cares which might have distracted her husband from his scientific work, undertaking the education of their children and the administration of all their possessions. Six children were born of the union, of whom four now survive. His mother still lived with them, and died at the age of eighty at the Palazzo Almagià in March 1916.

We must now proceed to an account of Volterra's scientific work. Instead of considering the individual papers one by one in chronological order, we shall group them according to subjects: and shall consider first those relating to functionals.

A *functional* may be introduced as a generalization of the idea of a function y of several independent variables $\phi_1, \phi_2, \ldots, \phi_n$, say $y(\phi_1, \phi_2, \ldots, \phi_n)$. Let us suppose that the set of variables, $\phi_1, \phi_2, \ldots, \phi_n$, is modified from being a finite set to being an

---

[1] The numbers in square brackets refer to the bibliography at the end of this notice.

44

enumerably-infinite set and finally to being a continuous set. To represent this analytically, we can regard ϕ_x as a function of its suffix x: then the functional y is a function of all the values that the function $\phi(x)$ takes when x lies in some interval $A \angle x \angle B$. The function $\phi(x)$ is arbitrary, and is, so to speak, the independent variable of which the functional y is a function. This definition may readily be extended to a functional y depending on several functions $\phi_1(x), \phi_2(x), \ldots$ and moreover on certain parameters $t_1, t_2, \ldots, y$ being a function in the ordinary sense of the t's. We may also introduce functions of several variables $\phi(x_1, x_2, \ldots)$ in place of the functions $\phi_r(x)$.

The transition from ordinary functions to functionals corresponds exactly to the transition from the theory of maxima and minima of functions of several variables, to the calculus of variations: and indeed the integrals, which in the calculus of variations are to be made maxima or minima by choosing the functions involved in them in a certain way, constitute a familiar and important example of functionals.

Volterra seems to have conceived the idea of creating a general theory of the functions which depend on a continuous set of values of another function, as early as 1883: but his first published work on the subject [17] did not appear until 1887. The name *functional* was introduced later by Hadamard and has now replaced Volterra's original nomenclature.

The first step in the theory must evidently be to extend to functionals the well-known fundamental concepts of the theory of functions: the continuity of a functional is first defined, and the derivative and the differential have their analogues. The partial derivative with respect to a particular variable passes into the derivative of the functional y with respect to ϕ at a certain point of the interval of definition of ϕ, say $x = \xi$. To the total differential, which is a linear form in the differentials of the independent variables with the partial derivatives as coefficients, there corresponds the total variation of the functional, which is an integral over the variation of the independent variables $\phi(\xi)$ with the derivatives of the functional y with respect to ϕ at the

point ξ as coefficients. By repeated application of these operations, higher differential coefficients and higher variations are
easily defined. The higher differential coefficients with respect
to ϕ at the points $\xi_1, \xi_2, \ldots$ are shown to be symmetrical
at these points (corresponding to $\dfrac{\partial^2 f(x, y)}{\partial x \partial y} = \dfrac{\partial^2 f(x, y)}{\partial y \partial x}$). In
some cases, however, a slightly generalized definition of the
variation is necessary in which, besides the integral, there appears
a finite or enumerably-infinite number of terms linear in the
variation of ϕ and the derivatives of ϕ at certain exceptional
points: the latter is more often than not the case in the Calculus
of Variations, where the exceptional points are usually the limits
of integration.

In two other notes of the same year, the concept of *functions
of a line* was introduced [18]. Let a closed curved line L in
space of n dimensions be specified by equations $x_i = \phi_i(t)$
$(i = 1, 2, \ldots, n)$ and suppose that to every such line L there
corresponds a definite value of a quantity y: then y is called a
function of the line L. Obviously y is a functional of the ϕ's:
but it is not the most general type of functional, since it is
invariant when the ϕ's are replaced by other functions ψ obtained
from them by a change of parameter $t = t(u)$, $\psi_i(u) = \phi_i(t)$:
y, in fact, depends on the line L but not on its mode of parametric
representation. Volterra defined the derivative of a function of a
line with respect to the line at a certain point of it, and then
defined the variation. He next introduced the idea of a *simple
function of a line*: let L_1 and L_2 be two contours having in common
an arc which is traversed in opposite directions in the two circuits,
and let $L_1 + L_2$ denote the contour obtained by deleting this
common arc: then a *simple function of a line* is defined to be a
function having the property $\phi(L_1 + L_2) = \phi(L_1) + \phi(L_2)$. Volterra
established several important theorems regarding these simple
functions.

A remarkable application of simple functions of a line is contained in a series of papers, the first of which appeared in the
same year [19]. Let L be a line, F and Φ two functions of this

line, $L + \Delta L$ a line which is identical with L except in the neighbourhood of a fixed point M, and let $F + \Delta F = F(L + \Delta L)$, $\Phi + \Delta \Phi = \Phi(L + \Delta L)$. Now let the deformation ΔL tend to zero, so ΔF and $\Delta \Phi$ tend to zero. If $\Delta F / \Delta \Phi$ tends to a limit which depends only on M and is independent of the sequence of diminishing deformations ΔL, then the functions F and Φ were said by Volterra to be *connected with each other in Riemann's sense*: he suggested that this corresponds to the relation between two complex variables z and w for which dw/dz is independent of the way in which the limit is approached, depending merely on the value of z. He developed the theory of these functions, showing that it depends on certain partial differential equations, which correspond to the Cauchy-Riemann equations.

In two more notes [21] he gave a theory of the differentiation and integration of connected functions of a line, defining first the 'connexion' between an ordinary function and a function of a line: the limit $d\Phi/dF$ introduced above is shown to be an ordinary function 'connected' with both F and Φ. If f (an ordinary function) and F (a function of a line) are 'connected' and without singularities inside a closed surface σ, then $\int_\sigma f \, dF = 0$: this corresponds to Cauchy's integral theorem in the theory of analytic functions. Morera's converse of Cauchy's theorem can also be extended to 'connected' functions of a line. Integration and differentiation as introduced in these papers are inverse operations. The theory has some connexions with the theory of analytic functions of two variables.

The whole theory of this generalization of the functions of a complex variable was systematically presented in a considerable memoir [25] in 1889: and at the same time in a series of notes [26, 27, 28] the idea of functions of a line was extended by considering, instead of a line, any sub-space S_r in a space S_n of any number n of dimensions: in particular, the notion of 'conjugate functions' (depending on an S_{r-1} and S_{n-r-1} respectively) was developed. Differential parameters, corresponding to the ∇ and Δ of ordinary theory, were introduced: as in ordinary

potential theory, the vanishing of the second differential parameter is a necessary condition for the existence of a conjugate
function (the second differential parameter of which also vanishes).

In the following year Volterra showed [30] that by means of
his functional calculus the Hamilton-Jacobi theory of the integration of the differential equations of dynamics might be
extended to general problems of mathematical physics. The idea
was, that whereas the equations of dynamics arise from variational problems relating to simple integrals, the equations of
physics arise from variational problems relating to multiple
integrals, which must be regarded as functionals of the boundary
of the field of integration.

After this, some years elapsed before the work on functionals
was continued. In 1892–1894 he published a number of papers
[35, 36, 39, 40, 41, 42, 44] on the Partial Differential Equations
of Mathematical Physics, especially the equation of cylindrical
waves

$$\frac{\partial^2 u}{\partial t^2} = \frac{\partial^2 u}{\partial x^2} + \frac{\partial^2 u}{\partial y^2}.$$

For this equation he obtained an expression of the solution in
terms of the initial values of u and $\partial u/\partial t$, which may be regarded
as the analogue in two dimensions of the Riemann-Green
formula for the propagation of waves in one dimension. He also
enquired how Kirchhoff's well-known expression of Huygens'
Principle in the wave-theory of light could be extended to space
of two dimensions, or of more than three dimensions: and he
obtained a formula which gives the value of a cylindrical wave
function at a point at the instant t, in terms of the disturbance
at points Q of a given curve at the instant $t - PQ/c$ and all
previous instants.[2] This formula was used later by Sommerfeld
in his work on the diffraction of X-rays.

It was after this that Volterra began his celebrated researches
in the theory of Integral Equations. He had first met with an
integral equation in a paper [11] of 1884, dealing with the

---

[2] These results are brought into relation with more recent work by Hadamard,
Marcel Riesz and others, in Baker and Copson, *Huygens' Principle* (Oxford, 1939).

distribution of electric charge on a segment of a sphere: the problem, as he showed, depends on the solution of what would to-day be called an integral equation of the first kind with a symmetric nucleus. It was, however, not until 1896 that he seriously took up work in this field, applying his theory of functionals to what was at that time called the 'inversion of definite integrals', and obtaining results [56, 57, 58] which were greatly admired. In these papers, he regarded the integral equation of the second kind with a variable limit of integration

$$\phi(y) = f(y) + \int_a^y f(x)\, S(x, y)\,dx$$

(where f is the unknown function), which is now generally called an *integral equation of Volterra's type*, as the limiting case of a system of linear algebraic equations; the nth equation of this system contains only the first n unknown quantities, so the system can be solved by recurrence. From the solution of the algebraic system, he passed to the corresponding solution of the integral equation, obtaining a formula

$$f(y) = \phi(y) + \int_a^y \phi(x)\, T(x, y)\,dx$$

where $T(x, y)$ is a function—the *resolvent nucleus*, as it would be called to-day—which can be constructed by a simple process from the given function $S(x, y)$. Unlike later writers—Fredholm, Hilbert, Schmidt—Volterra used the analogy with the linear algebraic equations only heuristically, the final results being proved independently.

In order to deal with integral equations of the first kind

$$\theta(y) - \theta(a) = \int_a^y f(x)\, H(x, y)\,dx$$

(where f is the unknown function), he differentiated this equation with respect to y, thus obtaining an integral equation of the second kind which could be solved by the method already found. A difficulty arises when $H(y, y)$ vanishes or becomes infinite at certain points: he discussed certain types of these 'singular' nuclei, and mastered them.

All these investigations were afterwards extended to the case of multiple integrals, and also to simultaneous systems of integral equations, involving several unknown functions. In 1897 he showed [65] that his method is applicable to integral equations with both limits variable: the range of integration actually considered was $ay \leq x \leq y$ where $-1 \leq a < 1$.

In a lecture [68] on the oscillations of liquids under the influence of gravitational forces (the problem of seiches), he advocated the application of infinite determinants to the theory of integral equations—a method which became of great importance later in the work of Fredholm.

In a paper written on the occasion of the centenary of Abel's birth [86], he applied his theory of 'the inversion of a definite integral' to a problem of stability. After pointing out that Abel was the first to solve an integral equation of the first kind of Volterra's type (namely, in the theory of the tautochrone), he discussed the problem of the stability of a fluid mass rotating about one of its principal axes and consisting of concentric, similar, and similarly situated layers: this configuration he proved to be unstable. A simpler proof, independent of the theory of integral equations, was also given: but it was pointed out that this latter proof is less general, as regards the conditions to be satisfied by the function representing the density of the liquid.

Meanwhile, a wide interest in integral equations had been awakened by Fredholm's theory, which was published in Swedish in 1900 and in French (in the *Acta Mathematica*) in 1903. Volterra [101] pointed out the connexion between Fredholm's theory and some problems of his own theory of functionals: the solution of an integral equation is, indeed, a simple case of the solution of a functional equation. In the same paper, and also in the seventh of his Stockholm lectures of the same year, he discussed certain transcendental integral equations originating in the 'Taylor expansion' of the theory of functionals.

To this period of his life belong some celebrated researches in the theory of elasticity, which were important not only on their own account but also because they suggested much of his subsequent work in Pure Mathematics. Perhaps the most notable of these was the theory of what he called *distorsioni*, a term for which Love introduced the English rendering *dislocations*. In elastic solids which occupy a multiply-connected region of space, the displacements may be many-valued functions, corresponding to deformations for which certain fundamental results of the ordinary theory of elasticity are untrue. As a simple example, suppose that a thin slice of material is cut out of an anchor-ring and the new surfaces thus formed are brought into contact and, after having been twisted, joined together. There is then an initial stress even in the absence of all external forces, and there are certain discontinuities in the displacements, although stress and strain are continuous. The joint may in fact be regarded either as a seat of discontinuities in the displacement, or else as a barrier (a branch-cut) for the many-valued functions which represent the components of the displacement.

Such systems had been introduced earlier, notably in Larmor's attempt to explain electrons as places of intrinsic strain in the aether: but Volterra was the first to develop in 1905–1906 [**100, 103**] a fairly general theory of these 'dislocations'. A comprehensive account of it, with some improvements by E. Cesàro, was published in 1907 [**107**]. He first determined the many-valued displacement in a multiply-connected region, corresponding to a given one-valued strain: by aid of the formulae thus obtained, he was able to discuss the type of discontinuity at the barrier: this he proved to be of the type of displacement of a *rigid* body. In particular, he discussed in some detail the possible displacements in a hollow cylinder, and also in a system of thin rods: in the former case he was able to compare his deductions with the results of experiment. Once during the war, when he was on a mission to England to discuss scientific questions of common interest to the Allies, he returned after a tiring day to the College where he was a guest, and found that his kind hosts had placed round the walls of his room a number of models of cylinders subjected to Volterra dislocations. He was deeply touched, and often recalled this incident in later life.

His work in elasticity was the origin of his theory of integro-differential equations, i.e. integral equations between the unknown function and its (partial) derivatives. In 1909 [**114**] he studied a particular type of such equations, and showed that the solution of this integro-differential equation was equivalent to the solution of a simultaneous system consisting of three linear integral equations and a partial differential equation of the second order.

Integro-differential equations occur in various branches of mathematical physics. Thus, for certain substances, the electric or magnetic polarization depends not only on the electro-magnetic field at the moment, but also on the history of the electromagnetic state of the matter at all previous instants (*hysteresis*). When the terms corresponding to this physical fact are introduced into the fundamental equations, these become integro-differential equations [**115**].

A similar situation is found in 'hereditary elasticity' (as Picard called it), to which two of Volterra's notes of the same year are devoted [**119, 120**]. He assumes 'linear heredity', i.e. that the strain is a linear functional of the stress; in this case the fundamental equations are systems of linear integro-differential equations, and he showed that the strain in a definite interval of time can be determined, provided that the forces in the body and the stress or strain on its surface are known for this time-interval.

In 1910 Volterra introduced into the theory of functionals the fruitful notions of *composition* and of *permutable functions* [**128**]. The *composition* of two functions $F(x, y)$ and $\Phi(x, y)$ is defined to be the formation of the integral

$$\int F(x, \xi)\Phi(\xi, y)d\xi$$

which is denoted by $F*\Phi$, the composition being said to be of the *first kind* if the limits of integration are x and y, and of the *second kind* if the limits are constants a and b: these two cases evidently correspond to Volterra's and Fredholm's integral equations respectively. Two functions F and Φ are said to be *permutable* if their composition is commutative. In the first note, he introduced permutability of the first kind, and in particular investgitaed functions Φ which are permutable (of the first kind) with a given function F. He transformed the defining equation $F*\phi=\phi*F$ into an integro-differential equation and by solving the latter showed that every function permutable (of the first kind) with F can be represented by a linear aggregate of F and compositions (of the first kind) of F with itself. Since the operation of composition is evidently an extension of the operation of matrix multiplication, to matrices in which the row-number and column-number take a continuous sequence of values, it is obvious that Volterra's result corresponds to the well-known theorem that every matrix which is permutable with a matrix F (whose latent roots are all simple) must be a polynomial in F. The application of Volterra's theorem to the case when F is a constant yielded the important result that the aggregate of all functions permutable of the first kind with a constant is identical with the aggregate of all functions of $(y-x)$.

In three notes of the same year [125, 127, 129] the theory of composition and permutable functions was applied to the theory of integral and integro-differential equations. Consider an algebraic relation $F(z, \zeta)=0$ between two variables. Let the variables be replaced by two functions f and ϕ and let all multiplications of z with itself or with ζ be replaced by compositions of the corresponding functions. We thus obtain an integral equation between f and ϕ. If it is possible to represent the solution of $F(z, \zeta)=0$ (where ζ is the unknown) by a power-series in z, then this representation, when z has been replaced by f and the multiplications by compositions, will yield the solution ϕ of the integral equation. Thus the solution of an integral equation has been made to depend on the solution of an algebraic equation. By a similar process the solution of an integro-differential equation may be deduced from the solution of a differential equation.

The results of these notes were applied to the problem of hereditary elasticity in two notes [124, 126] in which Volterra solved the fundamental integro-differential equations for the case of an isotropic sphere. In the second note he also solved a quadratic integral equation.

In 1911 he took up the problem of integro-differential equations with constant limits, i.e. of the Fredholm type [133]. As we have seen, the solution of an integro-differential equation can be reduced to the problem of finding the 'fundamental solution' of a certain associated differential equation. This 'fundamental solution' is represented by a series proceeding according to powers of the parameter which occurs in the integro-differential equation: when this parameter is zero, the integro-differential equation which Volterra was studying, reduced to Laplace's partial differential equation in n variables, and his 'fundamental solution' to the elementary solution of Laplace's equation. In order to solve the integro-differential equation, a series of compositions was derived from the fundamental series: this series of compositions is always convergent in the case of compositions of the first kind, but in the case of integro-differential equations of the Fredholm type, the compositions are of the second kind and the series formed of them do not always converge. Volterra therefore now gave another theory for linear integro-differential equations of Fredholm type, which, however, is applicable only when the coefficients of the integro-differential equation are permutable functions (of the second kind). The chief feature of the new theory is that the fundamental solution appears now as the solution of a linear integral equation of Fredholm's type (instead of being the solution of a differential equation).

In another note [134] he determined all the functions permutable (of the first kind) with a given function of order two (i.e. such that $\phi(x, x) = 0$ and $\left(\dfrac{\partial \phi}{\partial y}\right)_{y=x} = 0$) showing that they depend on an arbitrary function of one variable: he also solved the equation $\psi * \psi = \phi$ completely, on the assumption that ϕ is a function of order two. In a further note [135] he pointed out that for functions of the special form

$$F(x, y) = \sum_{i, \kappa = 1}^{n} a_{i\kappa} f_i(x) \phi_\kappa(y)$$

the question of permutability is identical with the question of commutability of products of the corresponding matrices $(a_{i\kappa})$. Thus from matrix-algebra he was able to find all functions permutable with F (i.e. all matrices permutable with $(a_{i\kappa})$). The theory was applied to the solution of an integral equation of the nth degree.

Later in 1911 he published [131] a general survey of his method of utilizing the theory of composition and of permutable functions in order to solve integral and integro-differential equations.

In 1912 he gave a more detailed exposition [136] of the theory of integro-differential equations with a variable upper limit and of elliptic type: and (in the same memoir) the theory of 'hereditary elasticity' and of electric and magnetic hysteresis. He also [139] extended his theory of hereditary elasticity by considering vibrations: this led him to integro-differential equations of hyperbolic type, and to some fundamental results regarding vibrations of hereditary type. In many simple cases the 'hereditary' solution may be obtained from the commonly known one by replacing the trigonometric functions of the known solutions by certain transcendental functions which he now defined.

In the following year (1913) he returned once more [147] to integro-differential equations of elliptic type and completed the investigations of 1911 by considering in greater detail the case of an odd number of (spatial) dimensions. In a lecture to the Fifth International Congress of Mathematicians at Cambridge [149] he dealt with transcendental integral equations.

In the same year appeared in book form his lectures at Rome on integral and integro-differential equations [145], and his lectures at the Sorbonne on functions of lines [146]; in these a full account is given of the theory of functionals, the basis being always the transition from a finite number of variables to a continuously infinite number. These works did much to make Volterra's ideas widely known. On the invitation of the Berlin Mathematical Society he also delivered a lecture [152] outlining the fundamental notions of the functional calculus and indicating applications to the calculus of variations, integral and integro-differential equations, the theory of quadratic forms in an infinite number of variables, hereditary elasticity, mechanics and electro-magnetism.

In 1914 he published a couple of notes [155, 156] dealing with functional derivative equations, i.e. equations between a functional and its derivative. After studying some simple types of such equations (linear functional derivative equations of the first and second order) he showed that these equations correspond to the differential equations of the ordinary theory. He also discussed a system of integro-differential equations corresponding to a canonical system of equations in dynamics, and obtained the functional derivative equation which corresponds to the Hamilton–Jacobi equation.

In a substantial memoir produced on the eve of the war—in fact, written for the Napier tercentenary in Edinburgh in July 1914—but not published until 1916 [162], he gave a systematic exposition of his theory of composition of the first kind, introducing some important new ideas. The most notable of these is the 'zero'th compositional power of a function' which plays the part of unity and is essentially identical with Dirac's ∂-function. By the aid of this unit, compositional fractions and hence negative compositional powers are easily defined. Proceeding further in the same direction, a definition of the 'logarithm by composition' is obtained, and the theory of 'functions by composition' developed. In particular, the fundamental notions of the Calculus are extended to the domain of functions by composition, such as: derivative of a function by composition which is proved to be a (new) function by composition, and the definite integral by composition of a function by composition.

This theory of functions by composition was the subject of three lectures delivered at the Rice Institute, Houston, Texas [174].

Before describing the scientific work of the last twenty-five years of Volterra's life, let us take up again the thread of his personal history. In March 1905 he was created a Senator of the Kingdom of Italy—a great honour for a man still comparatively young—and about this time he was appointed by the Government as Chairman of the Polytechnic School at Turin, and Royal Commissioner. The way was open for him to become a great figure in political and administrative life: but he preferred the career of a pure scientist, and took an active part in public affairs on only two occasions—the Great War of 1914–1918, and the struggle with Fascism.

In July 1914 he was, according to his custom at that time of year, at his country house at Ariccia, when the war broke out. Almost at once his mind was made up that Italy ought to join the Allies: and in concert with D'Annunzio, Bissolati, Barzilai and others, he organized meetings and propaganda which were crowned with success on the 24th of May in the following year, when Italy entered the war. As a Lieutenant in the Corps of Engineers he enlisted in the army, and, although now over fifty-five years of age, joined the Air Force. For more than two years he lived with youthful enthusiasm in the Italian skies, perfecting a new type of airship and studying the possibility of mounting guns on it. At last he inaugurated the system of firing from an airship, in spite of the general opinion that the airship would be set on fire or explode at the first shot. He also published some mathematical works relating to aerial warfare, and experimented with aeroplanes. At the end of these dangerous enterprises he was mentioned in dispatches, and decorated with the War Cross.

Some days after the capitulation of Gorizia he went to this town while it was still under the fire of Austrian guns in order to test the Italian instruments for the location of enemy batteries by sound. At the beginning of 1917 he established in Italy the Office for War Inventions, and became its Chairman, making many journeys to France and England in order to promote scientific and technical collaboration among the Allies. He went to Toulon and Harwich in order to study the submarine war, and in May and October 1917 took part in the London discussions regarding the International Research Committee, to the executive of which he was appointed. He was the first to propose the use of helium as a substitute for hydrogen, and organized its manufacture.

When in 1917 some political parties—especially the Socialists—wanted a separate peace for Italy, he strenuously opposed their proposals: after the disaster of Caporetto, he with Sonnino helped to create the parliamentary *bloc* which was resolved to carry on the war to ultimate victory.

On the conclusion of the Armistice in 1918, Volterra returned to his purely scientific studies and to his teaching work in the University. The most important discoveries of his life after the war were in the field of mathematical biology, and of these we must now give an account.

The title of his discourse [85], at the opening of the academic year following his election to the Roman Chair, shows that already in 1901 he was interested in the biological applications of mathematics. His own researches in this field, stimulated by conversations with the biologist Dr Umberto D'Ancona of the University of Sièna, began at the end of 1925: his first and fundamental memoir [189] (reprinted with modications and additions as [193]: summarized briefly in English [191]: and more fully [202]. Critical summary by J. Pérès, *Rev. gén. Sci. pur. appl.* 38, 285–300, 337–341 (1927)) on the subject appeared in the following year. The theory was developed in several further papers [194, 195, 196, 198] and in the winter 1928–1929 was made the subject of a course of lectures delivered by Volterra at the Institut Henri Poincaré in Paris. These lectures, together with a historical and bibliographical chapter compiled by D'Ancona, were published in 1931 [210].

The entities studied in these investigations were *biological associations*, i.e. systems of animal (or plant) populations of different species, living together in competition or alliance in a common environment: and the theory is concerned with the effects of interaction of these populations with one another and the environment as expressed in their numerical variations.

At the beginning of these researches, Volterra was unaware of

similar work already in existence. Some of this [3] referred only
to special problems. Other work, such as that of W. R. Thompson
on parasitology, though more general in character and allied to
Volterra's studies, differed in this respect: that it was necessary
to regard generations as distinct and the number of a population
as varying discontinuously in time. It will be obvious, however,
that there are many problems relating to species in which
generations overlap, where the number of a population can more
appropriately be regarded, with sufficient approximation, as a
continuously varying function of the time: and it is to problems
of this kind that Volterra's methods apply. Such problems had
been studied to some extent by A. J. Lotka, and some of Volterra's
simpler results referring to associations of two species had been
anticipated by him.

To understand the ideas on which Volterra's theory is based,
we may consider first the by no means trivial case of a single
species. Let the number of the population at the time t be $N(t)$.
The simplest assumption of constant birth and death rates leads
immediately to the 'Malthusian' equation

$$\frac{d\mathrm{N}}{dt} = \epsilon \mathrm{N} \tag{1}$$

giving on integration the geometric law of increase

$$\mathrm{N} = \mathrm{N}_0 e^{\epsilon t}. \tag{2}$$

Under more general conditions, there will be corrections to be
applied to the right-hand side of (1): and it is of course such
corrections which falsify the famous predictions of Malthus
regarding human populations.

If the environment will support only a limited number of
individuals, we must suppose that the 'coefficient of increase'
ϵ is no longer constant, but a decreasing function of N. It is
simplest to assume that this decrease is linear so that we have the
Verhulst equation

$$\frac{d\mathrm{N}}{dt} = (\epsilon - \lambda \mathrm{N})\mathrm{N} \tag{3}$$

[3] e.g. the analysis of the Ross-Martini malaria equations.

where ϵ, λ are constants, ϵ being called in Volterra's later papers the *coefficient of auto-increase* of the population, to distinguish it from the whole coefficient of increase $\epsilon - \lambda N$. Integrating (3), we obtain the well-known 'logistic curve'

$$N = \epsilon/(\lambda + ke^{-\epsilon t}) \qquad (4)$$

which is verified observationally in many contexts in biology and economics.

Under more general conditions further corrections must be allowed for; thus in a complicated case we might have to consider an integro-differential equation such as

$$\frac{dN}{dt} = \left\{ \epsilon - \lambda N + \kappa \sin \nu t - \int_o^t N(\tau) f(t - \tau) d\tau \right\} N + \alpha.$$

The total coefficient of increase here consists of ϵ, the coefficient of auto-increase of the population, together with three corrections: $-\lambda N$ representing the Verhulst effect of competition within the species: a periodic term arising perhaps from seasonal variations of the environment and generally producing forced oscillations in N: and an integral representing some delayed effect, such as the intoxication of a closed environment by the accumulation of waste products. The final constant α indicates immigration at a uniform rate.

Volterra's concern, however, is with associations of 2, 3, . . . or n species, and hence with 2, 3, . . . or n differential equations (or integro-differential equations if delayed effects are envisaged). He investigated in considerable detail the association of two species, one of which feeds on the other: we may call them *predators* and *prey*. For this case we have the Lotka-Volterra equations

$$\begin{aligned} (\text{prey}) \qquad & \frac{dN_1}{dt} = (\epsilon_1 - \gamma_1 N_2) N_1 \\ (\text{predators}) \quad & \frac{dN_2}{dt} = (-\epsilon_2 + \gamma_2 N_1) N_2 \end{aligned} \right\} \qquad (5)$$

By themselves, the prey would increase and the predators die out: hence the coefficients of auto-increase (ϵ_1, $-\epsilon_2$) are respec-

45

tively positive and negative. The competition terms, since they depend on encounters, are proportional to N_1N_2 and are positive for the predators, negative for the prey. By integrating these equations, Volterra deduced the existence of periodic fluctuations, whose period is independent of γ_1, γ_2 (so, e.g., it would not be affected by increased protection of the prey): the average values of N_1 and N_2 do not depend on their initial values, and are given by

$$\bar{N}_1 = \epsilon_2/\gamma_2, \quad \bar{N}_2 = \epsilon_1/\gamma_1. \tag{6}$$

The existence of periodic fluctuations in biological associations was already well known from observation: but ecologists had generally considered that it was necessary to seek an explanation of the fluctuations in some external cause, such as the seasons, or human interference. Partly as a result of the Lotka-Volterra analysis, it is now generally admitted that periodic fluctuations in a constant environment may under some circumstances be sufficiently explained by the mere fact of interaction.

If the populations considered in equations (5) are different species of fish, the effect of fishing (i.e. uniform destruction of both species of fish proportionately to their numbers) would be to increase ϵ_2 and to decrease ϵ_1. From (6) it follows that the mean number of predators ($\bar{N}_2$) is decreased, and that of their prey ($\bar{N}_1$) increased. Similarly, a cessation of fishing, such as occurs on a large scale in time of war, will be to the relative advantage of the predator species. This effect had already been observed by D'Ancona in his statistical study of the Adriatic fisheries over the period 1905–1923. He found, that is to say, a temporary increase in the mean relative frequency of the more voracious kinds of fish, as compared with the fish on which they preyed, during the years 1914–1918.

In the *Leçons sur la Théorie mathématique de la Lutte pour la Vie* [210] a general theory of *n* species is developed, and the suggestion of a dynamical analogy is introduced by the distinction between what he called *conservative* and *dissipative* associations. Without going into the details of the definition, it may be explained that the association (of one species) represented in equation (1) is conservative, that of equation (3) is dissipative. Equations (5) represent a conservative association: while if we modify them by taking account of competition within each species, just as (3) was obtained by modification of (1), we obtain the equations of a dissipative association, namely,

$$\begin{aligned}
\frac{dN_1}{dt} &= (\epsilon_1 - \lambda_1 N_1 - \gamma_1 N_2)N_1 \\
\frac{dN_2}{dt} &= (-\epsilon_2 - \lambda_2 N_2 + \gamma_2 N_1)N_2
\end{aligned} \right\} \tag{7}$$

The effect of this change is that the fluctuations of N_1 and N_2 are now damped: that is, their amplitudes diminish and the association tends with increasing time to its equilibrium state. This, Volterra showed, is a general property of the associations which he called dissipative, and gives an obvious point to the mechanical analogy. He regarded conservative associations as representing ideal situations not generally attained in nature, and supposed that actual associations are more often of the dissipative type.[4]

Finally, in chapter 4 of his book, Volterra extended the theory of two species to the cases, so important in many biological problems, where delayed effects occur. The two differential equations were now replaced by a pair of integro-differential equations. He solved them by a method of successive approximations, and discussed at length the analogy of this case with his previous studies of hereditary phenomena in elasticity and electromagnetism.

The extension of this study of delayed effects to an association of n species was tackled only in Volterra's last publication on mathematical biology [235], which appeared in 1939. In the

[4] G. F. Gause, *The Struggle for Existence* (Baltimore, 1934), described experiments which aimed at reproducing some of Volterra's mathematical models with simple biological models, using yeast cells and different species of Protozoa in competition. The attempts were not altogether successful, but it may be noted that the periodic oscillations eventually obtained in a predator-prey experiment were of diminishing amplitude.

intervening years, two very different developments of the theory occurred, in what may be called the 'applied' and 'pure' aspects of the subject.

Firstly [232], Volterra collaborated with D'Ancona in surveying the relevant biological literature for confirmation and further applications. The inherent difficulties of this task arose rather from the immense variety of special conditions occurring in different cases and the correct assessment of these conditions in any particular case, than in the mathematical analysis of the conditions once identified. The fact that the authors were able to find the appropriate mathematical model in many different cases is a vindication of Volterra's technique.

Secondly [222, 223, 224, 225, 226, 227], on the 'pure' side, he extended into the very core of classical dynamics the suggestion of an analogy which has been noticed above. This new development, which he described in a lecture [228] at the University of Geneva on 17 June 1937, may be explained simply by reference again to the predator-prey equations (5). With a change of notation, these can be written

$$\dot{N}_1 = \left(\epsilon_1 - \frac{N_2}{\alpha}\right)N_1, \quad \dot{N}_2 = \left(-\epsilon_2 + \frac{N_1}{\beta}\right)N_2.$$

We deduce easily

$$\alpha\dot{N}_1 + \beta\dot{N}_2 - \alpha\epsilon_1 N_1 + \beta\epsilon_2 N_2 = 0,$$

$$\alpha N_1 + \beta N_2 - \alpha\epsilon_1 \int_0^t N_1 dt + \beta\epsilon_2 \int_0^t N_2 dt = \text{a constant, say H.} \quad (8)$$

Introducing what he called the *quantities of life* of the two populations,

$$x = \int_0^t N_1 dt, \quad y = \int_0^t N_2 dt,$$

and regarding these as analogous to co-ordinates in dynamics, we have in (8) the result which Volterra called the *conservation of demographic energy*, namely,

$$T + V = H,$$

where $T = \alpha x + \beta \dot{y}$ $\quad$ = actual demographic energy
$\quad\quad V = -\alpha\epsilon_1 x + \beta\epsilon_2 y$ = potential demographic energy.

Moreover, if we introduce the function

$$\Phi = (\alpha x \log x + \beta y \log y) + \tfrac{1}{2}(\dot{x}y - x\dot{y}) - V,$$

it can be shown that the original fluctuation equations (5) are equivalent to the Lagrangian equations of motion

$$\frac{d}{dt}\left(\frac{\partial \Phi}{\partial \dot{x}}\right) - \frac{\partial \Phi}{\partial x} = 0, \qquad \frac{d}{dt}\left(\frac{\partial \Phi}{\partial \dot{y}}\right) - \frac{\partial \Phi}{\partial y} = 0.$$

It then follows as in classical dynamics that, as far as the variations of N_1, N_2 are concerned, the properties of the association may be summed up in a variational principle

$$\partial \int_{t_0}^{t} \Phi\, dt = 0. \tag{9}$$

The analogy has been described so far for the special association (5); but Volterra worked it out generally for a conservative association (and also for certain types of dissipative associations) of n species.

Yet one more analogy was described in a lecture [229, 230] to the Réunion Internationale des Mathématiciens. Generalizing the analysis of 'cessation of fishing' described above by means of equations (5), (6), it referred to the changes in the equilibrium state of a conservative association of n species, caused by variation of their coefficients of auto-increase; and it took the form of some principles of reciprocity not unlike those which appear in the theories of elasticity and electrostatics.

Biologists have been apt to criticize Volterra for preoccupying himself so elaborately with abstract mathematical models based on simplifying assumptions remote from the complexities of nature. Yet this, after all, is the procedure on which the triumphs of physical science have been founded. It would be rash to say whether the analogies with physical science which he unearthed will remain what they appear to be at first, and certainly are, *at least*, a clever and remarkable *tour de force*—or whether they will eventually be seen as the germs of a profound biodynamics, essential to the theoretical and economic biology of the future:

what is beyond dispute is that his contributions to pure mathematics will be in demand more and more inescapeably as mathematical biology develops.

While the researches which have last been described were Volterra's chief interest during the later years of his life, he still from time to time published contributions to pure analysis.

In 1924 appeared the well-known Volterra-Pérès book [185] on the theory of composition and permutable functions, marking the completion of his work in this field: and a course of lectures on the theory of functionals and of integral and integro-differential equations, given in Madrid in 1925 by invitation of the Faculty of Science of the University, were published in Spanish in 1927 [192] and in an English translation [204] in 1930.

Still more significant from the modern point of view was a work on the general theory of functionals written in collaboration with J. Pérès, of which the first of three projected volumes appeared in 1936 [233]. This first volume contains the general principles of the functional calculus and its applications to the theory of integral equations: the second volume was planned to contain the theories of composition, of permutable functions, of integro-differential equations and functional derivative equations, and of Volterra's generalizations of analytic functions: the third volume would deal with some subsidiary topics and with the applications of the functional calculus. Account was to be taken of the modern theory of functions and of abstract spaces, and the complete work would therefore have been of great importance.

In 1938 Volterra published, in the Borel series of monographs, a work [234] written in conjunction with B. Hostinský, concerning researches whose origins belong to his earlier period. In 1887–1888 he had written two notes [20, 23] and a substantial memoir [16], dealing with the theory of substitutions or matrices, the infinitesimal operations which can be performed upon these, and their applications to the theory of linear differential equations. He regarded the n^2 elements of a matrix of order n as functions of a variable x which was supposed to be real in these earlier papers, the extension to complex variables being given in a later memoir [75]: and he defined the *derivative* and the *integral* of a matrix with respect to x, showing that these two operations are inverse. The value of these concepts is shown when we consider a system of linear differential equations of the first order

$$\frac{dy_i}{dx} = \sum_{j=1}^{n} a_{ij}(x)y; \quad (i = 1, 2, \ldots, n).$$

Volterra showed that the elements of the 'integral' of the matrix of the $a_{ij}(x)$ yields a fundamental system of solutions of the differential equations. There is no difficulty in extending the formulae so as to include the case of non-homogeneous systems of differential equations, or the case of more than one independent variable. Theories of total differentials of matrices, and of double and curvilinear integrals were then developed: and the transformation of these latter into each other led to the introduction of differential parameters. When the variable x is supposed to be complex, the integral of a matrix along a closed contour can be defined, and a calculus of residues developed, the residues depending on the singularities of the elements of the matrix: in short, the main ideas of the theory of functions of a complex variable can be carried over into matrix theory: and an extension is thus obtained of the well-known results of Fuchs on the expansion of the solutions of a linear differential equation in the neighbourhood of one of its singular points. Volterra then went on to study matrices whose elements are one-valued and regular functions of position on a Riemann surface: these he called *algebraic* matrices, and their integrals he named *Abelian* matrices: the theory is an interesting analogue of the ordinary theory of algebraic functions and their integrals. The later chapters of the 1938 monograph contain many developments and applications of Volterra's original work, due chiefly to Hostinský: one of the most important relates to the solution of the celebrated functional equation which Professor S. Chapman introduced in 1928 (*Proc. Roy. Soc.*, **119**) in connexion with a problem in diffusion.

Volterra's scientific activity overflowed in many domains quite outside his more usual fields of research. The student of topology, for instance, who reads Professor Lefschetz's admirable monograph on *Analysis Situs* (Paris, 1924), finds therein the photographs of a number of ingenious models constructed by Volterra in order to show how two manifolds, defined in very different ways, may nevertheless be homeomorphous to each other.

There remains to be told the melancholy story of his later years. In 1922 Fascism seized the reins in Italy. Volterra was one of the very few who recognized from the beginning the danger to freedom of thought, and immediately opposed certain changes in the educational system, which deprived the Italian Middle Schools of their liberty. When the opponents of the Fascist Government in the House of Deputies withdrew altogether from the debates, a small group of Senators, headed by Volterra, Benedetto Croce and Francesco Ruffini, appeared, at great personal risk, at all the Senate meetings and voted steadily in opposition. At that time he was President of the Accademia dei Lincei and generally recognized as the most eminent man of science in Italy.

By 1930 the parliamentary system created by Cavour in the nineteenth century had been completely abolished. Volterra never again entered the Senate House. In 1931, having refused to take the oath of allegiance imposed by the Fascist Government, he was forced to leave the University of Rome, where he had taught for thirty years: and in 1932 he was compelled to resign from all Italian Scientific Academies.[5] From this time forth he lived chiefly abroad, returning occasionally to his country-house in Ariccia. Much of his time was spent in Paris, where he lectured every year at the Institut Henri Poincaré: he also gave lectures in Spain, in Roumania, and in Czechoslovakia. On all these journeys he was accompanied by his wife, who never left him, and learnt typewriting in order to copy his papers for him:

[5] He was however, on the nomination of Pope Pius XI, a member of the Pontifical Academy of Sciences, and this honour he retained until his death.

he was accustomed to say that the signature 'V. Volterra' in his later works represented not Vito but Virginia Volterra.

In the autumn of 1938, under German influence, the Italian Government promulgated racial laws, and his two sons were deprived of their University positions and their civil rights: at their father's suggestion, they left their native country to begin a new life abroad.

He had a remarkable power of inspiring affection. When in the last months of his life the new laws forbade him to have Italian servants, all of them refused to leave: a maid who had been with him for more than twenty years, and who was forced to leave, died of sorrow a week afterwards.

In December 1938 he was affected by phlebitis: the use of his limbs was never recovered, but his intellectual energy was unaffected, and it was after this that his two last papers [235, 236] were published by the Edinburgh Mathematical Society and the Pontifical Academy of Sciences respectively. On the morning of 11 October 1940 he died at his house in Rome. In accordance with his wishes, he was buried in the small cemetery of Ariccia, on a little hill, near the country-house which he loved so much and where he had passed the serenest hours of his noble and active life.

He had received countless honours. He was elected a Foreign Member of our Society in 1910, and had received a similar distinction from almost every National Academy and Mathematical Society in the world; and he was a doctor *honoris causa* of many Universities: in this country, of Cambridge, Oxford and Edinburgh. The photograph which is reproduced at the head of this memoir was taken when he received the honorary Sc.D. of Cambridge in 1900. In August 1938 he was offered the honorary doctorate of the University of St Andrews, and wished to travel to Scotland to receive it, but was forbidden by his medical attendant. In his native land he had the Gran Cordone della Corona d'Italia and the Croce di Guerra, and was a Senator of the Kingdom and a Knight of SS. Maurice and Lazarus. In France he was a Grand Officier de laLégion d'Honneur: he had also the orders of Leopold of Belgium, of S. Carlo of Monaco and of the Polar Star of Sweden. On 23 August 1921 he received from King George V the dignity of an honorary K.B.E.: and this was an immense gratification to him, for he was deeply attached to his many friends in this country.

In the words in which his death was announced to his fellow-members of the Pontifical Academy of Sciences, we believe him *ex hac vita in scientiarum profectum sedulo impensa ad sapientiae aeternitatem transisse.*

E. T. WHITTAKER

[Acknowledgment of help received in the preparation of this notice is gratefully made to Dr Enrico Volterra, Dr A. Erdélyi and Dr I. M. H. Etherington.]

LIST OF PUBLICATIONS OF VITO VOLTERRA

1. Sul potenziale di un' elissoide eterogenea sopra sè stessa. *Nuovo Cim.,* **9,** 221–229 (1881).
2. Alcune osservazioni sulle funzioni punteggiate discontinue. *G. Mat.,* **19,** 76–87 (1881).
3. Sui principii del calcolo integrale. *G. Mat.,* **19,** 333–372 (1882).
4. Sopra una legge di reciprocità nella distribuzione delle temperatura e delle correnti Galvaniche costanti in un corpo qualunque. *Nuouo Cim.,* **11,** 188–192 (1882).
5. Sopra alcuni problemi di idrodinamica. *Nuovo Cim.,* **12,** 65–96 (1882).
6. Sulla apparenze elettrochimiche alla superficie di un cilindro. *Atti Accad. Torino,* **18,** 147–168 (1882).
7. Sopra alcune condizioni caratteristiche delle funzioni di una variabile complessa. *Ann. Mat. pura appl.* (2), **11,** 1–55 (1883).
8. Sopra alcuni problemi della teoria del potenziale. *Ann. Scu. norm. sup. Pisa* (1883). (Tesi di abilitazione.)
9. Sull' equilibrio delle superficie flessibili ad inestendibili. *Atti Accad. Lincei,* **8,** 214–218 and 244–246 (1884).
10. Sopra un problema di elettrostatica. *Nuovo Cim.,* **16,** 49–57 (1884).
11. Sopra un problema di elettrostatica. *R. C. Accad. Lincei* (3), **8,** 315–318 (1884).

12. Sulla deformazione delle superficie flessibili ed inestendibili. *R. C. Accad. Lincei* (4), **1**, 274–278 (1885).

13. Integrazione di alcune equazioni differenziali del secondo ordine. *R. C. Accad. Lincei* (4), **1**, 303–306 (1885).

14. Sulle figure elettrochimiche di A. Guebhard. *Atti Accad. Torino*, **18**, 329–336 (1885).

15. Sopra uno proprietà di una classe di funzioni trascendenti. *R. C. Accad. Lincei* (4), **2**, 211–214 (1886).

16. Sui fundamenti della teoria delle equazioni differenziali lineari. Parte prima. *Mem. Soc. ital. Sci. nat.* (3), **6**, no. 8, 104 pp. (1887).

17. Sopra le funzioni che dipendono da altre funzioni. *R. C. Accad. Lincei* (4), **3**, 97–105, 141–146 and 153–158 (1887).

18. Sopra le funzioni dipendenti da linee. *R. C. Accad. Lincei* (4), **3**, 225–230 and 274–281 (1887).

19. Sopra una estensione della teoria di Riemann sulle funzioni di variabili complesse. I. *R. C. Accad. Lincei* (4), **3**, 281–287 (1887).

20. Sulle equazioni differenziali lineari. *R. C. Accad. Lincei* (4), **3**, 393–396 (1887).

21. Sopra una estensione della teoria di Riemann sulle funzioni di variabili complesse. II, III. *R. C. Accad. Lincei* (4), **4**, 107–115 and 196–202 (1888).

22. Sulle funzioni analitiche polidrome. *R. C. Accad. Lincei* (4), **4**, 355–362 (1888).

23. Sulla teoria delle equazioni differenziali lineari. *R. C. Circ. mat. Palermo*, **2**, 69–75 (1888).

24. Sulla integrazione di una sistema di equazioni differenziali a derivate parziali che si presenta nella teoria delle funzioni conjugate. *R. C. Circ. mat. Palermo*, **3**, 260–272 (1889).

25. Sur une généralisation de la théorie des fonctions d'une variable imaginaire. 1er Mémoire. *Acta Math., Stockh.*, **12**, 233–286 (1889).

26. Delle variabili complesse negli iperspazi. *R. C. Accad. Lincei* (4), **5**, 158–165 and 291–299 (1889).

27. Sulle funzioni conjugate. *R. C. Accad. Lincei* (4), **5**, 599–611 (1889).

28. Sulle funzioni di iperspazi e sui loro parametri differenziali. *R. C. Accad. Lincei* (4), **5**, 630–640 (1889).

29. Sulle equazioni differenziali che provengono da questioni delle variazioni. *R. C. Accad. Lincei* (4), **6**, 43–54 (1890).

30. Sopra une estensione della teoria Jacobi-Hamilton del calcolo delle variazioni. *R. C. Accad. Lincei* (4), **6**, 127–138 (1890).

31. Sulle variabili complesse negli iperspazi. *R. C. Accad. Lincei* (4), **6**, 241–252 (1890).

32. Sopra le equazioni di Hertz. *Nuovo Cim.* (3), **29**, 53–63 (1891).

33. Sopra le equazioni fondamentali della elettrodinamica. *R. C. Accad. Lincei* (4), **7**, 177–188 (1891) and *Nuovo Cim.* (3), **29**, 147–154 (1891).

34. Enrico Betti. *Nuovo Cim.* (3), **32**, 5–7 (1892).

35. Sulla vibrazioni luminose nei mezzi isotropi. *R. C. Accad. Lincei* (5), **1**, 161–170 (1892).

36. Sulle onde cilindriche nei mezzi isotropi. *R. C. Accad. Lincei* (5), **1**, 265–277 (1892).

37. Sul principio di Huygens. *Nuovo Cim.* (3), **31**, 244–255 (1892) and **32**, 59–65 (1892).

38. Sur les vibrations lumineuses dans les milieux biréfringents. *Acta Math. Stockh.*, **16**, 153–215 (1892).

39. Sulle vibrazioni dei corpi elastici. *R. C. Accad. Lincei* (5), **2**, 389–397 (1893).

40. Sulle integrazione delle equazioni differenziali del moto di un corpo elastico isotropo. *R. C. Accad. Lincei* (5), **2**, 549–558 (1893).

41, 42. Sul principio di Huygens. *Nuovo Cim.* (3), **33**, 32–36 and 71–77 (1893).

43. Eserzisi di fisica matematica. I. Sulle funzioni potenziali. *Riv. Mat.*, **4**, 1–14 (1894).

44. Sur les vibrations des corps élastiques isotropes. *Acta Math., Stockh.*, **18**, 161–232 (1894).

45. Sulla teoria dei movimenti del polo terrestre. *Astr. Nachr.*, **138**, 33–52 (1895).

46. Sulla teoria dei moti del polo terrestre. *Atti Accad. Torino*, **30**, 301–306 (1895).

47. Sul moto di un sistema nel quale sussistono moti interni stazionarii. *Atti Accad. Torino*, **30**, 372–384 (1895).

48. Sopra un sistema di equazioni differenziali. *Atti Accad. Torino*, **30**, 445–454 (1895).

49. Un teorema sulla rotazione dei corpi e sua applicazione al moto di un sistema nel quale sussistono moti interni stazionarii. *Atti Accad. Torino*, **30**, 524–541 (1895).

50. Sui moti periodici del polo terrestre. *Atti. Accad. Torino*, **30**, 547–561 (1895).

51. Sulla teoria del moti del polo nelle ipotesi della plasticità terrestre. *Atti Accad. Torino*, **30**, 729–743 (1895).

52. Sulla rotazione di un corpo in cui esistono sistemi ciclici. *R. C. Accad. Lincei* (5), **4**, 93–97 (1895).

53. Sul moto di un sistema nel quale sussistono moti interni variabili. *R. C. Accad. Lincei* (5), **4**, 107–110 (1895).

54. Sulle rotazioni permanenti stabili di un sistema in cui sussistono moti interni stazionarii. *Ann. Mat. pura. appl.* (2), **23**, 269–285 (1895).

55. Osservazioni sulla mia nota 'Sui moti periodici del polo terrestre'. *Atti Accad. Torino*, **30**, 817–820 (1895).

56. Sulla inversione degli integrali definiti. *R. C. Accad. Lincei* (5), **5**, 177–185 (1896).

57. Sulla inversione degli integrali multipli. *R. C. Accad. Lincei* (5), **5**, 289–300 (1896).

58. Sull' inversione degli integrali definiti. *Atti Accad. Torino*, **31**, 311–323, 400–408, 537–567 and 693–708 (1896).

59. Lettera al Presidenta Brioschi. *R. C. Accad. Lincei* (1896).

60. Osservazioni sulla nota precedente del Prof. Lauricella e sopra una nota di analogo argumenti dell' Ing. Almansi. *Atti Accad. Torino*, **31**, 1018–1021 (1896).

61. Lezioni di meccanica. Prime nozioni di cinematica. Livorno. Giusti 98 pp. (1896).

62. Sulla rotazione di un corpo in cui esistono sistemi policiclici. *Ann. Mat. pura appl.* (2), **24**, 29–58 (1896).

63. Un teorema sugli integrali multipli. *Atti Accad. Torino*, **32**, 859–868 (1897).

64. Sul principio di Dirichlet. *R. C. Circ. Mat. Palermo*, **11**, 83–86 (1897).

65. Sopra alcune questioni di inversione di integrali definiti. *Ann. Mat. pura appl.* (2), **25**, 139–178 (1897).

66. Sulle scarica elettrica nei gas e sopra alcune fenomeni di elettrolisi. *R. C. Accad. Lincei* (5), **6**, 389–401 (1897).

67. Sulla scarica elettrica nei gas. Editore R. Giusti. Roma 1897.

68. Sul fenomeno delle seiches. *Nuovo Cim.*, **8**, 270–272 (1898).

69. Sulle funzioni poliarmoniche. *Atti Ist. Veneto* (7), **10**, 233–235 (1898).

70. Sopra una classe di equazioni dinamiche. *Atti Torino*, **33**, 451–475 (1898).

71. Sulla integrazione di una classe di equazioni dinamiche. *Atti Torino*, **33**, 542–558 (1898).

72. Sur la théorie des variations des latitudes. *Acta Math., Stockh.*, **22**, 201–296 (1898).

73. Sur la théorie des variations des latitudes. *Astr. Gesellsch.*, **33**, 275–329 (1898).

74. Sulla scarica elettrica nei gas e sopra alcune fenomeni di elettrolisi. *Nuovo Cim.* (4), **7**, 53–57 (1898).

75. Sui fondamenti della teoria delle equazioni differenziali lineari. (Parte seconda). *Mem. Soc. Ital. Sci. Nat.* (3), **12**, 3–68 (1899).

76. Sopra una classe di moti permanenti stabili. *Atti Accad. Torino*, **34**, 247–255 (1899).

77. Sul flusso di energia meccanica. *Nuovo Cim.* (4), **10**, 337–359 (1899).

78. Sul flusso di energia meccanica. *Atti Accad. Torino*, **34**, 366–375 (1899).

79. Sopra alcuni applicazioni della rappresentazione analitica delle funzioni del Prof. Mittag-Leffler. *Atti Accad. Torino*, **34**, 492–494 (1899).

80. Sopra alcune applicazioni delle leggi del flusso di energia meccanica nel moto di corpi che si attraggono colla legge di Newton. *Atti Accad. Torino*, **34**, 805–817 (1899).

81. Necrologia del Prof. Eugenio Beltrami. *Annuario Univ. Roma* (1900).

82. Betti, Brioschi, Casorati, trois analystes italiens et trois manières d'envisager les questions d'analyse. *Congrès intern. Math.*, 43–57 (Paris 1900).

83. Sugli integrali lineari dei moti spontanei a caratteristiche independenti. *Atti Torino*, **35**, 186–192 (1900).

84. Sur les équations aux dérivées partielles. *Congr. intern. Math.* 377–378 (Paris 1900).

85. Sui tentativi di applicazione delle matematiche alle scienze biologiche e sociali. Discorso letto il 4 novembre 1901 alla inaugurazione dell' anno scolastico nella R. Università di Roma., 25 pp. (1901).

86. Sur la stratification d'une masse fluide en équilibre. *Acta Math., Stockh.*, **27**, 105–124 (1903).

87. Sul numero dei componenti indipendenti di un sistema. *R. C. Accad. Lincei* (5), **12**, 417–419 (1903).

88. Commemorazione del Socio Straniero G. G. Stokes. *R. C. Accad. Lincei* (1903).

89. Sur les équations différentielles du type parabolique. *C. R. Acad. Sci. Paris*, **139**, 956–959 (1904).

90. Relazione sur viaggio compiuto dal Prof. V. Volterra per incarico avato dalla Commissione nominata per il riordinamento del Politecnico di Torino (1904).

91. Un teorema sulla teoria della elasticità. *R. C. Accad. Lincei* (5), **14**, 127–137 (1905).

92. Sull' equilibro del corpi elastici più volte connessi. *R. C. Accad. Lincei* (5), **14**, 193–202 (1905) and *Nuovo Cim.* (5), **10**, 361–385 (1905).

93. Sulle distorsioni generate de tagli uniformi. *R. A. Accad. Lincei* (5), **14**, 329–342 (1905).

94. Sulle distorsioni dei solidi elasticiti più volte connessi. *R. C. Accad. Lincei* (5), **14**, 351–356 (1905).

95. Sulle distorsioni dei corpi elastici simmetrici. *R. C. Accad. Lincei* (5), **14**, 431–438 (1905).

96. Contributo allo studio delle distorsioni dei solidi elastici. *R. C. Accad. Lincei* (5), **14**, 641–654 (1905).

97. Note on the application of the method of images to problems of vibrations. *Proc. London Math. Soc.* (2), **2**, 327–331 (1905).

98. Opere del Prof. Alfredo Cornu. *Atti Accad. Torino* (1905).

99. Fondazione di un Politecnico nella Città di Torino. Discorso pronunciato in Senato, giugno 1906.

100. Sull' equilibrio dei corpi elastici più volte connessi. *Nuovo Cim.* (5), **10**, 361–385 (1905) ; (5), **11**, 5–20, 144–161, 205–221 and 338–347 (1906).

101. Sur des fonctions qui dépendent d'autres fonctions. *C. R. Acad. Sci. Paris*, **142**, 400–409 (1906).

102. L'economia matematica ed il nuovo manuale del Prof. Pareto. *G. Economisti* (1906).

103. Nuovo studii sulle distorsioni dei solidi elastici. *R. C. Accad. Lincei* (5), **15**, 519–525 (1906).

104. Sui tentativi di applicazione delle matematiche alle scienze biologiche e sociali. *Arch. Fisiol.* (1906).

105. Leçons sur l'intégration des équations différentielles aux dérivées partielles, professées à Stockholm (février-mars 1906) sur l'invitation de S. M. le Roi de Suède. Uppsala, pp. 82 (1906).

106. Les mathématiques dans les sciences biologiques et sociales. *Revue du mois* (1906).

107. Sur l'équilibre des corps élastiques multiplement connexes. *Ann. École norm.* (3), **24**, 401–517 (1907).

108. Parole pronunziate alle feste giubilari di Augusto Righi. Bologna 1907.

109. Il momento scientifico presente e la nuova Società Italiana per il Progresso delle Scienze. *Atti Soc. Ital. Progr. Sc.*, 3–14 (1908).

110. Parole pronunziate al Congresso della Società Italiana per il Progresso delle Scienze. *Atti Soc. Ital. Progr. Sc.* (1908).

111. La matematiche in Italia nella seconda metà del secolo XIX. *4th Math. Congr. Rome*, **1**, 55–65 (1909).

112. Sull' applicazione del metodo della imagini alle equazioni di tipo iperbolico. *4th Math. Congr. Rome*, **2**, 90–93 (1909).

113. Giovanni Vailati (Necroligia). *Period. Mat. Inseg. Sec.* (3), **6**, 289–292 (1909).

114. Sulle equazioni integro-differenziali. *R. C. Accad. Lincei* (5), **18**, 167–174 (1909).

115. Sulle equazioni della elettrodinamica. *R. C. Accad. Lincei* (5), **18**, 203–211 (1909).
116. Parole pronunziate al Congresso della Società Italiana per il Progresso delle Scienze. *Atti Soc. Ital. Progr. Sc.* (1909).
117. Parole del Preside della Facoltà di Scienze. Onoranze al Prof. Cremona. (1909).
118. Alcune osservazioni sopra proprietà atte ad individuare una funzione. *R. C. Accad. Lincei* (5), **18**, 263–266 (1909).
119. Sulle equazioni integro-differenziali della teoria dell' elasticità. *R. C. Accad. Lincei* (5), **18**, 296–301 (1909).
120. Equazioni integro-differenziali della elasticità nel caso della isotropia. *R. C. Accad. Lincei* (5), **18**, 577–586 (1909).
121. Commemorazione di Valentino Cerrati. Roma 1909.
122. Lectures delivered at Clark University ; Worcester, Mass., 1909.
123. Parole pronunziate avanti feretro di Stanislao Cannizzaro. *Nuovo Cim.* (5), **19**, 387–389 (1910).
124. Soluzione delle equazioni integro-differenziali dell' elasticità nel caso di una sfera isotropa. *R. C. Accad. Lincei* (5), **19**, 107–114 (1910).
125. Questioni generali sulle equazioni integrali ed integro-differenziali. *R. C. Accad. Lincei* (5), **19**, 169–180 (1910).
126. Deformazione di una sfera elastica, soggetta a date tensioni, nel caso ereditario. *R. C. Accad. Lincei* (5), **19**, 239–243 (1910).
127. Osservazioni sulle equazioni integro-differenziali ed integrali. *R. C. Accad. Lincei* (5), **19**, 361–363 (1910).
128. Sopra le funzioni permutabili. *R. C. Accad. Lincei* (5), **19**, 425–437 (1910).
129. Sulle equazioni permutabili. *R. C. Accad. Lincei* (1910).
130. Espacio, tiempo i massa según las ideas modernas. *An. Soc. cient. Argent.*, **70**, 223–283 (1911).
131. Sopra una proprietà generale delle equazioni integrali ed integro-differenziali. *R. C. Accad. Lincei* (5), **20**, 79–88 (1911).
132. Parole del Presidente della Facoltà. Onoranze al Prof. De Helgnero. Roma, 1911.
133. Equazioni integro-differenziali con limiti costanti. *R. C. Accad. Lincei* (5), **20**, 95–99 (1911).
134. Contributo allo studio delle funzioni permutabili. *R. C. Accad. Lincei* (5), **20**, 296–304 (1911).
135. Sopre le funzioni permutabili di 2a specie e le equazioni integrali. *R. C. Accad. Lincei* (5), **20**, 521–527 (1911).
136. Sur les équations intégro-différentielles et leurs applications. *Acta Math., Stockh.*, **35**, 295–356 (1912).

137. Leçons sur l'intégration des équations différentielles aux dérivées partielles. Paris, 1912. (Republication of 105.)

138. Lectures delivered at the celebration of the twentieth anniversary of the foundation of Clark University, under the auspices of the department of physics. New York and London, 1912.

139. Vibrazione elastiche nel caso della eredità. *R. C. Accad. Lincei* (5), **21**, 3–12 (1912).

140. Sulle temperature nell' interno delle montagne. *Nuovo Cim.*, **4**, 111–126 (1912).

141. L'évolution des idées fondamentales du calcul infinitesimal. *Revue du mois*, **13**, 257–274 (1912).

142. L'application du calcul aux phénomènes d'hérédité. *Revue du mois*, **13**, 556–574 (1912).

143. Onoranze al Prof. Valentino Cerrati. Roma, 1912.

144. Henri Poincaré : L'oeuvre mathématique. *Revue du mois*, **15**, 129–154 (1913).

145. Leçons sur les équations intégrales et les équations intégro-différentielles, professées à la Faculté des sciences de Rome en 1910, publiées par M. Tomassetti et F. S. Zarlatti. Collection Borel, pp. 164 (1913).

146. Leçons sur les fonctions de lignes, professées à la Sorbonne en 1912, recuellies et rédigées par J. Pérès. Collection Borel, pp. 230 (1913).

147. Sopra equazioni integro-differenziali aventi i limiti costanti. *R. C. Accad. Lincei* (5), **22**, 43–49 (1913).

148. Sui fenomeni ereditarii. *R. C. Accad. Lincei* (5), **22**, 529–539 (1913).

149. Sopra equazioni di tipo integrale. *Proc. 5th intern. Congr. Math.*, **1**, 403–406 (1913).

150. Some integral equations. *Bull. Amer. Math. Soc.* (2), **19**, 170–171 (1913).

151. Onoranze al Prof. Dott. G. B. Guccia. *R. C. Circ. Mat. Palermo* (1914).

152. Les problèmes qui ressortent du concept de fonctions de lignes. *S. B. berl. math. Ges.*, **13**, 130–150 (1914).

153. Osservazioni sui nuclei delle equazioni integrali. *R. C. Accad. Lincei* (5), **23**, 266–269 (1914).

154. Henri Poincaré. *Nouvelle Collection Scientifique*. Paris, 1914.

155. Sulle equazioni alle derivate funzionali. *R. C. Accad. Lincei* (5), **23**, 393–399 (1914).

156. Equazioni integro-differenziali ed equazioni alle derivate funzionali. *R. C. Accad. Lincei* (5), **23**, 551–557 (1914).

46

157. Drei Vorlesungen über neuere Fortschritte der mathematischen Physik, gehalten in September 1909 an der Clark Universität. Deutsch von Ernst Lamla. *Arch. Math. Phys., Lpz.* (3), **22**, 97–181 (1914). Also Leipzig, pp. 84 (1914).

158. The theory of permutable functions. Princeton, 1915.

159. Henri Poincaré. *Rice Inst. Pamphl.,* **1**, 133–162 (1915).

160. Sulle correnti elettriche in una lamina metallica sotto l'azione di un campo magnetico. *Nuovo Cim.* (1914) and *R. C. Accad. Lincei* (5), **24**, 220–234, 289–303, 378–390 and 533–543 (1915).

161. Metodi di calcolo degli elementi di tiro dell' artiglieria aeronautica. *R. C. Inst. Centr. Aero. Roma,* 1916.

162. Teoria delle potenze dei logaritmi e delle funzioni di composizione. *Mem. Accad. Lincei* (5), **11**, 167–269 (1916).

163. The generalisation of analytic functions. *Rice Inst. Pamphl.,* **4**, no. 1, 53–101 (1917).

164. Relazione sulla missione in Inghilterra ed in Francia. Roma, 1917.

165. On the theory of waves and Green's method. *Rice Inst. Pamphl.,* **4**, no. 1, 102–117 (1917).

166. Inaugurazione dell' Instituto Centrale di Biologia Marina in Messina. Venezia, 1917.

167. Pietro Blaserna. *R. C. Senato, Roma,* 1918.

168. Relazione della conferenza interalleata sulla organisazione scientifica. *R. C. Accad. Lincei* (1919).

169. L'entente scientifique. *Nouv. Revue Italie* (1919).

170. Le congrès de mathématiques de Strasbourg. *Nouv. Revue Italie* (1920).

171. Sur l'enseignement de la physique mathématique et quelques points d'analyse. *Enseign. Math.,* **21**, 200–202 (1920).

172. Sur l'enseignement de la physique mathématique et quelques points d'analyse (conférence générale). *C. R. congr. intern. math.,* 81–97 (1920).

173. Commemorazione di Augusto Righi. *Atti Parl. Senato Regno Roma,* 1920.

174. Functions of composition. Three lectures delivered at the Rice Institute in the autumn of 1919. *Rice Inst. Pamphl.,* **7**, no. 4, 181–251 (1920).

175. Saggi scientifici. Bologna, 1920.

176. Osservazioni sul metodo di determinare la velocità dei dirigibili. *Rassegna Marittima Aero. Roma,* 1920.

177. G. Lippmann (Necrologia). *R. C. Accad. Lincei* (5), **30**, 388–389 (1921).

178. A. Ròiti (Necrologia). *R. C. Accad. Lincei* (5), **30**, 477 (1921).

179. The flow of electricity in a magnetic field. Berkeley, pp. 72 (1921).

180. Funzioni di linee, equazioni integrali e integro-differenziali. *An. Soc. cient. Argent.* (1921).

181. Les équations aux dérivées fonctionelles et la théorie de la relativité. *Enseign. Math.*, **22**, 77–79 (1922).

182. Mouvement d'une fluide en contact avec un autre en surfaces de discontinuité. *C. R. Acad. Sci. Paris*, **177**, 569–571 (1923).

183. Sur les fonctions permutables. *Bull. Soc. Math. Fr.*, **52**, 548–568 (1923).

184. Da annuncio della morte del Socie Corrado Segre e ne rimpiange perdita. *R. C. Accad. Lincei* (5), **33**, 459–461 (1924).

185. Leçons sur la composition et les fonctions permutables. Paris, 1924.

186. Da annuncio della morte del Socio straniero Fusakichi Omori. *R. C. Accad. Lincei* (5), **33**, 43 (1924).

187. Arthur Gordon Webster. Worcester, Mass., 1924.

188. Commemorazione del Presidente F. d'Ovidio. *R. C. Accad. Lincei* (1925).

189. Variazioni e fluttuazioni del numero d'individui in specie animali conviventi. *Mem. Accad. Lincei*, **2**, 31–113 (1926).

190. Lois de fluctuation de la population de plusieurs espèces coexistent dans la même milieu. *Association Française Lyon*, 96–98 (1926).

191. Fluctuations in the abundance of a species considered mathematically. *Nature*, **118**, 558–560 (1926).

192. Teoria de los funcionales y de las ecuaciones integrales e integro-differenciales. Conferencias explicadas en la Facultad de la Ciencias de la Universitad, 1925, redactadas por L. Fantappié. Madrid, pp. 208 (1927).

193. Variazioni e fluttuazioni in specie animali conviventi. *R. Comit. Talass. Italiano*, Memoria cxxxi, Venezia, 1927.

194. Sulle fluttuazioni biologiche. *R. C. Accad. Lincei* (6), **5**, 3–10 (1927).

195. Leggi delle fluttuazioni biologiche. *R. C. Accad. Lincei* (6), **5**, 61–67 (1927).

196. Sulla periodicita delle fluttuazioni biologiche. *R. C. Accad. Lincei* (6), **5**, 463–470 (1927).

197. Essai mathématique sur les fluctuations biologiques. *Bull. de la Soc. d'Oceanographie de France*, 1927.

198. Una teoria matematica sulla lotta per l'esistenza. *Scientia*, **41**, 85–102 (1927) : French translation, 33–48.

199. Lois de fluctuations de la population de plusieurs espèces coexistant dans la même milieu. *Ass. Franç. Avanc. Sci.* (1927).

200. In memoria di H. A. Lorentz. *Nuovo Cim.* (2), **5**, 41–43 (1928).
201. Sur la théorie mathématique des phénomènes héréditaires. *J. Math. pur. appl.* (9), **7**, 249–298 (1928).
202. Variations and fluctuations of the number of individuals in animal species living together. *J. Conseil Int. Explor. Mer*, **3**, 1–51 (1928).
203. In memoria di Traiano Lalesco. *Revista Universitara Bucarest*, **1**, 213–215 (1929).
204. Theory of functionals and of integral and integro-differential equations. Edited by L. Fantappiè. Translated by M. Long. London, 1929.
205. La teoria dei funzionali applicata ai fenomeni ereditari. *Atti Congresso Bologna*, **1**, 215–232 (1929).
206. Alcune osservazioni sui fenomeni ereditarii. *R. C. Accad. Lincei* (6), **9**, 585–595 (1929).
207. La théorie des fonctionelles appliquée aux phénomènes héréditaires. *Rev. gén. Sci. pur. appl.*, **41**, 197–206 (1930).
208. Sulle fluttuazioni biologiche. *R. C. Semin. mat. fis. Milano*, **3**, 154–174 (1930).
209. Sulle meccanica ereditaria. *R. C. Accad. Lincei* (6), **11**, 619–625 (1930).
210. Leçons sur la théorie mathématique de la lutte pour la vie. Paris, 214 pp. (1931).
211. (With V. D'ANCONA) La concorrenza vitale tra le specie nell' ambiente marino. 7. *Congr. Int. Agriculture* (1931).
212. Ricerche matematiche sulle associazioni biologiche. *G. Ist. Ital. Attuari*, **2**, 295–355 (1931).
213. Italian physicists and Faraday's researches. *Nature*, **128**, 342–345 (1931).
214. Le calcul des variations, son évolution et ses progrès, son rôle dans la physique mathématique. (Conf. faites en 1931 à la fac. d. sci. univ. Charles, Praha et à la fac. d. sci. univ. Masaryk, Brno.) Praha and Brno (1932). Czech translation, *Časopis*, **62**, 93–116 and 201–227 (1933).
215. Sur les jets liquides. *J. Math. pur. appl.*, **11**, 1–35 (1932).
216. Équations aux dérivées partielles et théorie des fonctions. *Ann. Inst. H. Poincaré*, **4**, 273–352 (1933).
217. De Moivre's 'Miscellanea analytica'. *Nature*, **132**, 898 (1933).
218. Sur la théorie des ondes liquides et la méthode de Green. *J. Math. pur. appl.* (9), **13**, 1–18 (1934).
219. Représentations des fonctionelles analytiques déduites du théorème de Mittag-Leffler. *J. Math. pur. appl.* (9), **13**, 293–316 (1934).

220. Rémarques sur la note de M. Régnier et Mlle. Lambien. *C. R. Acad. Sci. Paris*, **199**, 1684–1686 (1934).

221. La théorie mathématique de la lutte pour la vie et l'expérience. *Scientia*, **59**, 169–174 (1936).

222. Les équations des fluctuations biologiques et la calcul des variations. *C. R. Acad. Sci. Paris*, **202**, 1953–1957 (1936).

223. Les équations canoniques des fluctuations biologiques. *C. R. Acad. Sci. Paris*, **202**, 2023–2026 (1936).

224. Sur l'intégration des équations des fluctuations biologiques. *C. R. Acad. Sci. Paris*, **202**, 2113–2116 (1936).

225. Le principe de la moindre action en biologie. *C. R. Acad. Sci. Paris*, **203**, 417–421 (1936).

226. Sur la moindre action vitale. *C. R. Acad. Sci. Paris*, **203**, 480–481 (1937).

227. Principes de biologie mathématique. *Acta biotheor.*, **3**, 1–36, Leyden (1937).

228. Applications des mathématiques à la biologie. *Enseign. math.*, **36**, 297–330 (1937).

229. Leggi delle fluttuazione e principii di reciprocità in biologia. *Riv. Biol.*, **22**, (1937).

230. Fluctuations dans la lutte pour la vie, leurs lois fondamentales et le reciprocité. *Conf. Réunion int. mathém.*, Paris, 1938.

231. Population growth, equilibria and extinction under specified breeding conditions : a development and extension of the theory of the logistic curve. *Human Biology* (Baltimore), **10**, 1–11 (1938).

232. (With V. D'ANCONA) Les associations biologiques au point de vue mathématique. Paris, 1935.

233. (With J. PÉRÈS) Théorie générale des fonctionelles. Tome 1 : Généralités sur les fonctionelles. Théorie des équations intégrales. Collection Borel, Paris, pp. 359 (1936).

234. (With B. HOSTINSKÝ) Opérations infinitésimales linéaires, applications aux équations différentielles et fonctionelles. Collection Borel, Paris, pp. 238 (1938).

235. The general equations of biological strife in the case of historical actions. *Proc. Edinburgh Math. Soc.*, **6**, 4–10 (1939).

236. Energia nei fenomeni elastici ereditarii. *Pontif. Acad. Scient. Acta*, **4**, 115–130 (1940).

APPENDIX B

On the Attempts to Apply Mathematics to the Biological and Social Sciences

The Inaugural address delivered at the formal opening of the academic year at the University of Rome in 1901, published in Annuario dell'Università *1901-1902 and reprinted in* Giornale degli Economisti, Serie II, vol. 23, *1901. It was also printed in French in* Revue du Mois I:1 (Paris, Soudier, *1906) and in* Archivio di fisiologia III:2 (Firenze, Gennaio, 1906).

Anatole France, that brilliant philosopher and novelist and delight of so many discerning readers, recounts the following anecdote. "Some years ago," he says, "I visited the museum of natural history in a great European city with one of its curators, who described to me with immense satisfaction the fossil animals. He instructed me very well up until the Pliocene, but when we found ourselves in front of the first traces of human beings, he turned away, and to my questions replied that this was not his display cabinet. I realized my indiscretion: Never ask a scientist about the secrets of the universe that cannot be found in his display cabinet."

Although a mind as keen and paradoxical as that of Anatole France was bound to conclude from this simple encounter that scientists are the least curious people in the world and uninterested in what lies outside their own display cabinets, we will be careful not to draw the same conclusion but will instead consider the episode an example of the natural and often justified reluctance of scholars to discuss ideas and make statements outside their own field.

Among scientists the curiosity to range far and wide is very strong, as is their desire to rummage through the display cabinets of their peers in order to better gauge the value of their own. Sometimes a collegial stocktaking overcomes whatever constraints Anatole France's friend felt in front of a stranger. And such curiosity is much greater in those devoted to the study of mathematics than in those engaged in other disciplines.

The mathematician possesses a wonderful and valuable tool, fashioned over the centuries by the cumulative efforts of the keenest minds that ever lived. He has, so to speak, the key that opens a passage to the hidden mysteries of the universe—the means to summarize, in a few symbols, a synthesis embracing and connecting the vast and disparate results of many sciences.

While he devotes his life and his mental powers to refining and perfecting his tools, to fit them for the subtlest inquiry and an ever greater understanding of the facts, he is continually badgered by a growing wave of scholars, who beg him for help and frequently for more than he can give. Only the rarest of spirits can roam the sphere of numbers and the abstractions of geometry and logic, remaining indifferent to and apart from all that lives and moves around them and laboring purely for the glory of human thought. Instead, it is more natural to break out of the circle of pure mathematical analysis—to gather information, to compare the results of the various methods at your disposal, to classify them in preparation for their application, to direct them toward their proper use, to improve the most useful, to reinforce the weakest, to create methods even more powerful.

But that curiosity is most intense in the biological and social sciences—which mathematics has only recently begun to penetrate—because of a strong desire to find out whether the classic methods that produced such wonderful results in the mechanical-physical sciences can be applied with equal success to those new and unexplored fields.

Succumbing to the desire to give the impression that a mathematician can entertain these matters, as compared with the classic applications of mathematics, I will allow myself to abandon the area of my own studies in order to touch on a field—in a very limited way—which is closely tied to the major problems of the philosophy and history of science and is itself extensive. Indeed, to trace and compare the paths, old and new, that mathematics has taken as it infiltrated the various branches of science, gauging its effects and counter effects, examining in depth the mutual relations thus engendered, while offering a grand panorama and superb synthesis of a great part of the completed work of human thought and providing a guide to future progress—all this would be a huge endeavor and quite beyond my powers.

First of all, it is important to clarify a subtle point concerning our subject. Some people expect too little from mathematics, others too much; this accounts for the suspicions of the one and the hyperenthusiasm of the other toward its new applications. If it is true, as the saying goes, that you get out of something only what you put into it, and that analysis adds nothing essential to the postulates forming the basis of mathematics, it is also true that mathematics is the royal road to arriving at general laws and the surest guide to forming new hypotheses—that is, to refining and perfecting those same postulates that form the basis for every single treatment—indeed, it offers the most precise way to test them, to bring them out of the realm of abstraction into that of reality. In truth, there is no better tool than the calculus with which to make accurate estimates of the long-term consequences of data obtained from observation and experiment.

But the history of science reveals a more direct contribution of mathematics to the perception and understanding of nature. When we use the calculus to establish the exact course of two apparently different phenomena, we often find an identity, or, as we say, we find they are governed by the same equations—it is a single step to conclude that the two constitute two representations of a single phenomenon. This was the procedure Maxwell followed to reach the conclusion that electromagnetic perturbations are the same as light—a momentous discovery opening the way to the work of Hertz which has so influenced modern physics and inspired the practical inventions of Ferraris and Marconi.

None can predict to what vast horizons the geometer will be led on the narrow and thorny path of the calculus. Did Lagrange suspect, when he conceived of analytical mechanics, not only that he was creating a powerful tool and sure guide for all the most difficult problems of the science of motion and equilibrium but that his formulas would one day become—in the hands of men of genius like Maxwell and Helmholtz—so comprehensive as to embrace and dominate all the phenomena of the physical world?

Nevertheless, while such is the importance of analysis, one must also consider its limitations. Unfortunately, professional mathematicians are separated from the rest of the world by a barrier of symbols, which lend an air of mystery to their work—so much so that those not privy to the secrets of algebra and the calculus sometimes labor under the delusion that their methods are of a different nature from those of everyday reasoning. Many people make a similar error with regard to the capability of machinery whose mechanisms are hidden.

Well, the gap is not as great as it might first appear between the crude reasoning enabling those ignorant of the calculus to determine the course of certain phenomena and the mechanism of the forces governing them and the subtle reasoning of the geometer, who derives in the same case a precise result from a convoluted set of algebraic symbols in a manner often astonishing even to those trained in and accustomed to analytical argument. On the contrary, if we examine the matter carefully, we see that the subtler process is nothing more than the crude process perfected and refined; moreover, in the mind of the geometer the crude reasoning preceded and guided the calculation, pointing the way to go and suggesting what to attempt. In a sense, it represents the framework from which the analytical edifice is built. But when we see the completed work, we are confronted with a magnificent monument stripped of all its scaffolding and supports. The props that served to support the dome during construction have vanished, and it appears to the astonished eye of the beholder as a miracle of construction.

Therefore, it is neither with undue hopes, nor often dangerous illusions, but also without indifference, that we ought to welcome new attempts to apply the calculus to whatever species of phenomena.

The transition of science from what I will call its pre-mathematical era to the mathematical era is characterized thus: Elements under study are examined in a quantitative rather than a qualitative way; in this transition, definitions that suggest an idea of the elements, in a more or less vague picture, give way to those definitions or principles that determine them, offering instead a way to measure them.

What importance in Newtonian mechanics, for example, can the primitive notion of force—as expressed by the definition "Force is the cause of motion"—have when compared with the first two laws, which essentially provide a way to measure it? So little that in some modern attempts to reformulate mechanics, this word—"force," the last verbal residue of personification in the inanimate world—can be done away with and replaced with the components that, combined, give its magnitude.

In light of this classic example, and many analogous ones I could easily cite, we joyfully salute Galton's attempts to measure quantitatively certain elements of the theory of evolution, such as heredity and variation.[1] Whereas Galton has taken only the first step on this path, and we may have to accept the criticisms of his results and revise much of what he has done, we also must recognize that a new day has dawned with the rise of the methods he pioneered.

However, to translate natural phenomena into arithmetical or geometrical language is to open a new avenue for mathematics rather than simply to apply the tools of analysis. To study the laws of the variation of measurable entities, to idealize these entities, to strip them of particular properties or attribute some property to them, to establish one or more elementary hypotheses that regulate their simultaneous and complex variation—all this marks the moment when we lay the foundations on which we can erect the entire analytical edifice. And it is then that we see the brilliance and power of the tools that mathematics offers in abundance to those who know how to use them.

Although political economics, for example, is only at the beginning of this path, its practitioners have experimented with how simply it represents —as in a picture—the mechanisms connecting the elements of the economic world, and how algebraic computation expresses the degree of change of each with the others, or according to the conditions in which they find themselves; whereas pre-mathematical economics never achieved the total picture because it was forced to examine each of those relations by itself, in isolation from the others.

To manipulate concepts so that we can introduce measurement; to measure them; to deduce laws; from those to work backward to hypotheses; from those to deduce, thanks to analysis, a science of idealized entities but still a rigorously logical one; to then compare this science to reality; to reject or modify, as contradictions are found between the results of calculations and the real world, the fundamental hypotheses; and thereby to arrive at new facts and analogies, or to argue from the present what the past was and

---

[1]*Natural Inheritance*, by Francis Galton, London 1889. In a short but interesting article by [Charles] Davenport in which the theory of these studies is explained (*Science*, N. S., XII: 310), he observes that Galton was led to his research mainly by the work of [Adolphe] Quetelet.

what the future will be—thus do I summarize as briefly as possible the birth and evolution of a science with a mathematical character.

The path is long and rugged and strewn with difficulties. Consider that whereas ancient traces of human civilization attest to astronomical measurements made by primitive people, celestial mechanics itself is only three centuries old. What a wonder, therefore, are the results—even if limited compared with our ambitious hopes and demands—that the calculus has been able to obtain in those sciences that only yesterday were in the premathematical period and are still struggling to escape it.

But let us see these analytical methods in action, and appreciate them as they tackle these new problems. Among the physical sciences, one discipline has always led, while the others have gradually followed, imitating it and taking it as example. This science is mechanics, and it constitutes, together with geometry, if not the most brilliant then surely the most dependable and secure body of knowledge in which the human mind glories. Not among the biological but among the social sciences, we can identify a discipline—pure economics—that is shaped by mechanics, has used its procedures and methods, and has achieved similar results.

Mechanics, like all the other physical sciences and like economics, owes its success to the methods of the infinitesimal [calculus], the most subtle and at the same time most powerful analytical tool ever conceived. The fundamentals of infinitesimal analysis are not be easily or briefly explained, even if we strip it of nonessentials and lay bare the skeleton of this superb and noble edifice that so many geniuses, from Archimedes to Newton, have constructed. So I will not try. I will say only that natural phenomena of whatever kind display an obvious complexity: What is happening now is the result of all that has occurred in the past; changes that take place at one point in space depend upon and are related to those that occur everywhere else.

To discover all those hidden ties whose consequences are so apparent—to take all this in at a glance, to master it—seems at first not only difficult but impossible, yet it is a necessary task if we want to form a complete idea of a phenomenon. How do the methods of the infinitesimal [calculus] extricate us from such a tangle, which besets us on all sides and seems to choke off any avenue of escape?

Let us imagine a sequence of events in an infinitely short time and an infinitely small space. It then becomes possible to distinguish the dominant changes of variables from those that are comparatively negligible, and if we can measure the former, or establish relations between them, we can then use those data to work backward from what happens at a certain instant in a particular place to what will happen, as time goes by, everywhere—or wherever the elementary laws obtain. Establishing those elementary laws is known as setting down differential equations; working through them step by step and calculating each element is known as integrating the equations. The geometer can perform the latter operation even if he ignores (as often happens) the specific problem to which his formulas will be applied; just as the obscure and humble miner, deep in the bowels of the earth, enriches humanity with a wealth of energy, unaware that the fuel he laboriously quarries from the ground will power a factory, illuminate our nights with a thousand flames, or propel a ship through distant seas.

Thanks to the infinitesimal calculus, we can, for instance, plot the motions of the heavenly bodies, expound the law of the harpstring's vibration, and calculate the effects of the most powerful machines, and it is with just such methods that the differential equations of economics were formulated. A comparison between mechanics and economics is easily made. Let's imagine what impressions a practitioner of mechanics might get from the study of economics.[2]

The concept of *Homo economicus*, which has prompted so much discussion and provoked such enormous difficulty that there are still those who refuse to accept it, comes so naturally to our mechanist that he is surprised at the suspicions aroused by this abstract, schematic being. He sees in *Homo economicus* a concept similar to those that, by long habit, have become familiar to him. He is used to idealizing surfaces as frictionless, wires as inextensible, solids as undeformable, and to substituting perfect liquids and gases for the natural kind. Not only has he made a habit of all this, but he also knows the advantages of doing so.

If the mechanist looks a little further, he realizes that in his science, as in economics, everything is reduced to an interplay of tendencies and constraints, the latter limiting the former, which react by generating tensions. Sometimes equilibrium arises, sometimes motion—so there are statics and dynamics in both sciences.

We have already mentioned the concept of force in mechanics; it has descended from the heights of metaphysics to the plain of measurable entities. Similarly, in economics, we no longer speak, as Jevons did, of the mathematical expression of unmeasurable quantities.[3] Pareto, instead of beginning directly from the notion of "ophelimity," as he did in his *Course of Political*

---

[2]Compare *Principi di economia pura* by Maffeo Pantaleoni, Firenze, 1894, and his *Scritti vari di economia*, Palermo 1904. See *Mathematical Investigations in the Theory of Value and Prices*, by Dr. Irving Fisher (*Trans. Connecticut Academy*, IX, July 1892). An exposition of the mechanical model imagined by Fisher was made by Colonel [Enrico] Barone in vol. VIII, Serie 2ª of *Giornale degli Economisti*. In the *Encyklopädie der Mathematischen Wissenschaften* (Leipzig, Teubner), [Vilfredo] Pareto published an interesting article in 1902: *Anwendungen der Mathematik auf Nationalökonomie*, in which the fundamental ideas, the different theories and the principal results on this subject are summarized and accompanied by a rich bibliography.

[3]*The Theory of Political Economy* by W. Stanley Jevons, London 1888. It is interesting to follow the evolution of Jevons' ideas, which can be linked to those of Laplace and Bernouilli and, according to Pantaleoni, to Jevons' studies under [Augustus] De Morgan's supervision (cf. *Contributo alla teoria del riparto delle spese pubbliche*, enclosed in *Scritti vari di Economia politica*, mentioned above).

Economy,[4] suggests that we begin from purely quantitative concepts, with his indifference curves, which correspond so well to the standard level curves and equipotential surfaces of mechanics.[5]

Molecular and atomic theories lead us to understand the innermost constitution of bodies as discontinuous: Lamé, Cauchy, and others who defined the mathematical theory of elasticity—which demonstrates its importance and practical applications daily—succeeded only when they passed, in a true stroke of genius, from the discontinuous to the continuous. Now, like them and like Fourier with the theory of heat, the economists assume that the quantities of goods at one's disposal, which are by nature discrete, may vary continuously.

Finally, our mechanist recognizes in the logic of economic equilibrium the same reasoning he uses to establish the principle of virtual work. When he is confronted by the differential equations of economics, he immediately wants to apply to them the proven integral methods he knows so well.[6] Thus we see a discipline belonging to the [social sciences] that, while preserving its originality, is progressively assimilating the tools of mathematics—and which in the short period from the appearance of works by Whewell, Cournot, Gossen and Walras to the present day, has attempted to apply the ideas of mathematics.[7]

The application of mathematics to the biological sciences appears to be in its initial stages, even though interest in it grows day by day. It is true that rather recently a so-called "biomechanical" school was founded,

---

[4]Pareto, Vilfredo. *Cours d'Economie Politique Professé a l'Université de Lausanne.* Lausanne, 1896.

[5]Summary of chapters of a new treatise on political economy by Prof. Pareto, *Giornale degli Economisti, II XI, XX.* – See also the *Encyklopädie der Math. Wiss* §§ *3, 4.* and the appendix to *Manuale di Economia politica* by Pareto (Milano, Società Editrice Libraria, 1906).

[6]See Amoroso, Luigi, *Sulle analogie fra l'equilibrio meccanico e l'equilibrio economico* (Modena, 1910) – *Contributo alla teoria matematica della dinamica economica.* (Roma, 1912).

[7]The oldest papers on political economics by these authors are: Whewell, William, "Mathematical exposition of some doctrines of political economics," Cambridge, *Phil. Trans.*, VIII, 1829 ; Cournot, Antoine Augustin, *Recherches sur les Principes Mathématiques de la Théorie des richesses*, 1838 ; Gossen, Hermann Heinrich, *Entwickelung der Gesetze des menschlichen Verkehrs und der darausfliessenden Regeln fürmenschlichen Handeln,* Brauschweig, 1854; Walras, Leon, *Elèments d'économie politique pure ou théorie de la richesse sociale*, Paris, 1874. Leon Walras is the son of Antoine Auguste Walras, also an economist and the author of *De la nature de la richesse et de l'origine de la valeur.* Paris, 1831. In order to find the oldest traces of the ideas and principles of mathematical economy it is necessary to go back to Giovanni Ceva (born in 1647 or 1648), mathematician and hydraulic engineer. The title of his economic work is *De re numeraria*, Mantova, 1711. Compare the article by Pantaleoni on Giovanni Ceva in the *Dictionary of Political Economy*, edited by R.H. Inglis Palgrave, London, 1894.

but it does not seem to exhibit characteristics that would indicate the true beginning of a mathematical era.[8]

There are also some branches of physiology, such as physiological optics and physiological acoustics, to which men such as Helmholtz have contributed their largely mathematical training.[9] There is also what might be called a physiological thermodynamics:[10] the classic studies of the circulation of the blood—that is, on the motion of fluids in the elastic and contractile vessels; the mechanical-physiological studies on walking, running and jumping;[11] and many others I will not bother to mention. In all of these, the use of the calculus is very advanced and has produced very useful results; but those admirable and mature investigations seem to belong to the various branches of mathematical physics and mechanics rather than constituting a new field in which mathematics has found an original application.

Leaving them aside for this reason, we come, of course, to those fledgling endeavors that are tackling new problems pertaining to biology. Their results have not yet reached the level of reliability of the investigations just mentioned; they still raise some doubts, but, if only for this reason, they arouse our curiosity all the more. They concern the problems of classification and evolution—problems, moreover, so closely related that in genetic theories they tend to depend on each other.

It is obvious on only cursory inspection that the mathematical studies begun in this field exhibit all the characteristics of the first stages of scientific research—that is, of a period of orientation; in the end, we find these studies to be dominated by the method of mathematical analogy and statistical methods based on the calculation of probabilities and on the theory of errors. Moreover, the researches of the so-called biometrical school are indistinguishable from the classic statistical studies typical of social phenomena.

The method of analogy in mathematical physics is certainly not new. Nowadays, we have abandoned many illusions about giving a mechanical

---

[8]See Roux, Vilhelm, *Gesammelte Abhandlungen über Entwickelungsmechanik der Organismen*, Leipzig, 1895.

[9]Helmholtz, Hermann, *Handbuch der phisiologischen Optik.* Hamburg, 1894. – *Die Lehre von den Tonempfidungenals physiol. Grundlage für die Theorie der Musik.* Braunschweig, 1877.

[10]Compare *Les transformations d'énergie dans l'organisme*, by André Broca (Rapports présentés au congrès international de Physique réuni a Paris en 1900) t. III, Paris, 1900.

[11]*Theorie der durch Wasser oder andere inkompressibele Flüssigkeiten in elastischen Röhren fortgepflansten Welle*, by Wilhelm Weber. (Berichte d.k. Sachs. Ges. D. Wiss. Math. Phys. Klasse XVIII, 1866). Compare the paper by E. H. Weber, *Ueber die Anwendung der Wellenlehre auf die Lehre von Kreislaufe des Blutes und insbesondere auf die Pulslehre. Ibid.*, 1850. *Mechanik der menschlichen Gehwerkzeuge. Eine anatomisch-physiologische Untersuchung* by W. WEBER and E. WEBER, 1836. These studies were preceded by a long series of works, among which the deep research by [Giovanni Alfonso] Borelli is especially memorable (*De motu animalium.* Roma, 1630). About the so-called school headed by Borelli, see for example *The History of Medicine*, by [Kurt] Sprengel. Venezia, 1814.

explanation of the universe. If we are no longer confident of explaining all physical phenomena by laws like that of universal gravitation, or by a single mechanism, we substitute mechanical models for those collapsed hopes—models that may not satisfy those who look for a new system of natural philosophy but do suffice for those who, more modestly, are satisfied by analogy, and especially mathematical analogy, that somewhat dissipates the darkness enshrouding so many phenomena of nature.

A mechanical model of a phenomenon is a simple device designed only to bear some sort of relationship to the phenomenon; the sole condition is that when it operates, its parts will move, or change, according to the same laws that govern change in the corresponding variable elements of the phenomenon: These elements are assumed to be its fundamental parameters. Experience shows that such models have been useful, and still are, to orient us in the newest and darkest scientific fields, where we are groping to find our way.

We therefore welcome, with the same interest that marked the acceptance of the mechanical models of electrical induction and the thermal cycle[12] by Maxwell and Boltzmann, the bold attempt of our celebrated astronomer Schiaparelli to build a geometrical model for the study of organic forms and their evolution;[13] all the more so, because it would not be too difficult to transform Schiaparelli's model from geometrical to mechanical, making it even more intuitive.

We must distinguish two parts in the work of the Italian astronomer in order to better understand it: One concerns the proper geometrical representation of variations in the organic world, the other relates to a hypothesis that, if not intrinsically new, is at least expressed in a new form, because the author has applied his model, putting it immediately to the test.

Even those scarcely acquainted with the most elementary notions of geometry know that lines are classified: We have, for example, the straight line, the circle, and the curves of the family of conical sections, comprised of the ellipse, the hyperbola, and the parabola. Schiaparelli tried to establish a parallel between the classification of curves in a family and any system of organic entities sharing certain characters and belonging to the same group, be it order, class, or kingdom. The curves and the entities both follow a law of correlation between their parts, such that each depends on the value of certain parameters that can be expressed as a point—so passage from one form to another can be characterized by the movement of this point.

If we accept that the nature of organic entities is described by analogous parameters, Darwin's speciation hypothesis finds a model in a similar movement, closely corresponding to the law of natural selection based on

---

[12]Compare Antonio Garbasso, *Fisica d'oggi, filosofia di domani* (Milano 1910) – *Vorlesungen über Maxwells Theorie der Elektricität und des lichtes*, by Dr. Ludwig Boltzmann, Leipzig, 1891.

[13]*Studio comparativo tra le forme organiche naturali e le forme geometriche pure*, by Prof. [Giovanni] Schiaparelli, Milano, Hoepli 1898.

the struggle for existence. However, Schiaparelli recognizes in both the inorganic and organic world a general law that leads him to modify the Darwinian structure by adding a new hypothesis, with which he arrives at what he calls the principle of controlled evolution, or fixed types. In the inorganic kingdom, he sees a marked tendency emerging from the general range of phenomena toward creation of specific, well-determined types quite distinct from one another, their series and classes proceeding by marked differences and not by insensible gradations; this tendency is even more apparent in the organic world. Therefore, in his geometrical scheme, he posits a discrete series of points corresponding to the forms predestined to be the types of those species that, because of a complex of circumstances unknown to us, are the only forms possible. According to this new hypothesis, evolution ceases to be unfettered, as in the pure Darwinian theory, but remains bound by these fixed points, departure from which would generate reactions comparable to elastic forces.

These considerations of one of the most crucial problems to occupy the human mind are so closely linked to the geometrical model that we cannot imagine any way to express them without recourse to the language it offers. This alone would be enough to render Schiaparelli's endeavor worthy of the highest consideration, because it is no small thing to offer a language to a science, especially when its source is the clear spring of geometry. So many theories have died, and most are buried in oblivion; still, a vestige of them remains, indicating that they did not pass useless into the earth. That they gave birth to a single term of our language is enough for us to declare that the remote gleam of their existence brightens, even today, the great torch of knowledge. Across the centuries, something of them still lives and is of use.

Schiaparelli's work, however, does not resolve matters but rather adds a new question to the many crowding the field of biology. Even the fiercest detractors of the biometrical school cannot deny that its aim is to respond to the innumerable questions and problems arising from the outsized conceptions of Lamarck, Geoffroy Saint-Hilaire, and Darwin—starting with observation and measurement and discussing [the problems] in light of known methods already noted or of new methods it is devising. Opposition to [this work] faults its applications as perhaps too specific, and also some of its results—though not the mathematical method itself on which they are founded. It is exactly that point which we want to emphasize today.[14]

---

[14]See: Georg Dumcker, *Die Methode der Variations-statistik (Archiv für Entwicklungsmechanik der Organismen* by W. Roux, VIII, 1899). – C.B. DAVENPORT: *Statistical Methods.* New York, 1899. Compare *Las Matématicas y la Biologia*, by Angel Gallardo (*Anales de la Sociedad Scientifica Argentina, t. ll).* Buenos Ayres, 1901. – *I metodi somatometrici in Zoologia* by G. Cattaneo (*Riv. Di biologia generale*, Aprile-Maggio 1901). – See the articles by Prof. CAMERANO in *Atti dell'Acc. Di Torino* 1900-01 and those by Prof. Andres in *Rend. Ist. Lomb.* 1897-901. – J. Ludwig published extensive bibliographies on biometrical studies in volumes XLIII and XLIX of the *Zeitschrift für Math. Und Physik.* The year 1901 marked the birth of the journal *Biometrika* with the aim of gathering and spreading biometrical research (*Biometrika. A Journal for the Statistical*

Better than anyone else, Pearson has shown why this new path was chosen, and nobody has more clearly outlined its scope and import.[15] According to Pearson, one must abandon—given the present state of our knowledge—the idea of a mechanism of inheritance, and renounce the hope of obtaining a mathematical relation between any one particular parent and any one particular descendant. The causes of natural inheritance in particular cases are so complex that they do not admit an exact treatment. One must therefore begin by examining a great number of cases en masse and work one's way down little by little to increasingly limited classes; one should never establish general rules from single examples. In other words, one must proceed by statistical methods, not by a consideration of typical cases. Nowadays this may perhaps seem discouraging to the practical man of medicine, who is more interested in, for example, the hereditary pathology of a particular family than in average probabilities pertaining to an entire class of people. On the other hand, it goes to show that in the study of heredity, as in that of variation, we find ourselves confronted by a huge number of small causes all acting simultaneously, so that it is impossible to isolate any one of them.

Thus there is no way to find our bearings other than to resort to those procedures so manifestly useful in attacking similar problems: that is, procedures based on the calculus of probabilities, that most singular and remarkable of mathematical branches. If we were to analyze any one of our conscious judgments, we would be sure to find, more or less concealed, the calculation of a probability. You might say that the simplest of men, awaiting the dawn and the rising sun, in some sense owes his faith in the coming of day to an unconscious application of Bernoulli's theorem of large numbers. However, the theory of probability is the only part of mathematics whose principles are not rigorously set and are still open to criticism and discussion. What, for instance, is the solid basis of the fundamental proposition of the theory of errors? Yet everybody believes it, because, as Lippmann once remarked to Poincaré, the experimentalists imagine that it is a mathematical theorem while the mathematicians consider it an experimental fact.

Whatever confidence we may have in its basis, the theory of probability certainly has produced and is producing incontestable and incalculable benefits to all the sciences. Even enumerating them, like discussing the aforementioned general problems and apparent contradictions, would take too long. Let us look instead, without going into particulars and just as an example, at how the new school treats one of the problems it has begun to examine.

Imagine a large number of individuals of a certain species. If their forms cluster around an average type, we will find, as we move away from that,

---

Study of Biological Problem. Founded by W. R. F. Weldon, Francis Galton and Karl Pearson. Edited by Karl Pearson, Cambridge University Press).

[15] "Mathematical Contributions to the Theory of Evolution. III Regression, Heredity and Panmixia," by Karl Pearson (*Phil. Transactions of the Royal. Society of London S.A.*, CLXXXIX), London, 1897.

fewer and fewer individuals. Galton represents this graphically, measuring a particular organ and constructing a curve that expresses the relation between its size and the greater or lesser number of corresponding individuals. We then find a line that geometers call the curve of error or frequency and statisticians call the Quetelet curve. Such a set of individuals is known as a monomorphic group.

However, it may be that when, for a certain set of individuals, we construct the curve as we said, we do not find a frequency line. This means that the individuals are clustering around two or more distinct types rather than one, in which case the curve can be decomposed into two or more frequency lines. The set is then known as dimorphic or polymorphic.[16]

The breakup of a polymorphic group into its elements then becomes a purely geometrical question, which Pearson has partly resolved, and corresponds to the division of a species into its varieties.[17] If we can follow such a division through time, and observe a group changing from monomorphic to polymorphic or vice versa—or also, quite simply, if we can discover a tendency to division or to recombination—we will have understood in detail a fundamental datum of evolution, shedding unexpected light on questions of variation and regression, continuity or discontinuity [in that species].

But there is more: A frequency curve while remaining such can assume different shapes, or, as we say, can change its characteristic parameters. Recognizing the variations of parameters that correspond to a group and its subgroups in successive generations and recognizing the correlations between parameters corresponding to the various organs, today constitute a complex and extensive chapter in which the subtle insights of Laplace and Bravais[18] on probabilities find important applications.

In this way, we can establish mathematical definitions for fundamental elements in the science of heredity and [natural] selection; so that these concepts seem to emerge from the fog in which they were shrouded and take precise shape in our minds.

Highly interesting results have already been obtained in this field, on various subjects. So, for example, Pearson has found that moral traits are inherited as readily as physical traits.[19] He has also discovered that civilized races exhibit more variability than savage ones.[20] Davenport studied

---

[16]Compare *Materials for the Study of Variation treated with special regard to Discontinuity in the Origin of Species* by William Bateson, London, 1894.

[17]"Contributions to the Mathematical Theory of Evolution" by Karl Pearson *(Phil. Trans. Roy. Soc. London (A)*, CLXXXV. London, 1895). – The solution given by Pearson is valid only for the division of a dimorphic group: Prof. De Helguero has given an interesting simplification of the Pearson method (*Biometrika* IV, 1, 2 June 1905).

[18]"Analyse mathématique sur les probabilités des erreurs de situation d'un point," by A. Bravais (*Memoires présentés par divers savants à l'Académie Royale des Sciences de l'Institut de Franc*, t. IX). Paris, 1846.

[19]"On the inheritance of the mental and moral characters in Man and its Comparison with the inheritance of the physical Characters." *The Huxley lecture for 1903*.

[20]*The Chances of Death and other Studies in evolution*, 2 vol., London, Arnold.

the phylogeny and geographical distribution of certain animals,[21] Duncker [studied] bilateral symmetry in animals,[22] De Vries vegetable hybrids and monstrosities,[23] Ludwig the specific characters of various vegetable species;[24] and we could cite a great many other notable investigations, for which I refer you to specialized bibliographies.[25]

Of the vast collection of facts before us, I tried to emphasize two, the most prominent, in particular: the great advances made by political economics in recent times, ever since the discipline that Descartes and Lagrange would not have hesitated to call analytical economics became a scientific field in its own right, and the even more recent beginnings of quantitative and statistical research in biology.

The new economic studies find their mathematical counterpart in infinitesimal procedures, which economists are already confident in using; and the new directions in biology find their counterpart in the laws of large numbers and in the calculation of probabilities, tools that a whole school has adapted. With the first of these powerful and wonderful instruments, our minds can penetrate deeply into the mysteries of the infinitely small; with the second, we take the long view, in an attempt to encompass the vast contours of an infinitely great mass of facts.

In the same way that the microscope and the telescope have revealed to the histologist and the astronomer two worlds that the naked eye cannot examine, these mathematical methods open new and unknown horizons to us; like those two optical devices, these two analytical instruments are in part different, in part quite alike. But there is something that renders their magic far more wonderful than that of any conceivable system of lenses: Both show

---

[21]"Quantitative Studies in the Evolution of Pecten." *Proc. Amer. Acad. Arts and Sci.* Comparison of some Pectens from the East and the West Coasts of the U.S. – Reprinted from the Mark Anniversary, 1903, ecc.

[22]"Symmetrie und Asymmetrie bei bilateralen Thieren" in *Arch. Entw.-mech*, XVII, 533-682.

[23]"Sur la loi de disjonction des hybrides." *Comptes rendus de l'Ac. des Sci. de Paris*, 26 Marzo 1900. – Sur l'origine expérimentale d'une nouvelle espèce végétale, ibid., 1900. – La loi de Mendel et les caractères constants des hybrides, ibid., 2 Feb. 1903. –Die Mutationslehre; Veit, Leipzig 1902, 2 vol. ecc.

[24]"Beiträge zur Phytarithmetik," *Bot. Centralbl.*, LXXI, 1897. – Ueber Variationskurven. Ibid., LXXV, 1898. – Variationstatistische Probleme und Materialen. *Biometrika*, I, 11-29, 316-8, ecc.

[25]In recent years, studies related to this matter have been very numerous and to mention them explicitly is outside the scope of this article. Besides the already mentioned magazine (*Biometrika*, Cambridge), see the *Journal of Genetics* (edited by W. Bateson and R. C. Punnett, Cambridge), *The Eugenics Review* (London) and others. The most complete source for bibliographical searches is the *International Catalogue of Scientific Literature* published by the Internat. Council at the Royal Society of London. See Volumes L. (general Biology), M. (Botany), N. (Zoology), P. (Anthropology), Q. (Physiology), and the chapters Variations, Evolution, Methods and Apparati (weights and measures, biometrics), etc.

us only what is useful to see; they serve to conceal the superfluities that would cloud our vision.

I could also mention the hopes, perhaps the dreams, for the future which other methods offer—such as energetics, for example; these things have not yet been successfully tested in the social and biological sciences. But it would take me far afield, so I will conclude my talk by alluding to the past.

If we take a quick look at the birth and development of the most original and fertile ideas that have transformed and revitalized human knowledge, we immediately see that a conspicuous number of them are due to the genius of Italians. Remaining within the scientific disciplines discussed today, I will note that Giovanni Ceva, in the seventeenth century, first conceived and proposed the concepts and principles which today inform economics; and the vestiges of probability theory can be found in the fourteenth century in a commentary about Dante.

And from those far-off epochs, across the centuries, to today continues unbroken the succession of Italians who have led us into the modern world, in which Italy now plays such a great part.

Science at the Present Moment and the New Italian Society for the Progress of the Sciences

This talk inaugurated the first congress of the Italian Society for the Progress of the Sciences, held in Parma on September 25th, 1907. It was published in the proceedings of the Society (Rome, 1908) and in Scientia, Year I, V. II: 4 (Bologna, 1907).

More than thirty years have elapsed since the last congress of Italian scientists was held, in Palermo. Today this noble institution reawakens and salutes the new sun shining before it.[1] Since the last congress, the material and social condition of Italy has undergone profound change, and scientific knowledge has rapidly developed and matured worldwide. The scientific advances appearing in this brief period of time have renewed not just our daily lives but also the general direction of our culture, breeding and strengthening a completely new, modern, original spirit—what I would call a scientific consciousness—that dominates our times just as other intellectual climates dominated earlier times. This scientific spirit, which has by now pervaded every level of our society, high and low, arose not just from the genius of our finest minds, to whom we owe the great discoveries and ideas, but also from the day-to-day activities of society as a whole, which makes constant use of them. Our healthiest and liveliest energies are today revived by this guiding spirit; heeding its call, our old association rises again.

We can be sure that the public's attitude toward science is very different today from what it was only fifty years ago. Indeed, the general public, as never before, has witnessed the birth and development of our generation's discoveries in the scientific laboratory, their transfer to the factory floor, and finally their invasion of everyday life.

Thus, at this moment in history we are presented with vast numbers of people who are fascinated by the inventions that have rapidly produced so much comfort and wealth and so fundamentally influenced their habits and their ways of thinking. They are trying to understand the truths of science, and to know them in detail; more important, they expect from science material and social progress. It is perhaps this state of expectation,

---

[1] See V. Volterra, "Proposta di una associazione italiana per il progresso delle scienze" (1906), *Saggi scientifici* (Bologna: Zanichelli, 1920; Zanichelli reprint, 1990, with an introduction by Raffaella Simili), 81-95.

so characteristic of our times, that has fostered the aforementioned scientific spirit.

I will try to illustrate my thesis with the familiar comparison between the development of steam and that of electric power:

Historically, the use of the former preceded the use of the latter; in fact the practical applications of electricity, along with its transmission, have taken hold in just the last thirty years, as we all know. [James] Watt and [George] Stephenson were practical-minded people, who rose by dint of their genius from their workshops to the Academy of Science and the highest levels of industry; they demonstrate that—at least in those heroic times when the steam engine was born—the workshops were the source of the most ingenious and celebrated inventions. Only then did science, investigating the operation of industrial engines, construct that wonderful monument of thermodynamics, whose principles govern all natural phenomena.

The contrary is true of electricity.

The [electric] battery and its immediate applications came straight out of the physics laboratory of the University of Pavia. With the induction principle, [Michael] Faraday laid the basis for all the applications of electricity, from the dynamo to the telephone. [Antonio] Pacinotti's [iron] ring, Galileo Ferraris's rotating field [motor], and the discovery of electrical waves are all the result of research carried out in scientific laboratories.

In summary, while the invention of the thermal engine was the starting point for much theoretical research, theoretical electrodynamics created, on its own, the many and wonderful applications of electricity. In this case, as in numerous others, the history of words summarizes and mirrors the history of a long, slow evolution of ideas. For instance, "temperature," originally a vague and approximate expression of atmospheric conditions, assumed, little by little, a more scientific import, until thermodynamics conceived of it as the integrating factor of a differential expression. By contrast, the notion of "potential"—originating with [Marquis Pierre Simon de] Laplace in celestial mechanics, enriched by [Carl Friedrich] Gauss, transplanted into electrostatics by [George] Green, and introduced into electrodynamics by [Gustav Robert] Kirchhoff—is in our day known as "voltage," a name on the lips of even the humblest laborers in the farthest, most isolated reaches, where just one electric light illuminates the night in a poverty-stricken village.

Everywhere apparent, among people of every stripe, benefiting us in every circumstance of our existence, enlivening and intensifying our lives, the applications of electricity are a demonstration, unparalleled in their effectiveness, of the power of scientific research and the usefulness of the most abstract thinking.

While there is thus a direct connection between the practical and the scientific life, science professionals, by a natural correspondence, are also drawn to the multitude of their fellow men; their existence is not confined to their laboratories and offices; they feel compelled by their daily, intimate contact with society, to participate in the life of that society. The demeanor of the modern scientist is thus very much different from that of the savant of a few years ago. In comparing these two distinctly opposite types, one thinks of two great men whose genius encompasses the world of physics: Gauss and Lord Kelvin.

The former worked alone for fifty years in modest Göttingen, neither approached nor approachable, making public only what he felt was perfected, while jealously guarding, or confiding only to close friends, the newest and most original of his thoughts—those that would later provoke such excitement and such an intellectual revolution. The latter, today our greatest living scientist, brought his fertile and protean efforts to bear in both worlds, bravely tackling and popularizing the most original and singular theories his genius could devise and playing his part in the grand modernization of England, to universal admiration.

Nevertheless, how many points of contact there are between the two scientists! If Lord Kelvin joined Europe and America with the transatlantic telegraph, Gauss was the first to imagine an electric telegraph linking his observatory to the physics laboratory of his friend [Wilhelm Eduard] Weber. The crystalline, geometrical elegance of Lord Kelvin's theory of images is alone comparable to the harmonious divine beauty of the properties of numbers discovered by Gauss. Not the form of their genius, then, but their character—and even more, the dissimilarity of their circumstances—was the source of the marked difference between them.

The intimate connection of science with practical life has not, moreover, diminished its majestic and solemn character—a character nourished and enlivened by what I have called the scientific spirit. Dynamos, the gigantic monuments of our present era, accomplish their work silently and at high speed in marvelous, modern edifices—not smoking and roaring, like the factories of old, but bright and quiet; such places recalling in their solemn and austere grandeur the monuments of another era, the ancient cathedrals that raise their wonderful spires to the sky. Beneath those airy arcades raised by the artisans of the Middle Ages, the soul fills with a solemn emotion, an empathy with the hopes and dreams of long-ago centuries. An emotion just as profound adheres to the sacred places of our modern industries; our hearts are overcome with pride and satisfaction and a serene faith in a secure future.

In the recent movement of science toward practical applications, Italy may well have benefited more than any other country, so the aforementioned scientific spirit—although it developed later here than elsewhere—is making increasingly rapid and gratifying progress before our very eyes. There was, not so many years ago, much gloom over the state of our economy, but by virtue both of people and circumstances it has miraculously and unexpectedly improved: An abundant source of wealth sprang from industry, a source we had thought denied to our country by Nature itself.

When the 1884 Turin Exposition saw the appearance of the first electric transformers—which would be compared to the lever, the fundamental mechanical device—the seed that would produce so much wealth was sown: The potential energy contained in our mountains and our rivers flowed onto the plain, powering thousands of our factories and lighting up our cities. The wires we see spreading like a net above our houses and stretching far away are the most eloquent evidence of our economic prosperity. In the lonely Roman countryside, they run parallel to the magnificent aqueducts. Lit by a fiery sunset, they once spoke to Lord Kelvin in a language as solemn as the majestic ruins of ancient Rome.

I have up to now tried to describe the effects of recent scientific developments on the modern world—and I mentioned briefly the evolution of the scientist—as best I could within the narrow scope allowed me. I brought to light, and only fleetingly, just one aspect of the grand landscape of science. This aspect we may consider its exterior; the interior, which is doubtless of greater interest, has remained concealed.

Nevertheless, the value of science does not lie only in its practical utility, nor do its strength and its support rest solely with the public who make use of its results and admire its producers. The value of science, which inspired in [Henri] Poincaré so many deep thoughts and so many eloquent pages, is revealed in another way—a nobler, higher way: It is manifest in the intimate character of the work itself, in the satisfaction it provides. In the pure, disinterested search for the truth, which is the supreme goal, the researcher's greatest joy is in learning, not in knowing.

But it is not my task to speak to you about these interior aspects of science. That will be made clear in the plenary talks given by illustrious scientists on the three great branches of science—the physical-chemical, biological, and social sciences; in the opening addresses of the presidents of the various sections; in the progress reports of the chapters of different disciplines; in original communications and discussions. In short, only the entire proceedings of this congress can present the spectacle of all that lives and breathes within the scientific world; it will show you the mysteries we are feverishly trying to solve, the victories we have achieved, and the disappointments we have suffered; however cruel, these must not be ignored.

The present moment offers no more than a chance for a fleeting overview of the various disciplines. Too many fundamental discoveries are rapidly accumulating and waiting to be classified and organized, while a deep, intense, and, I would say, almost ruthless critique scrutinizes and dissects every thought, every speculation, undermining systematic constructions that only yesterday seemed poised to defy the centuries. Today, they are nothing but scattered ruins upon which a few are already trying to rebuild.

But I must not fail to mention what every careful observer already knows from personal experience—that almost all scientific disciplines are today in great crisis, one in the conditions under which they work and one in the philosophical thought that informs the work.

The first of these presents a singular contrast: While, on the one hand, the acquisition of technical skill requires specialization and the division of scientific labor (so much so that an entire lifetime is often barely enough to acquire the abilities essential to progress), on the other hand, the various disciplines have so interpenetrated each other that it is not at all clear how we can move forward in one without a deep understanding of many others—and not only those generally thought to be closely related but also new fields now seen to be related as well. Collaborative efforts—common and more intense in the established sciences, such as astronomy, and in the great schools that arise around men of genius, as in the most advanced countries—tend to coordinate and discipline individual energies, but we are still far from achieving the balance that alone can generate the economy of effort we all aspire to.

But this is not the only crisis: The other, of interest to the philosophical thinker, is even more striking.

It is an ancient belief that, in science, hypotheses are means and not ends, that tomorrow we can abandon what today we embrace—so ancient a belief, in fact, that the Greek astronomers found all cosmic hypotheses acceptable, provided they were able to use them to calculate the positions of the stars.

But the present period differs from those preceding, because suddenly not just single hypotheses but also great principles, some of which had been unquestioned, universally accepted, and taught almost as dogma, have become the subject of debate, while systems that had long seemed buried for good have been unexpectedly resurrected. Perhaps posterity will regard the present moment as we regard the Renaissance, when the very foundations of our worldview were laid.

At the center of the modern critical movement that led to this period of unrest is undoubtedly mathematics.

Scarcely a century has passed—as [Gősta] Mittag-Leffler acutely observes in the beautiful pages dedicated to the memory of [Niels Henrik] Abel[2]—since that great analyst proclaimed that mathematics is an end in itself, and finds its model in itself. And yet, we can add, in this century more than in any other, mathematics moved beyond the limits of its intrinsic activity and enriched fields far from its own, giving rise to a new and flourishing philosophy.

As Abel reasoned, mathematics fell back upon itself, first to build, and then to consolidate the theory of functions and the geometry that would provide the foundation for recent research. This has led, by a profound and assiduous study of its own methods and concepts, the analysis of thought to such perfection, such sharpness, flexibility, and power, that [mathematics now] penetrates and affects all scientific and philosophical speculation. This

---

[2]G. Mittag-Leffler, *Niels Henrik Abel* (*Revue du Mois*, t. VI, 10 July -10 August, 1907).

is how, to mention a single famous example, a purely geometrical paper by [Eugenio] Beltrami,[3] derived from research on non-Euclidean geometry by Gauss, [Nicolai Ivanovich] Lobatschewski, and [Bernhard] Riemann, became of such universal importance that it cast new light on the fundamentals of logic itself. The modern critique initiated by mathematicians has successfully invaded the physical sciences and produced new currents of thought. Mechanics—not for the first time—was the road by which this new direction arrived.

But an important development has come along which has tended to change the very position of this discipline within the physical sciences.

Everyone in our generation (I think I can say openly) was taught those principles that we nowadays call mechanistic; and in fact the idea that all phenomena, at least those studied in physics, could be reduced to motions and all fell into the orbit of classical mechanics was a dogma that all schools used to bow to, its origins lost in the mists of long-ago Cartesian philosophy.

But little by little the mechanistic theories have changed; difficulties have accumulated; ideas beginning with those of [William] Rankine and strenuously supported by [Ernst] Mach (who however holds a unique position in the philosophy of science), and developed by others, have made their way. Many have fought beneath the flag bearing that celebrated inscription: "War Against the Mechanical Mythology."

Energetics came along and classified mechanics as just another physical science, no longer the basis for all of them. A new common basis was established, founded on broader and more comprehensive principles.

This framework of ideas, the subject of so much discussion and debate among mathematicians and natural scientists, does not, however, represent the extreme limit that may be arrived at. The critique developed by mathematical physicists, comparable to ultramicroscopic observation as opposed to ordinary microscopy, is now scrutinizing and casting doubt on these very principles.

In truth, the modern concepts of the electrical constitution of matter, while related to atomic and kinetic ideas, reintroduce principles similar to those of the old physical mechanics; on the other hand, they engender a profound revolution in the concepts of mass and inertia established by Newton as the basis of all natural philosophy—a revolution made even greater by the new theories of relativity being developed today. We can well understand that the dominant principles of half a century ago scarcely survive the storm that seemingly overwhelms them.

This crisis reverberates throughout the natural sciences; and in the meantime, in the heavens as on the earth, a thousand things are revealed that philosophy could not even dream of, from the effects of light on the movement of stars to new sources of terrestrial heat.

---

[3]Eugenio Beltrami, *Saggio di interpretazione della geometria non-euclidea. Giornale di Matematiche*, vol. VI. See also *Opere matematiche* by Eugenio Beltrami, v. I, p. 374, Milano 1902.

Moreover, when we realize that little by little the theories based on [particle] emission seem on the rise again, while a few years ago the wave theory was the sole victor in the field of phenomena propagating over distance, our surprise increases—to see new and unexpected speculation joined by the no-less-unexpected ancient ideas that, like ghosts, rise from graves thought forever closed.

Perhaps more than in physics itself, this revolution in ideas shows up in its sister science, chemistry, where new ideas of atomic structure have shaken up the old classical theories; where the alchemical dream is reborn, full of mystery and promise; where a new and flourishing field, physical chemistry, rich in results and in hopes, has arisen.

In the virgin field of physical chemistry we encounter opposing trends, and it is very difficult to decide to which of them the most interesting results are due. In fact, if, on the one hand, the purely kinetic theories have produced the discoveries of [J. D.] Van der Waals, Arrhenius, and many others, energetics, on the other, has found nothing to destroy or change but much to build on profitably; in this field, its beneficial influence has been deeply felt.

There is a characteristic kind of reasoning that I would not hesitate to call energetics. Powerful and fertile, its origins go back to [Sadi] Carnot and his memorable cycle; it dominates all of theoretical physical chemistry, and the doctrines it has inspired have had the most important consequences. But the applications of this kind of reasoning have extended much further, into different sciences: I must not fail to mention that what I consider the most characteristic and stimulating result of mathematical economics—the general demonstration that the differential equation of economic equilibrium is infinitely integrable—can be based on it.[4] Let us presume and hope that many other results can be so obtained.

This notion naturally leads me to the subject of the influence of mathematics and the innovations it has produced in the social sciences, but I would be overstepping the boundaries I set for myself. These limits allow me to note only in passing that the infant physical chemistry, of which we have just spoken, has imparted new data and thus a new direction to physiology.

With regard to the biological sciences, I will mention only briefly the crisis affecting fundamental concepts of life, evolution, and heredity that has so disrupted Darwinian doctrine. After guiding minds for half a century, Darwinism, in the wake of recent zoological and botanical research, seems to be losing, if not its importance, perhaps the predominance it once enjoyed.

Several factors have combined, and are combining, to produce this transformation of thought, and I cannot mention all of them, but they certainly include observations arising from science and technology, new experimental methods in physiological chemistry, and not least those of biometry (always

---

[4]Compare V. Volterra, "L'economia matematica e il nuovo manuale del Prof. Pareto" (*Giornale degli Economisti*, April 1906).

a much appreciated source of data and definitive laws), [all of which] herald a new era for biology.[5] It seems to me that we are seeing the outline of methods that one day may have broad application.

There exists, as [Émile] Picard observes, a mechanics of heredity that contrasts with the classical one, according to which the present state of a system alone determines its future. The concept of function that has dominated mathematics in the last century has been extended, and this extension raises new questions, which have led to useful results. It is easy to foresee that this might constitute a hereditary mechanics capable of representing with greater precision the elastic, magnetic, and other such phenomena in which hysteresis is of great importance.[6] I wonder what the future holds for this mechanics, if it succeeds in penetrating the field of biological phenomena.

But it is not safe to make any kind of prophecy. The history of science teaches us that sometimes all it takes is the discovery of one trivial fact to overturn the most well-founded predictions. Extrapolation in a field whose laws are uncertain or unknown is a danger I intend to avoid.

But it is time for me to conclude my talk and summarize my thoughts.

I wanted to lay before you two facts: the growing proximity between the public and scientists, due to the predominantly scientific spirit of the present age, and the great crisis that today afflicts so many branches of knowledge. Correspondingly, human society has developed new needs—needs that every civilized country must meet, lest its intellectual life languish or come to a halt and the source of its prosperity dry up.

The internal crisis threatening so many disciplines requires widespread and open discussion among scholars, in which they can express the thoughts that occupy them, the doubts that trouble them, the difficulties that thwart them, the hopes that drive them on. Books and papers do not and never will serve this function; the need is precisely to talk about and listen to ideas that nobody dares yet to publish or that will never be published.

The old academies are too closed; the teaching institutions have other goals; the individual scientific societies are too narrow to be useful for these purposes, which can be fulfilled only within a society that includes experts in all disciplines, like the association we inaugurate today.

On the other hand, every day we see the proliferation of scientific works and publications that engage the larger public, who flock to conferences and lectures in ever increasing numbers and with ever increasing curiosity. But no book will ever show how a scientific thought is born and formed, how an initially vague idea takes shape in the mind of the researcher; no lecture will ever illuminate it, just as no display in a zoological museum will ever fully represent life itself.

---

[5]See V. Volterra, "Sui tentativi di applicazione delle matematiche alle scienze biologiche e sociali" (1901), in *Saggi scientifici*, 1-35.

[6]See V. Volterra, "L'applicazione del calcolo ai fenomeni di eredità" (1912), in *Saggi scientifici*, 189-218.

Well, whatever the public cannot learn from books and lectures will become clear when they attend the discussions among the scientists, since it is in spontaneous and lively debate that the development of ideas artificially popularized by those who know too much appear in their truest light.

Not this alone does our country ask of its rising institution—not only the satisfaction of intellectual curiosity but also the useful promotion and encouragement of fruitful study and new and vital research. Every day, industrialists, businessmen, and those in the professions turn to science, continually urged by a growing multitude who hope that science will provide a solution to the new, complex, and pressing problems they face and who invoke a science that will prove victorious over ever rising difficulties.

These questions, of such interest to science and technology, can be effectively posed only before an open and liberal association like our own, which welcomes men in diverse fields, because even to formulate them requires the cooperation of different disciplines. It will then be up to the laboratories and the scientific institutes to develop and resolve them. This is why the committee has warmly and enthusiastically promoted the new Society, why it has convened all of you and now rejoices in seeing how many of you have gathered here from your schools, laboratories, and practical occupations.

We are all warmed by an equal passion for this newborn Society, which our votes consecrate to a great and noble end; its destiny smiles on us with equal hope; its future seems linked to the future of our country itself, as it moves on toward its high destiny.

In conclusion, my thoughts return to the comparison I drew a little while ago between the present era and the Renaissance. In that time of the wonderful restoration of intellectual life, Italy became the very center of universal scientific thought. Today, I venture to wish that the destiny reserved for us not be a lesser one, as the pure and authentic Italian soul rises and takes shape, reviving our thought and restoring to us our ancient country.

Acknowledgments

Asked once why she wrote about families, the Italian novelist Natalia Ginzburg replied, "because that is where everything starts, where the germs grow." In the case of Vito Volterra's "germs," I have benefited greatly from the generous help offered over the years by family members, including Vito Volterra Jr.; Edoardo and Nella Volterra and their daughter Virginia Volterra; Silvia D'Ancona, Edoardo Achille Almagià, Luisa Almagià, Ginetta Montecorboli, Fiorenza Almagià, and Massa Mori. I owe a special debt of thanks to Libera Levi-Civita, who welcomed me into her home in Rome in the summer of 1971, and put her husband's extensive collection of letters and personal documents at my disposal. Pier Vittorio and Susanna Silberstein Ceccherini, and Francesco and Teresa Scaramuzzi have been faithful companions on my long voyage to tell the story of the golden age of Italian mathematics through the eyes of Volterra and his circle of mathematicians. For permission to quote from the Vito Volterra Collection and the Levi-Civita Papers at the Accademia Nazionale dei Lincei, I thank Edoardo Vesentini, past president of the Lincei, and Giovanni Conso, the current president; and William Roberts, University Archives, UC Berkeley, for permission to quote from the Griffith Evans papers.

Thanks also go to the many archivists, librarians, and historians in Italy who patiently answered my questions, and found and supplied copies of documents, located out-of-print books, and in some cases, photographs. In Florence, Teresa Porcella, at the Archivio dello Stato Civile del Comune di Firenze; in Rome, Marco Guardo and Enrica Schettini, at the Accademia Nazionale dei Lincei, and Emilia Campochiaro, Cristina Cannizzo, at the Archivio Storico del Senato della Repubblica, and Federico Pommier, at the Biblioteca del Senato della Repubblica; in Pisa, Sandra di Majo, Rosangaela Cingottini, Milletta Sbrilli, and Stefano Pieroni, at the Scuola Normale, and Giovanna Tanti, at the Archivio di Stato; in Turin, Livia Giacardi, Sandro Caparrini, Clara Silvia Roero, and Paola Novaria, at the University of Turin; in Bologna, Fulvio Cammarano and Mariangela Mafessanti, at the University of Bologna; and in Crema, Mauro De Zan.

Giovanna Bergonzi and Anna Maria Galoppini provided information about anarchists, women, and higher education in late 19th century Tuscany; Vittoria Amighetti used her pharmaceutical training to diagnose Volterra's heat rash at Pisa; Bruno Di Porto, Riccardo Di Segni, and Lionella Viterbo educated me about Bar Mitzvahs and other Jewish ceremonial traditions in nineteenth-century Italy; Franca Foà Ascoli, president of the Comunità Ebraica di Ancona; Barbara Martinelli at the Comunità Ebraica di Pisa, and the Archivio Storico della Comunità Ebraica di Roma graciously searched their records on my behalf.

Closer to home, I wish to thank Lev Ginzburg for a mini-tutorial on mathematical biology; Judith Nollar and Shady Peyvan, and the entire interlibrary loan office in Millikan Library, at the California Institute of Technology, who never flinched at my bibliographic questions and requests; Bonnie Ludt, who dealt with the endnotes and much else with her usual efficiency; Charlotte Erwin, for her assistance with translating German documents; and all the staff of the Caltech Archives, who kept the place running smoothly while I was off doing research. I am also grateful for the hospitality of Jed Buchwald and the Burndy Library and the Dibner Institute for the History of Science and Technology on the Massachusetts Institute of Technology campus, where the Vito Volterra Collection was housed before it moved in 2006 to the Huntington Library in San Marino, California. I also benefited from reading Tim Sluckin's unpublished manuscript, "Mathematical Appreciation of Vito Volterra."

In addition to interviewing members of the Volterra family, Carlotta Scaramuzzi, a 1994 Caltech Summer Undergraduate Research Fellow, transcribed and translated many Italian letters; Elisa Piccio, who joined the project in 2002, picked up the reins from Carlotta and has done an equally fine job as a translator, indefatigable Internet fact checker, and jack-of-all trades research assistant.

Finally, I have had the help of two consummate editors, Sara Lippincott and Heidi Aspaturian, in shaping the narrative backbone of this biography. In some ways, *The Volterra Chronicles* is their story too, and I am grateful not only for their keen eyes and appreciation of the English language, but also for their collegiality and warm friendship. Donald Babbitt deserves a round of thanks also for serving as my mathematics mentor and for introducing me to Sergei Gelfand, the editor at the American Mathematical Society, who kept faith with the book and the author.

Selected Bibliography

The starting point for this biography is the collection of Vito Volterra's scientific papers and correspondence at the Accademia Nazionale dei Lincei, in Rome, and Giovanni Paoloni's important documentary history, *Vito Volterra e il suo tempo*, published on the occasion of an international conference at the Lincei in 1990 in memory of the mathematician. Family papers collected by Luisa Almagià and Edoardo Achille Almagià provided further details about the Almagià-Volterra family.

In Rome, the Besso Foundation houses an important collection of contemporary journals and books relating to Rome around the turn of the twentieth century. The Italian State Archives, situated on the grounds of the Esposizione Universale Rome (the intended site of an aborted 1942 World's Fair) contain the records of the Accademia d'Italia, the Consiglio Nazionale delle Ricerche, and other government agencies—records that reveal how the Fascist state viewed its scientific community.

Oral histories provided me with another important source of information about Vito Volterra, his family, and conditions in Italy in the 1930s. Interviews with his personal librarian, Gualda Massimi; two of his granddaughters, Dr. Silvia D'Ancona and Professor Virginia Volterra; and various other family members, including his cousins Edoardo A. Almagià, Ginetta Montecorboli, and Luisa Almagià, figure in the book.

Archives in the United States that contain information relevant to the life and times of Volterra include the Griffith Evans Papers at the Bancroft Library in Berkeley, California, the General Education Board files in the Rockefeller Archive Center in North Tarrytown, New York, the George E. Hale Papers at the California Institute of Technology, in Pasadena, and the Vito Volterra collection at the Huntington Library, Art Collections, and Botanical Gardens, in San Marino, California. Secondary sources in English on Volterra and on the history of Italian mathematics are scarce. In Italian, there is a considerable list of books about the development of mathematics in Italy between 1860 and 1940. These books are rarely translated into English. The Italian mathematician Giorgio Israel has written extensively on Vito Volterra; his most recent book, *The Biology of Numbers* (Birkhäuser, 2002), written in collaboration with Ana Millán Gasca, explores the development of Volterra's ideas in the field of mathematical biology; another recent book, *Scienza e razza nell'Italia fascista* (Società editrice il Mulino, 1998), written in collaboration with Pietro Nastasi, deals with issues of race and anti-Semitism. Other important works to appear in recent years include Umberto Bottazzini's *Va' Pensiero* (Società editrice il Mulino, 1994); Raffaella Simili, ed., *Vito Volterra, Saggi Scientifici* (*Zanichelli*, 1990); and *La Matematica Italiana dopo L'Unità* (Marcos y Marcos, 1998), an important collection of essays by some of the leading Italian historians of mathematics working today, edited by S. Di Sieno, A. Guerraggio, and P. Nastasi. In general, these books are aimed at working mathematicians and specialized historians of mathematics.

There is a small number of books and articles dealing with Volterra's circle of mathematicians. Emma Castelnuovo has written an article "Federigo Enriques e Guido Castelnuovo" (1997, *Bollettino U.M I.*) about her father, Guido Castelnuovo; Giovanni Enriques offers memories of his father, Federigo Enriques, in his book *Via d'Azeglio 57* (*Zanichelli*, 1983); the mathematician and historian of science Dirk Struik has touched on his association with Tullio Levi-Civita in Rome in the 1920s in an unpublished autobiography he shared with me. Emilio Segrè (no relation to Beniamino) writes about the Rome mathematicians in his own autobiography, *A Mind Always in Motion* (University of California Press, 1993). In 1959 Dover republished the first English translation of Volterra's *Theory of Functionals and of integral and Integro-Differential Equations*. The Dover edition includes a preface by Griffith C. Evans and the Royal Society biography and bibliography prepared by Sir Edmund Whittaker. Evans supplemented Whittaker's bibliography, but the biography is abridged.

The standard political history of Italy during the period of Volterra's life is Denis Mack Smith's *Modern Italy* (University of Michigan Press, 1997). The standard history of Italian Jewry is Attilio Milano's *Storia degli ebrei in Italia* (Turin, 1963); Renzo De Felice's account of Italian Jews under Mussolini, *The Jews in Fascist Italy: a history* (Enigma Books, 2001) remains the authoritative work on the subject. A small selection of other works consulted in the preparation of this biography is appended.

Accademia Nazionale dei Lincei. *Convegno internazionale in memoria di Vito Volterra: Roma, 8-11 ottobre 1990.* Rome, 1992.

Artusi, Luciano and V. Giannetti. *A vita nuova: ricordi e vicende della grande operazione urbanistica che distrusse il centro storico di Firenze.* Florence, 1995.

Artusi, Pellegrino. *La scienza in cucina e l'arte di mangiar bene.* Turin, 1974.

Berengo, Marino. "Italian Historical Scholarship since the Fascist Era." In *Daedalus* 100 (1971): 469-484.

Boatti, Giorgio. *Preferirei di no: le storie dei dodici professori che si opposero a Mussolini.* Turin, 2001.

Bôcher, Maxime. *An Introduction to the Study of Integral Equations*. New York, 1971.

Borioni, F. *Le feste anconitane nel settembre dell'anno 1841 per la faustissima venuta e dimora del N.S. Gregorio XVI regnante felicemente*. Ancona, 1841.

Bottazzini, Umberto, Alberto Conte, and Paola Gario, eds. *Riposte Armonie: lettere di Federigo Enriques a Guido Castelnuovo*. Turin, 1996.

Brizzi, Gian Paolo. *Bologna 1938: Silence and Remembering. The Racial Laws and the Foreign Jewish Students at the University of Bologna*. Bologna, 2002.

Capristo, Annalisa. "L'esclusione degli ebrei dall'Accademia d'Italia." 67 (2001): 1-36.

———. *L'espulsione degli ebrei dalle accademie italiane*. Turin, 2002.

———. "Tullio Levi-Civita e l'Accademia d'Italia." *Rassegna Mensile di Israel 69* (2003): 237-254.

Casella, Antonia, Alessandra Ferraresi, Giuseppe Giuliani, and Elisa Signore. *Una difficile modernità*. *Tradizioni di ricerca e comunità scientifiche in Italia, 1890-1940*. Pavia, 2000.

Castelnuovo, Guido. "Obituary." *Accademia Nazionale dei XL* 25 (1945): 87-95.

Compte rendu du deuxième Congrès international des mathématiciens, tenu à Paris du 6 au 12 août 1900. Procès-verbaux et communications, publiés par E. Duporcq, secrétaire général du congrès. Paris, 1902.

"Conseguenze culturali delle leggi razziali in Italia." *Atti dei convegni Lincei* 84 (1990) 5-133.

Cooke, Roger. *The Mathematics of Sonya Kovalevskaya*. New York, 1984.

Corbino, Epicarmo. *Racconto di una vita*. Naples, 1972.

D'Ancona, Alessandro. *Ricordi ed affetti*, Milano, 1902.

De Zan, Mauro. "Sul carteggio tra Vito Volterra e Giovanni Vailati." Annuario del Centro Studi Giovanni Vailati (2003): 79-90.

Dragoni, Giorgio. *Instrumenta*. Proceedings from a conference held on March 10-11 in Bologna, Italy. 1990.

Foà, Salvatore. *Gli ebrei nel Risorgimento italiano*. Assisi, 1978.

Giacardi, Livia. "Corrado Segre maestro a Torino. La nascita della scuola italiana di geometria algebrica." *Annali di storia delle università italiane* 5 (2001): 139-163.

——— and Marcella B. Bagnasco, eds. *I due volti del sapere: centocinquant'anni delle Facoltà di Scienze e di Lettere a Torino*. Turin, 1999.

———— and Clara Silvia Roero, eds. *Bibliotheca mathematica: Documenti per la storia della matematica nelle biblioteche torinesi.* Turin, 1987.

Giuliani, Giuseppe. *Il Nuovo Cimento:novant'anni di fisica in Italia, 1855-1944.* Pavia, 1996.

Giusti, Enrico and Luigi Pepe, eds. *La matematica in Italia 1800-1950.* Florence, 2001.

Goodstein, Judith R. "The Italian Mathematicians of Relativity." *Centaurus* 26 (1983): 241-261.

————. "Levi-Civita, Einstein, and Relativity in Italy," *Atti Rendiconti Accademia dei Lincei* 8 (1975): 43-51.

————. "The Rise and Fall of Vito Volterra's World." *Journal of the History of Ideas* 45 (1984): 607-617.

————. "Theodore von Karman and Applied Mathematics in America." With John L. Greenberg. *Science* 222 (1983): 1300-1304.

Guerraggio, Angelo. *La matematica italiana tra le due guerre mondiali.* Bologna, 1987.

———— and Pietro Nastasi. *Italian Mathematics between the Two World Wars.* Basel, 2005.

Gunzberg, Lynn M. *Strangers at Home: Jews in the Italian Literary Imagination.* Berkeley, 1992.

Hadamard, Jacques. *An Essay on The Psychology of Invention in the Mathematical Field.* Princeton, 1945; Dover reprint, 1954.

Hearder, Harry. *Italy in the Age of the Risorgimento, 1790-1870.* London, 1983.

Holmes, George, ed. *The Oxford History of Italy.* Oxford, 1997.

Impresa Almagià. Rome, ca. 1954.

International Congress of Mathematicians. *Verhandlungen des ersten Internationalen mathematikerkongresses in Zürich vom 9. bis 11. August 1897,* ed. by Ferdinand Rudio. Leipzig, 1898.

Israel, Giorgio and Pietro Nastasi. *Scienza e razza nell'Italia fascista.* Bologna, 1998.

Keen, Linda, ed. *The Legacy of Sonya Kovalevskaya: Proceedings of a Symposium,* Vol. 64. Providence, 1987.

Kertzer, David I. *The Kidnapping of Edgardo Mortara.* New York, 1997.

Kingsland, Sharon E. *Modeling Nature: Episodes in the History of Population Ecology.* Chicago, 1985.

Koblitz, Ann, H. *A Convergence of Lives. Sofia Kovalevskaia: Scientist, Writer, Revolutionary.* Boston, 1963.

Laras, Giuseppe. "Intorno al 'ius cazacà' nella storia del ghetto di Ancona" *Quaderni storici delle Marche* 7 (1968): 27-55.

Linguerri, Sandra. *Vito Volterra e il Comitato Talassografico Italiano: imprese per aria e per mare nell'Italia unita (1883-1930).* Florence, 2005.

Maiocchi, Roberto, "Il ruolo delle scienze nello sviluppo industriale italiano." In *Storia d'Italia: scienze e tecnica nella cultura e nella società dal Rinascimento a oggi,* edited by Gianni Micheli, 865-999. Turin, 1980.

"Matériaux pour une biographie du mathématicien Vito Volterra." *Archeion* 23 (1941): 325-359.

Maz'ya, Vladimir and Tatyana Shaposhnikova. *Jacques Hadamard, A Universal Mathematician.* Providence, 1998.

Minazzi, Fabio, ed. *Giovanni Vailati intellettuale europeo.* Milan, 2006.

Monna, A. F. *Functional Analysis in Historical Perspective.* Utrecht, 1973.

Moscati Benigni, Maria Luisa. *Marche: itinerari ebraici: i luoghi, la storia, l'arte.* Venice, 1996.

Nastasi, Pietro and Rossana Tazzioli. *Per l'archivio della corrispondenza dei matematici italiani: aspetti scientifici e umani nella corrispondenza di Tullio Levi-Civita (1873-1941).* Palermo, 2000.

Neider, Charles, ed. *The Complete Travel Books of Mark Twain.* Garden City, 1966.

Paoloni, Giovanni. "Ricerca e istituzioni nell'Italia liberale." In *Ricerca e istituzioni scientifiche in Italia,* edited by Raffaella Simili, 93-117. Rome-Bari, 1998.

———. "La ricerca fuori dall'Università: il quadro istituzionale." In *Una difficile modernità, 1890-1940. Tradizioni di ricerca e comunità scientifiche in Italia 1890-1940,* edited by Casella, A., A. Ferraresi, G. Giuliani, E. Signori, 389-404. Pavia, 2000.

———. "Vito Volterra and the International Board of Education. A correspondence (1924-1930)." In *The "Unacceptables." American Foundations and Refugee Scholars between the two Wars and after,* edited by G. Gemelli, 273-298. Brussels, 2000.

Papa, Emilio. *Storia di due manifesti: il fascismo e la cultura italiana.* Milan, 1958.

Pavia, Rosario and E. Sori, eds. *Ancona.* Bari, 1990.

Pérès, Joseph. "Cenni biografici." *Opere matematiche* I (1954): xxvii-xxxiii.

Pugno, Giuseppe Maria. *Storia del Politecnico di Torino: dalle origini alla vigilia della seconda guerra mondiale.* Turin, 1959.

Ravà, Luigi. "Le laureate in Italia." *In Bollettino ufficiale del Ministero dell'IstruzionePubblica, April 3, 1902*, 634-654.

Reid, Constance. *Hilbert.* New York, 1970.

Sacerdoti, Annie and L. Fiorentino. *Guida all'Italia ebraica.* Genoa, 1988.

Schram, Albert. *Railways and the Formation of the Italian State in the Nineteenth Century.* New York, 1997.

Segrè, Emilio. *Enrico Fermi: Physicist.* Chicago, 1970.

———. *A Mind Always in Motion: The Autobiography of Emilio Segrè.* Berkeley, 1993.

Segre, Michael. "Peano's Axioms in their Historical Context." *Archives for History of Exact Sciences* 48 (1994): 201-342.

Siegmund-Schultze, Reinhard. *Rockefeller and the Internationalization of Mathematics between the Two World Wars: Documents and Studies for the Social History of Mathematics in the 20th Century.* Basel, 2001.

Simili, Raffaella and Giovanni Paoloni, eds. *Per una storia del Consiglio Nazionale delle Ricerche.* 2 Vols. Rome-Bari, 2001.

Somigliana, Carlo. "L'opera scientifica di Vito Volterra." *Opere matematiche* 1(1954): xv-xxvi.

———. "Tullio Levi-Civita e Vito Volterra." *Seminario Mat. e Fis. di Milano* 17 (1946): 1-61.

———. "Vito Volterra." *Acta Pontificia Academia Scientiarum* 6 (1942): 57-85.

Sori, Ercole, ed. *La Comunità Ebraica ad Ancona: la storia, le tradizioni, l'evoluzione sociale, i personaggi.* Ancona, 1995.

———. "Una 'comunità crepuscolare': Ancona tra otto e novecento." In *La presenza ebraica nelle Marche secoli XIII-XX*, edited by Sergio Anselmi and Viviana Bonazzoli, 189-266. Quaderno monografico, 14, 1993.

Struik, Dirk, J. *Matematica: un profilo storico*, trans. Umberto Bottazzini. Bologna, 1981, with an appendix by U. Bottazzini. Originally published as *A Concise History of Mathematics*, 2 vols. New York, 1948.

Tomasi, Tina and Nella Sistoli Paoli. *La Scuola normale di Pisa dal 1813 al 1945: cronache di un'istituzione.* Pisa, 1990.

Toscano, Fabio. *Il genio e il gentiluomo: Einstein e il matematico italiano che salvò la teoria della relatività generale.* Milan, 2004.

Traina, Richard P. *Changing the World: Clark University's Pioneering People, 1887-2000.* Worcester, 2005.

Tricomi, Francesco, G. *Integral Equations.* New York, 1985.

————. *La mia vita di matematico attraverso la cronistoria dei miei lavori.* Padua, 1967.

Truesdell, Clifford. "Functionals in the Modern Mechanics of Continua," in *Convegno Internazionale in Memoria di Vito Volterra*, Rome, 1992.

Volterra, Enrico. "Vito Volterra." *Dictionary of Scientific Biography*, XIV: 85-8.

Volterra, Vito. *Opere matematiche. Memorie e note*, 5 vols. Rome, 1954-1962.

————. *Theory of Functionals and of Integral and Integro-Differential Equations*, edited by Luigi Fantappiè, transl. M. Long, London and Glasgow, 1930.

Weintraub, Roy, E. *How Economics Became a Mathematical Science.* Durham, 2002.

Yandell, Benjamin, H. *The Honors Class: Hilbert's Problems and their Solvers.* Natick, 2002.

Notes

Notes to Introduction

[1] Albert Einstein to Michele Besso, *The Collected Papers of Albert Einstein*, trans. A. Beck, vol. 5 (Princeton: Princeton University Press, 1987), 374.

[2] Einstein to Tullio Levi-Civita, *Collected Papers*, trans. A. M. Hentschel, vol. 8 (1998), 89.

[3] George Ellery Hale to John J. Carty, July 3, 1918, George Ellery Hale Papers, Box 10, Archives, California Institute of Technology.

[4] G. E. Hale to Vito Volterra, June 30, 1932 and V. Volterra to G. E. Hale, Dec. 24, 1934, Hale Papers, Box 41.

[5] Vito Volterra, "Trois analystes italiens: Betti, Brioschi, Casorati," *Bull. Am. Math. Soc.* 7 (1900), 60.

[6] Dirk J. Struik, "Some Mathematicians I Have Met," Proc. AMS Special Session, Joint Mathematics Meeting, Cincinnati, Jan. 1994, in Bettyne A. Case, ed., *A Century of Mathematical Meetings* (Providence: American Mathematical Society, 1996), 253.

[7] Francesco G. Tricomi, "Matematici italiani del primo secolo dello stato unitario," *Mem. Acc. Sci. Torino*, cl. sci., fis., 4th ser. 1 (1962): 3

[8] Ferdinand Gregorovius, *The Ghetto and Jews of Rome*, trans. M. Hadas (New York: Schocken, 1948), 83-86.

[9] Ibid., 90.

[10] David I. Kertzer, *The Kidnapping of Edgardo Mortara* (New York: Knopf, 1997).

[11] Ercole Sori, ed., *La Comunità ebraica ad Ancona: la storia, le tradizioni, l'evoluzione sociale, i personaggi* (Ancona: Comune di Ancona, 1995), 46.

[12] P. Levi, Preface, in Vivian B. Mann, ed., *Gardens and Ghettos: The Art of Jewish Life in Italy* (Berkeley: Univ. of California Press, 1989), xvi.

Notes to Chapter 1

[1]Quoted in Mario Natalucci, *Ancona attraverso i secoli, vol. 3* (Città di Castello: Unione Arti Grafiche, 1960), 252-253.

[2]Ercole Sori, "Una 'comunità crepuscolare': Ancona tra otto e novecento," in Sergio Anselmi and Viviana Bonazzoli, eds., *La presenza ebraica nelle Marche secoli XIII-XX* (Ancona: University of Ancona, 1993), 196-197.

[3]Ibid., 199.

[4]Ibid., 201.

[5]F. Borioni, *Le feste anconitane nel settembre dell' anno 1841 per la faustissima venuta e dimora del N.S. Gregorio XVI Regnante felicemente* (Ancona, 1841), 51.

[6]Edoardo Almagià to his sons, May 23, 1907, The Volterra papers, private collection (VP hereafter).

[7]E. Almagià diary, VP.

[8]Mosè Volterra, "Al suo affettuoso fratello Abramo Volterra ed egregia cognata Angelica Almagià," [Mar. 16], 1859, VP.

[9]Saul Almagià to Roberto Almagià, October 4, 1860. I am indebted to Luisa Almagià for a copy of this letter.

[10]Edoardo Volterra to Virginia Volterra, May 24, 1947 and Vito Terni, same date, Vito Volterra files, Institute Archives, California Institute of Technology.

Notes to Chapter 2

[1]Angelica Volterra to Virginia Camiz, Sept. 18, 1864, in The Volterra papers, private collection (VP hereafter).

[2]A. Volterra to V. Camiz, Mar. 5, 1865, and A. Volterra to Alfonso Almagià, Jan. 22, 1865; A. Volterra to V. Camiz, Sept. 18, 1864, VP.

[3]A. Volterra to Alfonso Almagià, Mar. 5, 1865, VP.

[4]Charles Neider, ed., *The Complete Travel Books of Mark Twain* (Garden City: Doubleday, 1966), 163-166.

[5]A. Volterra to V. Camiz, Aug. 20, 1866, VP.

[6]Ibid., Oct, 7, 1866, VP.

[7]A. Volterra to A. Almagià, Aug. 15, 1869, VP.

[8]Ibid., Aug. 15, 1869, VP.

[9]Vito Volterra to A. Almagià, July 29, 1869, VP.

[10]Ibid., Aug. 15, 1869, VP.

[11]Ibid., July 29, 1869, VP.

[12]Ibid., Aug. 22, 1869, VP.

[13]A. Volterra to A. Almagià, Aug. 15, 1869, VP.

[14]V. Camiz to A. Almagià, Aug. 22, 1869, VP.

[15]V. Volterra to A. Almagià and Fortunata Almagià to A. Almagià, [Aug.] 1869; A. Volterra to A. Almagià [Sept.] 1869, VP.

[16]V. Volterra to A. Almagià, Sept. 5, 1869; V. Volterra to A. Volterra, same date, VP.

Notes to Chapter 3

[1]Angelica Almagià to Edoardo Almagià, Aug. 16, 1875, in The Volterra papers, private collection (VP hereafter).

[2][Joseph Pérès?], "Matériaux pour une biographie du mathématicien Vito Volterra," *Archeion* 23 (1941): 326.

[3]Vito Volterra, "L'Évolution des idées fondementales du calcul infinitésimal," in Joseph Pérès, ed., *Leçons sur les fonctions des lignes* (Paris: Gauthier-Villars, 1913), 8-9.

[4]Edoardo Almagià to Vito Volterra, Oct. 1876, VP.

[5]"Matériaux pour une biographie," 328.

[6]Alfonso Almagià to E. Almagià. Nov. 13, 1876, VP.

[7]Ibid., VP.

[8]E. Almagià to A. Almagià, Nov. 20, 1876, VP.

[9]A. Almagià to E. Almagià, Nov. 26, 1876, VP.

[10]The exchange is related in Alessandro Faedo, "Discorsi pronunciati nella cerimonia tenuta presso la Scuola Normale Superiore di Pisa il giorno 20 novembre 1960," in *Vito Volterra nel primo centenario della nascita* (Rome: Accademia Nazionale dei Lincei, 1961), 43.

[11]E. Almagià to A. Almagià, June 13, 1877, VP.

[12]Ibid. [Aug.] 1877, VP.

[13]E. Almagià, "Dal libro di memorie," Mar. 12, 1867, VP.

[14]E. Almagià to Alfonso Almagià, [Aug.] 1877, VP.

[15]A. Almagià to E. Almagià, Aug. 3, 1877, VP.

[16]Ibid., Aug. 25, 1877 and Sept. 8, 1877, VP; Antonio Ròiti to A. Almagià, Sept. 1, 1877, in Giovanni Paolini, ed., *Vito Volterra e il suo tempo* (Rome: Accademia Nazionale dei Lincei, 1990), 4.

[17]A. Almagià to E. Almagià, Oct. 7, 1877, VP.

[18]A. Volterra to E. Almagià, Oct. 7, 1877, and E. Almagià to A. Almagià, Oct. 11, 1877, VP.

[19]E. Almagià to Vito Volterra, Mar. 15, 1889, VP.

Notes to Chapter 4

[1]Vito Volterra to Angelica Volterra, Nov. 5, 1878, in The Volterra papers, private collection (VP hereafter).

[2]Ibid., Dec. 9, 1878, VP.

[3]A. Volterra to V. Volterra, Nov. 4, 1878, VP.

[4]Ibid., Nov. 6, 1878, VP.

[5]Alfonso Almagià to V. Volterra, Nov. 6, 1878.

[6]V. Volterra to A. Volterra, Nov. 15, 1878, VP.

[7]Ibid., Dec. 7, 1878, VP.

[8]Ibid., Jan. 10, 1879, VP.

[9]Ibid., Nov. 10, 1878, VP.

[10]Giovanni Paoloni, ed., *Vito Volterra e il suo tempo* (Rome: Accademia Nazionale dei Lincei, 1990), 6-7.

[11]V. Volterra to A. Volterra, Nov. 10, 1878, VP.

[12]A. Almagià to V. Volterra, Jan. 11, 1879, VP.

[13]V. Volterra to A. Volterra, Jan. 12, 1879, VP.

[14]A. Almagià to V. Volterra, Nov. 6, 1878, VP.

[15]V. Volterra to A. Volterra, Nov. 12, 1878, VP.

[16]Paoloni, Vito Volterra, 5-6.

[17]V. Volterra to A. Volterra, Nov. 8, 1878, VP.

[18]Ibid., Nov. 17, 1878, VP.

[19]Ibid., Feb, 1, 1879, VP.

[20]Ibid., Jan. 14, 1879, VP.

[21]Fedele Romani, "I miei ricordi di Pisa," *La Lettura* 8 (1908): 120-121.

[22]John A. Davis, "Italy 1796-1870: The Risorgimento," in George Holmes, ed., *The Oxford History of Italy* (Oxford: Oxford Press, 1997), 208.

[23]V. Volterra to A. Volterra, Nov. 9, 1878, VP.

[24]Oscar Browning, *Memories of Sixty Years at Eton, Cambridge, and Elsewhere* (London: John Lane, 1910), 170.

[25]V. Volterra to A. Volterra, Nov. 21, 1878, VP.

[26]Alessandro D'Ancona, *Ricordi ed affetti* (Milan: Fratelli Treves, 1901), 243.

[27]V. Volterra to A. Volterra, Dec. 11, 1878, VP.

[28]Romani, "I miei ricordi," 122.

[29]V. Volterra to A. Volterra, [Jan.], 1879, VP.

[30]Ibid.

[31]Ibid., Jan. 14, 1879, VP.

[32]Ibid., Apr. 1, 1879, VP.

[33]Ibid., Mar. 8, 1879, VP.

[34]Translation supplied by Professor George Pigman, private communication.

[35]Carlo Somigliana, "Vito Volterra," *Pont. Acad. Sci. Acta* 6 (1942): 58.

[36]Tina Tomasi and Nella Sistoli Paola, La scuola normale di Pisa dal 1813 al 1945 (Pisa: ETS Editrice, 1990), 100.

Notes to Chapter 5

[1]Carlo Somigliana, "Tullio Levi-Civita e Vito Volterra," *Rendiconti del seminario matematico e fisico di Milano*, 17 (1946): 8.

[2]Ernesto Padova, "Commemorazione di Enrico Betti," *Atti. R. Ist. Veneto di Scienze, Lettere ed Arti* 4 (1892-93): 614.

[3]V. Volterra, "Alcune osservazioni sulle funzioni punteggiate discontinue," in *Opere matematiche*, vol. 1 (Rome: Accademia Nazionale dei Lincei, 1954-62), 7-15.

[4]Enrico Giusti and Luigi Pepe, eds., *La matematica in Italia, 1800-1950* (Florence: Polistampa, 2001), 128.

[5]Vito Volterra to Angelica Volterra, Nov. 11, 1880, in The Volterra papers, private collection (VP hereafter).

[6]V. Volterra, "Sui principi del calcolo integrale," *Opere matematiche*, vol. 1, 16-48.

[7]C. Somigliana, "L'opera scientifica di Vito Volterra," ibid., xv.

[8]The French mathematician Henri Léon Lebesgue later showed how to generalize Riemann integrability to include these pathological functions. (Dr. Michele Vallisneri, private communication.)

[9]C. Somigliana to V. Volterra, n.d., 1881, Vito Volterra Collection, Accademia Nazionale dei Lincei, Rome.

[10]Ibid., May 4, 1882.

[11]Ibid., Dec. 9, 1883.

[12]Ibid., Sept. 19, 1885.

[13]Ibid., Oct. 4, 1885.

[14]Angelica Volterra to V. Volterra, May 8, 1881, VP.

[15]Alfonso Almagià to V. Volterra, Jan 17, 1880, VP.

[16]Ibid., June 12, 1880, VP.

[17]Ibid., June 2, 1881, VP.

[18]Ibid., June 8, 1881, VP.

[19]V. Volterra to Angelica Volterra, April 29, 1881, VP.

[20]V. Volterra, "Sul potenziale di un'ellissoide eterogenea sopra sè stessa," in Opere matematiche, vol. 1, 1-6.

[21]Alfonso Almagià to V. Volterra, June 8, 1881, VP.

[22]Angelica Volterra to V. Volterra, June 9, 1881, VP.

[23]Edoardo Almagià to Angelica Volterra, Aug. 18, 1881, VP.

[24]V. Volterra, "Enrico Betti," Opere matematiche, vol. 1, 600.

[25]Enrico Betti to V. Volterra, Oct 31, 1882, in Giovanni Paoloni, ed., Vito Volterra e il suo tempo (Rome: Accademia Nazionale dei Lincei, 1990), 7-8.

[26]V. Volterra to E. Betti, n.d., but Nov. 1882, in Paoloni, ed. Vito Volterra, 8.

[27]Alfonso Almagià to V. Volterra, Dec. 18, 1882, VP.

[28]Ibid., Mar. 5, 1883, VP.

[29]Ibid., Mar. 7, 1883, VP.

[30]Antonio Ròiti to V. Volterra, n.d., but Mar. 17, 1883, in Paoloni, ed., Vito Volterra, 9.

[31]Alfonso Almagià to V. Volterra, Mar. 7, 1883, VP.

[32]Corrado Padoa to Enrico Stracciati, April 1883, VP.

Notes to Chapter 6

[1]H. Stuart Hughes, *Prisoners of Hope: The Silver Age of Italian Jews, 1924-1974* (Cambridge: Harvard Univ. Press, 1983), 18.

[2]Leonardo Sciascia, "Il quarantotto," in *Gli zii di Sicilia* (Turin: Einaudi, 1958), 161.

[3]Angelica Volterra to Vito Volterra, Mar. 1, 1885, in The Volterra papers, private collection (VP hereafter).

[4]L.S.D., *Trip to the Sunny South in March 1885* (Birkenhead, U.K.: E. Griffith & Son, 1885), 48-9.

[5]Alfonso Volterra to V. Volterra, with postscript by A. Almagià, July 5, 1885, VP.

[6]Epicarmo Corbino, *Racconto di una vita* (Naples: Edizioni scientifiche Italiane, 1972), 28.

[7]Alfonso Volterra to V. Volterra, July 14, 1885, VP.

[8]Angelica Volterra to V. Volterra, Nov. 4, 1884, VP.

[9]Ibid., Mar. 30, 1884, VP.

[10]Ibid., Nov. 9, 1884, VP.

[11]Alfonso Almagià to V. Volterra, Mar. 11, 1880, VP.

[12]Angelica Volterra to V. Volterra, June 11, 1884, with postscript by A. Almagià, VP.

[13]Ibid., Dec. 18, 1884, VP.

[14]Ibid., Apr. 25, 1887, VP.

[15]Ibid., July 6, 1887, VP.

[16]Eugenio Beltrami to V. Volterra, July 19, 1883, Vito Volterra Collection, Accademia Nazionale dei Lincei, Rome.

[17]Enrico Betti to V. Volterra, Sept. 15, 1887, in Giovanni Paoloni, ed., *Vito Volterra e il suo tempo* (Rome: Accademia Nazionale dei Lincei, 1990), 15.

[18]V. Volterra to E. Betti, undated draft, but probably Sept. 1887, in Vito Volterra Collection.

[19]V. Volterra, *Theory of Functionals,* ed. Luigi Fantappiè, trans. M. Long (London: Blackie & Son, 1930), 22.

[20]———. *Opere matematiche,* vol. 1 (Rome: Accademia Nazionale dei Lincei, 1954-62), 294.

[21]Jacques Hadamard, *The Psychology of Invention in the Mathematical Field* (Princeton: Princeton Univ. Press, 1945; Dover reprint, 1954), 129-30.

[22]Carlo Somigliana to V. Volterra, May 4, 1882, Vito Volterra Collection.

Notes to Chapter 7

[1]Giovanni Battista Guccia to Vito Volterra, Dec. 4, 1887, in Giovanni Paoloni, ed., *Vito Volterra e il suo tempo* (Rome: Accademia Nazionale dei Lincei, 1990), 16.

[2]Dirk J. Struik, *A Concise History of Mathematics*, vol. 2 (New York: Dover Publications, 1948), 276.

[3]V. Volterra to Enrico Betti, Aug. 1, 1888, Betti Archive, Scuola Normale Superiore di Pisa.

[4]Ann Hibner Koblitz, *A Convergence of Lives: Sofia Kovalevskaia: Scientist, Writer, Revolutionary* (Basel: Birkhäuser, 1983), 136.

[5]Ibid., 186.

[6]Roger Cooke, *The Mathematics of Sonya Kovalevskaya* (New York: Springer-Verlag, 1984), 110.

[7]V. Volterra to E. Betti, Aug. 21, 1888, Betti Archive.

[8]Sophie Kowalevski, "Sur le problème de la rotation d'un corps solide autour d'un point fixe," *Acta Mathematica* 12 (1889): 171-232.

[9]Struik, *A Concise History*, 237.

[10]V. Volterra to E. Betti, Aug. 21, 1888, Betti Archive.

[11]Ibid., Aug. 1, 1888.

[12]Salvatore Pincherle, "Ricerche sopra una classe importante di funzioni monodrome," *Giornale di Matematiche* 18 (1880): 92-136.

[13]Eugenio Beltrami to Jules Hoüel, Feb. 10, 1881, in L. Boi, L. Giacardi, and R. Tazzioli, eds., *La découverte de la géométrie non euclidienne sur la pseudosphère: Les lettres d'Eugenio Beltrami à Jules Hoüel* (Paris: A. Blanchard, 1998), 192.

[14]Carlo Somigliana to V. Volterra, Jan.12, 1890, Vito Volterra Collection, Accademia Nazionale dei Lincei, Rome.

[15]Cooke, *The Mathematics of Sonya Kovalevskaya*, 173.

[16]Gabriel Lamé, *Leçons sur la théorie mathématique de l'élasticité des corps solides* (Paris: Bachelier, 1852).

[17]Sofia Kovalevskaya, "Über die Brechung des Lichtes in crystallinischen Mitteln," *Acta Mathematica* 6 (1885): 249-304.

[18]Cooke, *The Mathematics of Sonya Kovalevskaya*, 173.

[19]V. Volterra, "Sur les vibrations lumineuses dans les milieux biréfringents," *Acta Mathematica* 16 (1891): 153-215.

[20]Roger Cooke, "Sonya Kovalevskaya's Place in Nineteenth-Century Mathematics," in Linda Keen, ed., *The Legacy of Sonya Kovalevskaya* 64 (Providence: American Mathematical Society, 1987), 36.

[21]Paoloni, ed., *Vito Volterra e il suo tempo*, 75.

[22]Constance Reid, *Hilbert* (New York: Springer-Verlag, 1970), 47.

[23]Struik, *A Concise History*, 271.

[24]David E. Rowe, "'Jewish Mathematics' at Göttingen in the Era of Felix Klein," *Isis* 77 (1986): 432.

[25]V. Volterra to Angelica Volterra, July 25, 1891, in The Volterra papers, private collection (VP hereafter).

[26]Ibid., July 16 and 25, 1891, VP.

[27]Ibid., July 16, July 6, July 25 and July 27, 1891, VP.

[28]Ibid., July 27, 1891, VP.

Notes to Chapter 8

[1]Vito Volterra, "Enrico Betti," *Il Nuovo Cimento* 32 (1893): 5-7.

[2]Federigo Enriques to Guido Castelnuovo, Oct. 3, 1893, in U. Bottazzini, A. Conte, P. Gario, eds., *Riposte armonie* (Turin: Bollati Boringhieri, 1996), 28.

[3]Vito Volterra to Ulisse Dini, n.d. in Giovanni Paoloni, ed., *Vito Volterra e il suo tempo* (Rome: Accademia Nazionale dei Lincei, 1990), 23.

[4]Antonio Ròiti to V. Volterra, telegram, n.d. but probably ca. July 1893, Vito Volterra Collection, Accademia Nazionale dei Lincei, Rome.

[5]Riccardo Felici to V. Volterra, Aug. 9, 1893, Paoloni, ed., *Vito Volterra e il suo tempo*, 24.

[6]A. Ròiti to V. Volterra, Sept. 21, 1893, ibid., 26.

[7]Enrico D'Ovidio to V. Volterra, Sept. 23, 1893, ibid., 26.

[8]V. Volterra to E. D'Ovidio, Sept. 30, 1893, Vito Volterra Collection.

[9]E. D'Ovidio to V. Volterra, Oct. 5, 1893, ibid.

[10]V. Volterra to Luigi Cremona, Oct. 13, 1893, in Paoloni, ed., *Vito Volterra e il suo tempo*, 27.

[11]L. Cremona to V. Volterra, Oct. 17, 1893, ibid., 28.

[12]V. Volterra to Andrea Naccari, Oct. 22, 1893, ibid., 28.

[13]V. Volterra to A. Ròiti, Oct. 24, 1893, Vito Volterra Collection.

[14]Quoted in Angelo d'Orsi, "Un profilo culturale," in Valerio Castronovo, *Torino* (Bari: Laterza, 1987), 523.

[15]Primo Levi, in Vivian B. Mann, ed., *Gardens and Ghettos: The Art of Jewish Life in Italy* (Berkeley: Univ. of California Press, 1989), xv.

[16]Quoted in Susan Zuccotti, *The Italians and the Holocaust* (New York: Basic Books, 1987), 19.

[17]Roberto Maiocchi, "Il ruolo delle scienze nello sviluppo industriale italiano," in *Scienze e tecnica nella cultura e nella società dal rinascimento a oggi*, Storia d'Italia, vol. 3 (Turin, Giulio Einaudi, 1980), 877.

[18]Adrian Lyttelton, "Politics and Society, 1870-1915," in George Holmes, ed., *The Oxford History of Italy* (Oxford: Oxford University Press, 1997), 251.

[19]Livia Giacardi, "Corrado Segre maestro a Torino. La nascita della scuola italiana di geometria algebrica," in *Annali di storia delle Università italiane* 5 (2001): 162-63.

[20]Beppo Levi, "La personalidad de Vito Volterra," Publ. Inst. Mat. Univ. Nac. Litoral 3 (1941): 25.

[21]Quoted in Hubert C. Kennedy, *Peano* (Boston: D. Reidel, 1980), 54.

Notes to Chapter 9

[1]Costantino Botto, quoted in Hubert C. Kennedy, *Peano: Life and Works of Giuseppe Peano* (Dordrecht: D. Reidel, 1980), 101.

[2]Vito Volterra, "Replica ad una nota del Prof. Peano," *Opere matematiche*, vol. 2 (Rome: Accademia Nazionale dei Lincei, 1954-62), 213-215.

[3]V. Volterra to Luigi Bianchi, May 18, 1895, quoted in Angelo Guerraggio, "Le memorie di Volterra e Peano sul movimento dei poli," *Archive for History of Exact Sciences* 31 (Nov. 1984): 112.

[4]Giuseppe Peano, "Sopra lo spostamento del polo sulla terra," *Atti Accad. sci. Torino* 30 (May, 1895): 515.

[5]Guerraggio, "Le memorie di Volterra e Peano," 112-113.

[6]Quoted in Hermann Brunner, "1896-1996: One Hundred Years of Volterra Integral Equations of the First Kind," *Applied Numerical Mathematics* 24 (1997): 83.

[7]V. Volterra, "Sui moti periodici del polo terrestre," in *Opere matematiche*, vol. 2, 141-151.

[8]Guido Castelnuovo and Carlo Somigliana, "Vito Volterra e la sua opera scientifica," in *Atti della Accademia nazionale dei Lincei*, Oct. 17, 1946 (Rome, 1947), 14-15.

[9]G. Peano to V. Volterra, May 30, 1895, in Giovanni Paoloni, ed., *Vito Volterra e il suo tempo* (Rome: Accademia Nazionale dei Lincei, 1990), 33-34.

[10]V. Volterra to G. Peano, June 2, 1895, ibid., 34.

[11]V. Volterra to C. Segre, June 2, 1895, Vito Volterra Collection, Accademia Nazionale dei Lincei, Rome.

[12]Guerraggio, "Le memorie di Volterra e Peano," 113-114.

[13]C. Segre to V. Volterra, May 30, 1895, Vito Volterra Collection.

[14]Kennedy, *Peano*, 39.

[15]Ibid., 59.

[16]V. Volterra, *Opere matematiche*, vol. 2, 170-172; also quoted in Kennedy, *Peano*, 60.

[17]Kennedy, *Peano*, 59.

[18]Ibid., 60.

[19]V. Volterra, "Replica ad una nota del Prof. Peano," *Opere matematiche*, vol. 2, 213-215.

[20]V. Volterra to Tullio Levi-Civita, Jan. 15, 1896, Tullio Levi-Civita Collection, Accademia Nazionale dei Lincei, Rome.

[21]V. Volterra, "Sur la théorie des variations des latitudes," in *Opere matematiche*, vol. 2, 452-573.

[22]V. Volterra to C. Somigliana, April 19, 1896 and C. Somigliana to V. Volterra, April 22, 1896, Vito Volterra Collection.

[23]Francesco Tricomi, *La mia vita di matematico attraverso la cronistoria dei miei lavori* (Padova: CEDAM, 1967), 18.

[24]V. Volterra, "Sulla inversione degli integrali definiti," *Opere matematiche*, vol. 2, 216.

[25]V. Volterra to T. Levi-Civita, Feb. 29, 1896, Tullio Levi-Civita Collection.

[26]W.F. Osgood, "The International Congress of Mathematicians at Zurich," *Bull. Am. Math. Soc.* 4 (1898): 47.

[27]V. Volterra to Giovanni Vailati, Aug. 25, 1897, Giovanni Vailati Collection, University of Milan.

[28]Ibid., Aug. 14, 1898, Vito Volterra Collection.

[29]Ibid., July 24, 1898, Giovanni Vailati Collection, University of Milan. A copy of the letter supplied through the courtesy of Dr. Mauro De Zan.

Notes to Chapter 10

[1]Oscar Wilde, *The Importance of Being Earnest* (New York, Dover Publications, 1990), 12.

[2]Angelica Volterra to Vito Volterra, Nov. 10, 1899, in The Volterra papers, private collection (VP hereafter).

[3]Ibid., Nov. 19, 1899, VP.

[4]Ibid., Nov. 10, 1899, VP.

[5]Ibid.

[6]Antonio Ròiti to V. Volterra, Dec. 11, 1899, VP.

[7]A. Volterra to V. Volterra, Dec. 19, 1899, VP.

[8]Guido Castelnuovo to V. Volterra, Mar. 2, 1900, Vito Volterra Collection, Accademia Nazionale dei Lincei, Rome.

[9]A. Ròiti to V. Volterra, Mar. 8, 1900, Vito Volterra Collection.

[10]V. Volterra to Valentino Cerruti, Mar. 11, 1900, VP.

[11]V. Cerruti to V. Volterra, Mar. 12, 1900, in Giovanni Paoloni, ed., *Vito Volterra e il suo tempo* (Rome: Accademia Nazionale dei Lincei, 1990), 41-42.

[12]V. Volterra to A. Ròiti, Mar. 14, 1900, Vito Volterra Collection.

[13]A. Ròiti to V. Volterra, Mar. 15, 1900, Vito Volterra Collection.

[14]Ibid., Mar. 22, 1900.

[15]V. Volterra to A. Ròiti, Mar. 25, 1900, Vito Volterra Collection.

[16]V. Volterra to A. Volterra, Mar. 27, 1900, VP.

[17]A. Volterra to V. Volterra, May 10, 1900, VP.

[18]V. Volterra to A. Volterra, May 10, 1900, VP.

[19]Ibid., May 12, 1900, VP.

[20]A. Ròiti to V. Volterra, May 6, 1900, Vito Volterra Collection.

[21]V. Volterra to A. Volterra, May 13, 1900, VP.

[22]G. Castelnuovo to V. Volterra, May 17, 1900, Vito Volterra Collection.

[23]V. Volterra to A. Volterra, May 21, 1900, VP.

[24]A. Ròiti to V. Volterra, May 24, 1900, Vito Volterra Collection.

[25]Edoardo Almagià to Virginia Almagià, June 2, 1900, VP.

[26]A. Ròiti to V. Volterra, June 1900, undated, but ca. June 19, Vito Volterra Collection.

[27]G. Castelnuovo to V. Volterra, June 17, 1900, Vito Volterra Collection.

[28]G. Castelnuovo to V. Volterra, June 19, 1900, Vito Volterra Collection.

[29]V. Volterra, "Appunti di Vito per cattedra Roma," undated, but written after June 28, 1900, VP.

[30]Ibid.

[31]V. Volterra to Virginia Almagià, June 28, 1900, VP.

[32]Corrado Segre to V. Volterra, June 19, 1900, Vito Volterra Collection.

[33]V. Volterra to V. Almagià, July 1, 1900, VP.

[34]Interview of Ginetta Montecorboli, by Carlotta Scaramuzzi, 1995, p. 4, Institute Archives, California Institute of Technology.

[35]V. Volterra to A. Volterra, July 12, 1900, VP.

[36]Pietro Blaserna to V. Volterra, Sept. 8, 1900, Vito Volterra Collection.

Notes to Chapter 11

[1]Virginia and Vito Volterra to Angelica Volterra, July 28, 1900, The Volterra papers, private collection (VP hereafter).

[2]A. Volterra to Vito Volterra, July 26, 1900, VP.

[3]Virginia Volterra to Edoardo Almagià, July 31, 1900, VP.

[4]Virginia Volterra to A. Volterra, July 28, 1900, VP.

[5]Ibid., Aug. 6, 1900, VP.

[6]V. Volterra, "Betti, Brioschi, Casorati: Tre analisti e tre modi di considerare le questioni d'analisi," in *Saggi scientifici* (1920; reprint, with an introduction by Raffaella Simili, Bologna: Zanichelli, 1990), 37-38.

[7]Virginia Volterra to A. Volterra, Aug. 6, 1900, VP.

[8]Corrado Segre to V. Volterra, Sept. 11, 1900, Vito Volterra Collection, Accademia Nazionale dei Lincei, Rome.

[9]David Hilbert, "Mathematical Problems," trans. Dr. Mary Winton Newson, *Bull. Am. Math. Soc.* 8 (1902): 437-45, 478-89.

[10]V. Volterra, *Theory of Functionals*, Luigi Fantappiè, ed., trans. M. Long (London: Blackie & Son, 1930), vii-viii.

[11]Quoted in Benjamin H. Yandell, *The Honors Class: Hilbert's Problems and Their Solvers* (Natick, MA: A. K. Peters, 2002), 424.

[12]C. A. Scott, "The International Congress of Mathematicians in Paris," *Bull. Am. Math. Soc.* 7 (1901): 76.

[13]Vito Volterra to Virginia Volterra, Sept. 25, 1900, VP.

[14]Emilio Segrè, *A Mind Always in Motion* (Berkeley and Los Angeles: Univ. of California Press, 1993), 52-3.

[15]Julius Weingarten to V. Volterra, Nov. 11, 1900, Vito Volterra Collection, Accademia Nazionale dei Lincei, Rome.

[16]V. Volterra to J. Weingarten, Nov. 30, 1900, ibid.

[17]J. Weingarten to V. Volterra, Dec. 4, 1900, ibid.

[18]V. Volterra, "Un teorema sulla teoria della elasticità," *Rend. R. Accad. Lincei* 14 (1905): 127-37.

[19]V. Volterra to J. Weingarten, Feb. 27, 1905, Vito Volterra Collection.

[20]V. Volterra, "Henri Poincarè," in *The Book of the Opening of the Rice Institute*, vol. 3 (Houston, Texas: The De Vinne Press, 1912), 912-13.

[21]V. Volterra, "Sulle equazioni integro-differenziali," *Rend. R. Accad. Lincei* 18 (1909): 167-74.

[22]V. Volterra to Giovanni Vailati, Dec. 7, 1900, Vito Volterra Collection.

[23]Interview of Luisa Almagià by Carlotta Scaramuzzi, 1996, p. 5, Institute Archives, California Institute of Technology.

[24]Personal communication from Edoardo's daughter Dr. Virginia Volterra, Dec. 30, 1996.

[25]V. Volterra to G. Vailati, July 1, 1901, Giovanni Vailati Collection, University of Milan.

[26]G. Vailati to V. Volterra, July 3, 1901, Vito Volterra Collection.

[27]V. Volterra to Virginia Volterra, Aug. 2, 1901, VP.

[28]Ibid., Aug. 8, 1901.

[29]Ibid., Aug. 14, 1901.

[30]Ibid., Aug. 15, 1901.

[31]Ibid., Aug. 17, 1901.

[32]Ibid., Aug. 15, 1902.

[33]Edoardo Volterra to Virginia Volterra, undated, but summer 1902, VP.

[34]V. Volterra to Virginia Volterra, Aug. 14, 1902, VP.

[35]Ibid., Aug. 20, 1902.

[36]Quintino Sella to Rosa Sella, Nov. 17, 1879, Vito Volterra files, Institute Archives, California Institute of Technology.

[37]V. Volterra to Alfonso Sella, Aug. 29, 1902, ibid.

[38]V. Volterra to Virginia Volterra, Sept. 6, 1902, VP.

Notes to Chapter 12

[1]Vito Volterra to Virginia Volterra, Oct. 5, 1902, in The Volterra papers, private collection (VP hereafter).

[2]Vito Volterra to Nunzio Nasi, Feb. 17, 1902, draft, in Giovanni Paoloni, ed., *Vito Volterra e il suo tempo* (Rome: Accademia Nazionale dei Lincei, 1990), 43-44.

[3]V. Volterra to Tullio Levi-Civita, Mar. 13, 1905, Tullio Levi-Civita Collection, Accademia Nazionale dei Lincei, Rome.

[4]V. Volterra to Virginia Volterra, Aug. 9, 1903, VP.

[5]V. Volterra to Virginia Volterra, August 18, 1903, VP.

[6]V. Volterra to Virginia Volterra, Aug. 13, 1903, VP.

[7]V. Volterra to Virginia Volterra, August 18, 1903, VP.

[8]Ibid.

[9]V. Volterra to Virginia Volterra, August 21, 1903, VP.

[10]V. Volterra to Virginia Volterra, Feb. 13, 1904, VP.

[11]Hermann Weyl, "David Hilbert: 1862-1943," Obituary Notices of Fellows of the Royal Society of London 4 (1944), 547.

[12]V. Volterra to Virginia Volterra, Feb. 19, 1904, VP.

[13]V. Volterra to Virginia Volterra, Feb. 21, 1904, VP.

[14]V. Volterra to Virginia Volterra, Feb. 20, 1904, VP.

[15]V. Volterra to Virginia Volterra, Feb. 25, 1904, VP.

[16]V. Volterra to Virginia Volterra, Feb. 29, 1904, VP.

[17]V. Volterra to Virginia Volterra, Feb. 18, 1904, VP.

[18]Stanislao Cannizzaro to V. Volterra, June 10, 1904, Vito Volterra files, Institute Archives, California Institute of Technology.

[19]Atti Parlamentari, Legislatura XXII, 1st session, 1904-906, June 19, 1906.

[20]V. Volterra to Angelica Almagià, Aug. 20, 1904, VP.

[21]Vito and Virginia Volterra to the Almagià family, Aug. 28, 1904, VP.

[22]"L'Illustrazione Italiana," XXXII, 1905, 243.

[23]Guido Castelnuovo, "Vito Volterra," *Opere matematiche*, vol. 1, ix-x.

Notes to Chapter 13

[1]Cesare Arzelà to Vito Volterra, July 15, 1905, Veronica Gavagna, "Dalla teoria delle funzioni all'analisi funzionale: il carteggio Arzelà-Volterra," *Bollettino di storia delle scienze matematiche* 14 (1994), 70.

[2]Guido Castelnuovo to Vito Volterra, July 29, 1905, Vito Volterra Collection, Accademia Nazionale dei Lincei, Rome.

[3]Vito Volterra to Virginia Volterra, July 18, 1905, The Volterra Papers, private collection (hereafter, VP).

[4]V. Volterra to Virginia Volterra, July 19, 1905, VP.

[5]V. Volterra to Virginia Volterra, July 26, 1905, VP.

[6]V. Volterra to Virginia Volterra, Aug. 4, 1905, VP.

[7]V. Volterra to Virginia Volterra, Aug. 7, 1905, VP.

[8]V. Volterra to Virginia Volterra, Aug. 8, 1905, VP.

[9]Edoardo Almagià to Virginia Volterra, Aug. 15, 1905, VP.

[10]V. Volterra to Virginia Volterra, Sept. 3, 1905, VP.

[11]V. Volterra to Virginia Volterra, Aug. 29, 1905, VP.

[12]V. Volterra to Virginia Volterra, Sept. 25, 1905 and Sept. 26, 1905 and Sept. 28, 1905, VP.

[13]V. Volterra to Virginia Volterra, Sept. 28, 1905, VP.

[14]Vito Volterra, "Leçons sur l'intégration des équations différentielles aux dérivées partielles," *Opere matematiche*, vol. 3 (Rome: Accademia Nazionale dei Lincei, 1954-62) 64.

[15]V. Volterra to Virginia Volterra, Feb. 15, 1906, VP.

[16]V. Volterra to Virginia Volterra, Feb. 20, 1906, VP.

[17]V. Volterra to Virginia Volterra, Feb. 14, 1906 and Mar. 9, 1906, VP.

[18]Edoardo Almagià to Vito Volterra, Feb. 23, 1906, VP.

[19]V. Volterra to Virginia Volterra, Mar. 5, 1906, VP.

[20]V. Volterra to Virginia Volterra, Mar. 10, 1906, VP.

[21]V. Volterra, "Proposta di una associazione italiana per il progresso delle scienze," *Opere matematiche*, vol. 3, 148.

[22]Volterra, "Proposta," 150.

[23]Ibid., 149.

[24]Volterra, "Proposta," 151.

[25]Giovanni B. Guccia to V. Volterra, June 30, 1906, in Giovanni Paoloni, ed., *Vito Volterra e il suo tempo* (Rome: Accademia Nazionale dei Lincei, 1990), Fig. II. 3.

[26]Gösta Mittag-Leffler to Vito Volterra, Aug. 17, 1906, Vito Volterra Collection.

[27]V. Volterra to Virginia Volterra, Sept. 17, 1906, VP.

[28]V. Volterra to Virginia Volterra, Sept. 21, 1906, VP.

[29]V. Volterra to Virginia Volterra, Sept. 23, 1906, VP.

[30]E. Almagià to Ettore Marchiafava, undated but end of Oct. 1906, VP.

[31]V. Volterra to Virginia Volterra, Sept. 24, 1907, VP.

[32]Quoted in Giuseppe Giuliani, *Il Nuovo Cimento: novant'anni di fisica in Italia, 1855-1944* (Pavia: La Goliardica Pavese, 1996), 29.

[33]Orso Mario Corbino to Tullio Levi-Civita, Sept. 12, 1909, Tullio Levi-Civita Papers, Accademia Nazionale dei Lincei, Rome.

[34]Ibid.

[35]Arthur Gordon Webster to Vito Volterra, July 12, 1909, Vito Volterra Collection.

[36]G. Castelnuovo to V. Volterra, July 19, 1909, Vito Volterra Collection.

Notes to Chapter 14

[1]Quoted in J. J. O'Connor and E. F. Robertson, "Maxine Bôcher," [Internet], St. Andrews, MacTutor History of Mathematics, August 2005 [cited 4 Aug. 2006], available from: `http://www-history.mcs.st-andrews.ac.uk/Biographies/Bocher.html`.

[2]Maxime Bôcher, *An Introduction to the Study of Integral Equations* (New York: Hafner Publishing, 1971), 24.

[3]Vito Volterra to Griffith Evans, Aug. 13, 1913, Griffith C. Evans Papers, Carton 13, Bancroft Library, UC Berkeley.

[4]Vito Volterra to Virginia Volterra, August 19, 1909 (enclosure), in The Volterra papers, private collection (hereafter, VP).

[5]Vito Volterra to V. Volterra, Aug. 19, 1909, VP.

[6]V. Volterra to Virginia Volterra, Aug. 20, 1909, VP.

[7]V. Volterra to Virginia Volterra, Aug. 23, 1909, VP.

[8]V. Volterra to Virginia Volterra, ibid., VP.

[9]V. Volterra to Virginia Volterra, Aug. 21, 1909, VP.

[10]V. Volterra to Virginia Volterra, Aug. 29, 1909, VP.

[11]V. Volterra to Virginia Volterra, Aug. 31, 1909, VP.

[12]V. Volterra to Virginia Volterra, Sept. 3, 1909, VP.

[13]V. Volterra to Virginia Volterra, Sept. 3, 1909, VP.

[14]V. Volterra, *Lectures delivered at the Celebration of the Twentieth Anniversary of the Foundation of Clark University* (Worcester: Clark University, 1912), 1-2.

[15]V. Volterra to Virginia Volterra, Sept. 7, 1909, VP.

[16]V. Volterra to Virginia Volterra, Sept. 9, 1909, VP.

[17]V. Volterra, "Lectures," 41.

[18]V. Volterra to Virginia Volterra, Sept. 10, 1912, VP.

[19]V. Volterra to Virginia Volterra, Sept. 13, 1909, VP.

[20]V. Volterra to Virginia Volterra, Feb. 1, 1912, VP.

[21]V. Volterra to Virginia Volterra, Sept 19, 1909, VP.

[22]Federigo Enriques to Vito Volterra, Dec. 17, 1909, Vito Volterra Collection, Accademia Nazionale dei Lincei, Rome.

[23]V. Volterra to Virginia Volterra, June 18, 1910, VP.

[24]V. Volterra to Virginia Volterra, June 26, 1910, VP.

[25]V. Volterra to Virginia Volterra, July 1, 1910, VP.

[26]V. Volterra to Virginia Volterra, July 8, 1910, VP.

[27]V. Volterra to Virginia Volterra, July 14 and July 18, 1910, VP.

[28]V. Volterra to Virginia Volterra, Aug. 3, 1910, VP.

[29]Griffith Evans to Julian [surname unknown], Griffith Evans Papers, July 29, 1968, Carton 1, UC Berkeley.

[30]E. Roy Weintraub, *How Economics Became a Mathematical Science* (Durham: Duke University Press, 2002), 42-43.

[31]V. Volterra to Virginia Volterra, Jan. 18, and Jan. 19, 1912, VP.

[32]V. Volterra to Virginia Volterra, Feb. 11, 1912, VP.

[33]V. Volterra to Virginia Volterra, Feb. 18, 1912, VP.

[34]Edoardo Almagià to Vito and Virginia Volterra, Feb. 20, 1912, VP.

[35]V. Volterra to Virginia Volterra, Jan. 26, 1912, VP.

[36]V. Volterra to Virginia Volterra, Feb. 21, 1912, VP.

[37][V. Volterra, *Leçons sur les fonctions de lignes* (Paris: Gauthier-Villars, 1913)]

[38]Quoted in Denis Mack Smith, Modern Italy: A Political History (Ann Arbor: University of Michigan Press, 1997), 225.

[39]V. Volterra to Virginia Volterra, Jan. 30, and Feb. 1, 1912, VP.

[40]V. Volterra to Virginia Volterra, Feb. 6, VP.

[41]V. Volterra to Virginia Volterra, Feb. 7, 1912, VP.

[42]Orso M. Corbino to Vito Volterra, Feb. 24, 1912, in Giovanni Paoloni, ed., *Vito Volterra e il suo tempo* (Rome: Accademia Nazionale dei Lincei, 1990), 63.

[43]V. Volterra to Virginia Volterra, Feb. 13, 1912, VP.

[44]V. Volterra to Virginia Volterra, Oct. 17, 1912, VP.

Notes to Chapter 15

Portions of chapter 15 appeared in different form in *Journal of the History of Ideas.*

[1]Gaetano Arturo Crocco, in "Vito Volterra nel I centenario della nascita: 1860-1960," Accademia Nazionale dei Lincei 51 (1961), 26.

[2]V. Volterra to Gaston Darboux, Sept. 7, 1914, Giovanni Paoloni, ed., *Vito Volterra e il suo tempo* (Rome: Accademia Nazionale dei Lincei, 1990), 94.

[3]Quoted in Denis Mack Smith, *Modern Italy: A Political History* (Ann Arbor: University of Michigan Press, 1997), 264.

[4]Vito Volterra to Virginia Volterra, May 5, 1915, in The Volterra papers, private collection (hereafter, VP).

[5]Carlo Somigliana to V. Volterra, May 19, 1915, Vito Volterra Collection, Accademia Nazionale dei Lincei, Rome.

[6]V. Volterra to Virginia Volterra, Jan. 25, 1916, VP.

[7]V. Volterra to Virginia Volterra, April 1, 1916, VP.

[8]V. Volterra to Virginia Volterra, April 12, 1916, VP.

[9]V. Volterra to Virginia Volterra, May 31, 1916, VP.

[10]V. Volterra to Virginia Volterra, June 2, 1916, VP.

[11]G.A. Crocco, "Vito Volterra," 27.

[12]Paolo Morrone, "Report," Feb. 12, 1920, cartella XVII, fasc. 3, Vito Volterra Collection, Rome; V. Volterra to Virginia Volterra, Sept. 16, 1916, VP.

[13]V. Volterra to Virginia Volterra, July 21, 1915, and Feb. 21, 1916, VP.

[14]Emilio Lussu, *Sardinian Brigade*, trans. by Marion Rawson (New York: Grove Press, 1970), 3.

[15]V. Volterra to Virginia Volterra, June 9, 1916, VP.

[16]V. Volterra to Virginia Volterra, June 12, 1916, VP.

[17]V. Volterra to Virginia Volterra, Nov. 22, 1916, VP.

[18]V. Volterra to Virginia Volterra, Nov. 15, 1916, VP.

[19]V. Volterra to P. Morrone, Jan. 27, 1917, Vito Volterra Collection, Accademia Nazionale dei Lincei, Rome.

[20]V. Volterra to Virginia Volterra, April 15, 1917, VP.

[21]S.L.G. Knox to George Ellery Hale, Oct. 1, 1918, George Ellery Hale Papers, Box 26, Institute Archives, California Institute of Technology.

[22]Knox to Hale, ibid.

[23]Judith R. Goodstein, *Millikan's School: A History of the California Institute of Technology* (New York: W.W. Norton, 1991), 84.

[24]Judith R. Goodstein, "A Conversation with Franco Rasetti," in *Physics in Perspective* 3 (2001), 294.

[25]Emilio Segrè, *Enrico Fermi, Physicist* (Chicago: University of Chicago Press, 1970), 30.

[26]Laura Fermi, *Atoms in the Family: My Life with Enrico Fermi* (Chicago: University of Chicago Press, 1954), 30.

[27]V. Volterra to Virginia Volterra, July 14, 1921, VP.

[28]V. Volterra to Virginia Volterra, Jan. 18, 1922, VP.

[29]V. Volterra to Virginia Volterra, Oct. 26, 1922, VP.

[30]V. Volterra to G. Pestalozzi, April 25, 1923, Vito Volterra Collection, Accademia Nazionale dei Lincei, Rome.

[31]"Verbale delle riunione tenuta il 25 maggio 1923 in una sala del Senato da un gruppo di parlamentari universitari contrari alla riforma Gentile," Vito Volterra Collection, Rome.

[32]Giorgio Abetti to George Ellery Hale, Nov. 1, 1922, George Ellery Hale Papers, Box 1, Institute Archives, California Institute of Technology.

[33]J.R. Goodstein, "Conversation with Rasetti," p. 297.

[34]L. Fermi, *Atoms in the Family*, 42.

[35]Vito Volterra, "Variazioni e fluttuazioni del numero d'individui in specie animali conviventi," in *Opere matematiche*, V, 1.

[36]Ibid.

[37]Sharon E. Kingsland, *Modeling Nature: Episodes in the History of Population Ecology* (University of Chicago Press: Chicago, 1995), 109.

[38]V. Volterra, "The General Equations of Biological Strife in the Case of Historical Actions, in *Opere matematiche*, V, 496.

[39]V. Volterra, "Discorso presidenziale del 1925," in *Opere matematiche*, IV, 529-531.

[40]Dirk Jan Struik, "Viaggio in Italia," *Lettera Matematica Pristem* 12 (June, 1996), 25.

[41]Carlo Somigliana to V. Volterra, Feb. 7, 1926, Vito Volterra Collection, Lincei.

[42]V. Volterra to Luigi Errera, in Giovanni Paoloni, ed., *Vito Volterra*.

[43]V. Volterra, *Leçons sur la théorie mathématique de la lutte pour la vie*, ed. Marcel Brelot (Paris: Gauthier-Villars et Cie., 1931).

[44]Tullio Levi-Civita to V. Volterra, Feb. 27, 1929, in Paoloni, *Vito Volterra*, 176.

[45]The document, dated December 19, 1928, is reproduced in G. Paoloni, *Vito Volterra*, Fig. VII. 6.

[46]Quoted in Judith R. Goodstein, "The Rise and Fall of Vito Volterra's World," *Journal of the History of Ideas* 45 (1984): 614.

[47]Vito Volterra to Griffith Evans, Jan. 7, 1932, Box 1, Evans Papers, UC Berkeley.

[48]David M. Smith, *Modern Italy: A Political History*, 361.

[49]Guido Corbellini, "Per il centenario della nascita di Vito Volterra, Senato della Repubblica, Seduta del 12 Maggio 1960."

[50]Quoted in "Gualda Massimi," interview by Carlotta Scaramuzzi, Nov. 25, 1994, p. 3, Institute Archives, California Institute of Technology.

[51]Renzo de Felice, *The Jews in Fascist Italy: A History*, with a preface by Michael A. Leeden, trans. Robert L. Miller (New York, Enigma, 2001), 264.

[52]G. Paoloni, *Vito Volterra*, Dec. 19, 1938, Figure 10.1.

[53]Quoted in André Weil, *The Apprenticeship of a Mathematician*, trans. Jennifer Gage (Basel: Birkhäuser Verlag, 1992).

[54]Virginia Volterra to [name unknown] Volterra, May 20, 1940, VP.

[55]G. Castelnuovo, "Vito Volterra," *Opere matematiche*, I, xiii.

Index